Engineering Properties of Marine Soils and Offshore Foundations

Engineering Properties of Marine Soils and Offshore Foundations

Guest Editors

Youkou Dong
Dengfeng Fu
Xiaowei Feng

MDPI

Basel • Beijing • Wuhan • Barcelona • Belgrade • Novi Sad • Cluj • Manchester

Guest Editors
Youkou Dong
College of Marine Science and Technology
China University of Geosciences
Wuhan
China

Dengfeng Fu
Key Laboratory of Marine Environment and Ecology
Ocean University of China
Qingdao
China

Xiaowei Feng
State Key Laboratory of Costal and Offshore Engineering
Dalian University of Technology
Dalian
China

Editorial Office
MDPI AG
Grosspeteranlage 5
4052 Basel, Switzerland

This is a reprint of the Special Issue, published open access by the journal *Journal of Marine Science and Engineering* (ISSN 2077-1312), freely accessible at: www.mdpi.com/journal/jmse/special_issues/Q52TLA2473.

For citation purposes, cite each article independently as indicated on the article page online and using the guide below:

Lastname, A.A.; Lastname, B.B. Article Title. *Journal Name* **Year**, *Volume Number*, Page Range.

ISBN 978-3-7258-3124-1 (Hbk)
ISBN 978-3-7258-3123-4 (PDF)
https://doi.org/10.3390/books978-3-7258-3123-4

Contents

About the Editors

Youkou Dong

Dong Youkou, male, born in 1987, is a professor at the School of Marine Science of China University of Geosciences (Wuhan), as well as a doctoral supervisor and the head of the Department of Ocean Engineering and Technology. He is recognized as a young talent in Hubei Province. He has conducted extensive research focused on offshore wind power foundations, reef engineering, and the prevention and control of underwater landslide disasters. His research findings were incorporated into the internationally recognized standard ISO/DIS 19901-4 and applied in site investigations and geological stability assessments of reef engineering in various marine resource extraction areas both domestically and internationally, yielding significant economic and social benefits. As the first or corresponding author, he published 32 SCI papers, which were cited a total of 680 times. He led youth and general projects funded by the National Natural Science Foundation, as well as key research and development projects in Hubei Province and sub-projects of the Ministry of Science and Technology. He serves as the chair of the TC-10 subcommittee of the International Association of Geological Disasters and Disaster Reduction, a youth editorial board member for the SCI journal *The Innovation* (IF = 31.7), and a guest editor for the *Journal of Marine Science and Engineering* (Chinese Academy of Sciences Zone 2). In 2023, he was selected as a "Top Young Talent" in Hubei Province and received the second prize for scientific and technological progress in Shanxi Province.

Dengfeng Fu

Fu Dengfeng is a professor and doctoral supervisor in the College of Environmental Science and Engineering at Ocean University of China. He entered Tianjin University in 2017 and joined Ocean University of China in 2023. His research interests are mainly focused on theoretical innovation and technological development in the field of marine resource exploitation and marine soil investigations. So far, he has led two projects funded by the National Natural Science Foundation of China and 16 research projects commissioned by the industry, and received funding support from the Young Expert of Mount Taishan Scholarship in Shandong Province. He has published 46 papers, including 35 SCI papers (22 from top and authoritative journals), and had 13 patents (inventions, utility models, and software copyrights) authorized. His research achievements were applied to the world's first large-scale floating wind power development (suction anchor), Sicily offshore oil and gas development (finned suction anchor), Guangdong Sanda Three Gorges offshore wind power construction (bucket foundation), Tianjin Nangang land reclamation (foundation instability evaluation method), and other projects, one of which was adopted by the international authoritative standard ISO19901-4. He serves as a member of the Youth Working Committee of the Soil Mechanics and Geotechnical Engineering Branch of the Chinese Civil Engineering Society, a Committee of the Engineering Risk and Insurance Research Branch of the Chinese Civil Engineering Society, and a member of the Academic Committee of the Tunnel and Underground Engineering Branch of the Chinese Civil Engineering Society on Tunnel and Underground Space Operation Safety and Energy Conservation and Environmental Protection Technology Forum.

Xiaowei Feng

Xiaowei Feng is a Professor and Doctoral Supervisor at Dalian University of Technology. He was selected as the National Overseas Young Talents. He is mainly engaged in research and consulting in offshore geotechnical engineering. He is a member of the Technical Committee on Offshore Geomechanics of the International Society of Soil Mechanics and Geotechnical Engineering (TC209) and the Youth Committee of the Soil Mechanics and Geotechnical Engineering, Branch of the Chinese Civil Engineering Society. He has been granted one National Overseas Young Talents Project and one General Project and participated in one Major Project of the National Natural Science Foundation of China. He has completed more than 10 international consulting projects and published more than 40 journal papers, 7 of which have been published in *Géotechnique*, the leading journal of geotechnical engineering. He established a toolbox approach "Optimization Method for Geotechnical Engineering Design of Deepwater Engineering Subsea Foundation". Many aspects of his research, involving a total of nine papers, were adopted by the ISO 19901-4.

Preface

Dear Colleagues,

To reduce greenhouse gas emissions and enhance energy security, the role of renewable energy in the global energy structure is becoming increasingly significant. Within this context, marine clean energy holds an important position, encompassing tidal energy, wind energy, solar energy, and combustible ice found in the ocean. To achieve the extraction and utilization of these energy sources, the development of marine engineering research has become one of the areas of growing interest. In the field of marine engineering, it is crucial to thoroughly understand the engineering characteristics of marine soils, foundation systems, and offshore structures to better guide the work during the design, construction, and maintenance phases. Professors Dong Youkou from China University of Geosciences (Wuhan), Fu Dengfeng from Ocean University of China, and Feng Xiaowei from Dalian University of Technology selected 15 cutting-edge papers in the field of marine engineering to inform our colleagues about the current state of research in this area and to inspire their research endeavors, thereby advancing progress in the field. Among these, six papers focus on foundation treatment and foundation engineering properties, five papers are related to offshore engineering foundations, two papers address the inspection and maintenance of offshore structures, and two papers pertain to the exploitation of marine resources.

As a Guest Editor for this Special Issue on "Engineering Characteristics of Marine Soil and Offshore Foundations", I would like to express my heartfelt gratitude to all the authors who contributed their valuable work, as their contributions have been instrumental to the success of this Special Issue.

Youkou Dong, Dengfeng Fu, and Xiaowei Feng
Guest Editors

Journal of
Marine Science and Engineering

MDPI

Editorial

Engineering Properties of Marine Soils and Offshore Foundations

Youkou Dong [1], Dengfeng Fu [2] and Xiaowei Feng [3,*]

1 College of Marine Science and Technology, China University of Geosciences, 388 Lumo Road, Wuhan 430074, China; dongyk@cug.edu.cn
2 Key Laboratory of Marine Environment and Ecology, Ministry of Education, Ocean University of China, Qingdao 266100, China; dengfeng.fu@ouc.edu.cn
3 State Key Laboratory of Costal and Offshore Engineering, Dalian University of Technology, Dalian 116024, China
* Correspondence: xfeng@dlut.edu.cn

To reduce greenhouse gas emissions and enhance energy security, the role of renewable energy in the global energy structure is becoming increasingly significant. Within this context, marine clean energy holds an important position, encompassing tidal energy, wind energy, solar energy and combustible ice found in the ocean. To achieve the extraction and utilization of these energy sources, the development of marine engineering research has become one of the areas of growing interest.

In the field of marine engineering, it is crucial to thoroughly understand the engineering characteristics of marine soils, foundation systems and offshore structures to better guide the work during design, construction, and maintenance phases. This Special Issue includes a total of 15 relevant papers to facilitate a broader understanding of the latest advancements in this research area and to promote its development. Among these, six papers focus on foundation treatment and foundation engineering properties, five papers are related to offshore engineering foundations, two papers address the inspection and maintenance of offshore structures, and two papers pertain to the exploitation of marine resources.

In offshore engineering, subsea pipelines are particularly vulnerable to fatigue cracks due to the intricacies of their manufacturing processes and the harsh operating environments they endure under prolonged cyclic loads. A comprehensive and high-precision detection method for subsea pipeline damage can effectively mitigate the potential safety hazards. Peng et al. [1] proposed a three-dimensional ultrasonic imaging detection method modified for cylindrical coordinates, which adeptly addresses the complex three-dimensional wave fields involved in the inspection of subsea cylindrical pipelines. The study generalizes the three-dimensional staggered-grid finite-difference method from Cartesian coordinates to cylindrical ones and simulates the full wave field in the three-dimensional space of the pipeline. After processing the ultrasonic recording signals, the method utilizes reverse time migration and cross-correlation imaging conditions to achieve three-dimensional reverse time migration imaging of defects in subsea pipelines.

Accurate surveying is essential for coastal engineering. However, conventional seismic exploration methods have become inadequate due to the challenges posed by seismic waves traversing weathered zones and reaching the necessary depths. To address the limitations of existing cross-well imaging technologies, Peng et al. [2] investigated the application of three-dimensional (3D) cross-well elastic reverse time migration (RTM) imaging, utilizing multi-wave and multi-component techniques in coastal engineering surveys. They achieved precise decomposition of vector compressional waves (P-waves) and shear waves (S-waves) through two wavefield decoupling algorithms, without any amplitude or phase distortion. Furthermore, the pressure component of the compressional wave was extracted, enabling subsequent independent imaging. This method effectively leverages the elastic properties of the soil media through multi-wave and multi-component

Citation: Dong, Y.; Fu, D.; Feng, X. Engineering Properties of Marine Soils and Offshore Foundations. *J. Mar. Sci. Eng.* **2024**, *12*, 2077. https://doi.org/10.3390/jmse12112077

Received: 5 November 2024
Accepted: 11 November 2024
Published: 18 November 2024

elastic wave RTM imaging technology, providing valuable insights into subsurface rock layers, interfaces, and other structural distributions for coastal engineering projects without incurring additional costs. During the construction and operation of nearshore engineering structures, numerous engineering challenges emerge that necessitate resolution. This Special Issue comprises papers addressing the pile driving in calcareous sand foundations, the influence of structures on siliceous sands, studies on the permeability of back fill materials with a high clay content, research on the improvement and reuse of marine mud and coral sand, and methods for ground improvement. Among these contributions, Gao et al. [3] employed model experiments in conjunction with pressure sensors and close-range photogrammetry techniques to investigate the mechanical response and deformation characteristics of calcareous sands during pile driving and installation. Various types of piles were analyzed, including pipe piles, square piles, and partially closed ended steel pipe piles. The experimental results revealed that during pile driving, the tip resistance of the different piles increased with penetration depth; however, significant fluctuations in tip resistance were also observed due to particle breakage and energy dissipation in the calcareous sand. The extent of particle breakage and the variations in internal stress differed among the pile types, subsequently influencing the tip resistance. Chen et al. [4] examined the particle breakage characteristics of marine siliceous sand under high-pressure conditions. A series of conventional triaxial tests were conducted on siliceous sand subjected to confining pressures ranging from 2 MPa to 8 MPa. The results demonstrated that as the particle breakage index increased, the fractal dimension exhibited an upward convex hyperbolic trend. The boundary radius at which siliceous sand particles displayed fractal characteristics was determined to be approximately 0.4 mm. These experimental findings provide reference and supplementary data for analyzing particle breakage in sandy soils. Shan et al. [5] utilized a stepwise vacuum preloading method in the laboratory to reinforce back fill materials with a high clay content derived from dredging vessels. They assessed the pore structure and permeability characteristics of the dredged fill under varying vacuum pressures. Correlation analysis indicated a strong relationship between a large pore content and the permeability coefficient, which can be employed to describe the permeability characteristics of the soil. The findings of this study offer valuable insights for enhancing reinforcement methods and evaluating the effectiveness of dredged fill in engineering practice. Han et al. [6] proposed a method that integrates flocculation solidification with high-pressure filtration for the effective disposal of marine mud, demonstrating advantages in dewatering performance, material savings, and the shear strength of the treated mud. Yao et al. [7] explored the effects of vibration flotation and impact compaction on the densification and load-carrying capacity of coral sand foundations. Field tests were conducted in four different areas, encompassing plate load tests, California Bearing Ratio (CBR) tests, density measurements, dynamic penetration tests (DPT), and settlement monitoring. The results indicated that coral sand possesses favorable characteristics for foundation construction. The seepage and self-weight consolidation following land reclamation significantly enhanced the compaction of the coral sand, thereby meeting the requirements for areas with lower load-carrying capacities. Both vibration flotation and impact compaction techniques can substantially improve the load-carrying capacity of the foundation, with minimal differences between the two methods. Due to its simplicity and rapid construction speed, the impact compaction method is regarded as the most effective approach for improving coral sands. The DPT results suggested that the reinforcement effects of both vibration flotation and impact compaction methods are less pronounced in deeper foundations compared to surface layers. This study provides valuable insights for optimizing foundation treatment in coral reef reclamation projects. Based on the life cycle assessment (LCA) method, Yu et al. [8] developed an ontological framework for the sustainable assessment of sustainable cementitious systems (SCSs) and evaluated the effects of fineness, carbonation degree, and substitution rate of steel slag on the sustainability of SCS. The results indicated that, compared to pure cement stabilized soil (S-C), using 10% and 20% finely ground steel slag carbonated for 18 h (FSS-C-18h) as a cement substitute can significantly reduce carbon emissions and costs while achieving strength performance comparable to that of S-C. This demonstrates the feasibility of utilizing steel slag as a sustainable supplementary cementitious material for stabilizing soft soils.

In offshore structures, the stability of the foundation is a critical determinant of the overall structural integrity. This Special Issue encompasses studies on anchor rod foundations and bucket foundations. Notably, Yuan et al. [9] and Li et al. [10] investigated anchor rod foundations. Yuan et al. [9] proposed a novel unified method for estimating the uplift capacity of both deep single- and multi-helix anchors, based on the investigation of failure mechanisms. This method utilizes a modified Mohr–Coulomb model that incorporates strain softening in sand, alongside Euler–Lagrange coupling techniques for finite element analysis to ascertain deep failure modes. A simplified rupture surface was introduced, and an equation for estimating uplift capacity was derived using the limit equilibrium method. Two critical factors were examined including the lateral earth pressure coefficient and the average internal friction angle. The validity of this method was confirmed through comparisons with the results of centrifuge tests. In contrast, Li et al. [10] conducted model tests on torpedo anchors to investigate the effects of the uplift angle and bearing plate radius on the load-carrying behavior of T-type, EN3-type, EN4-type, and EC-type torpedo anchors. Research by Jiang et al. [11], Wang et al. [12], and Lian et al. [13] has concentrated on bucket foundations. Jiang et al. [11] conducted centrifuge tests to investigate the response of a single bucket foundation subjected to monotonic and symmetric cyclic loads in over-consolidated clay. The results indicated that the discrepancy between the monotonic vertical bearing capacity measured from the centrifuge tests and that obtained from the finite element results was less than 6%. This study examined the effects of the cyclic load amplitude (ranging from 37% to 64% of the vertical bearing capacity) and the number of cycles on the accumulation of vertical displacement and the evolution of stiffness. Simplified methods were proposed for predicting both dimensional and non-dimensional stiffness evolution. Wang et al. [12] explored the load-carrying behavior of tripod-bucket foundations through a series of physical modelling tests and numerical simulations. They observed that the center of rotation of the foundation diminished as the aspect ratio (L/D) decreased, leading to the transition of the failure mechanism from rotation to uplift. The impact of soil pressure on the bucket was studied through finite element analysis and model testing, which elucidated the failure mechanisms of tripod-bucket foundations with varying L/D ratios. A revised method is proposed to estimate the moment bearing capacity of the tripod bucket foundation under horizontal and moment loads. This method is thought to be more convenient and applicable in practice. Lian et al. [13] investigated composite bucket foundations (CBFs) with significant transition sections, initially developing a finite element method (FEM) to characterize the rigid deformation performance of these transition components. To clarify the impact of the transition section on wind turbines equipped with CBFs, the transition section was simplified as a rigid body, and a three-DOF theoretical model was established. This model accounted for horizontal and rotational foundation stiffness to illustrate the constraint effects below the mud line. A sensitivity analysis was conducted on the parameters of the transition section, including mass, moment of inertia, and center of mass height. Furthermore, the vibration characteristics of the CBF structure under various operational load conditions were compared between the theoretical model and field data. The results indicated that the relative error between the theoretical model and finite element model ranged from 3.78% to 5.03%, meeting accuracy requirements. The parameters of the transition section significantly influenced the natural frequency, foundation stiffness, and vibration response of wind turbines with CBF. Compared to wind loads and 1P loads, the effect of 3P loads was more pronounced when the 3P frequency approached the natural frequency of the wind turbine. This Special Issue also encompasses research focused on maintaining the safety and stability of offshore structures and resource extraction. Dong et al. [14] examined submarine pipelines subjected to the impacts of giant waves generated by natural disasters, such as hurricanes and tsunamis. The influence of various factors, including insulation layer configuration, pipeline structure, and marine environment, were investigated on the protective performance of the insulation layer. Wei et al. [15], utilizing critical state theory, proposed a predictive model for the drained shear strength of hydrate-bearing fine-grained sediments. They conducted multiple consolidated undrained triaxial tests on hydrate-bearing fine samples from the Shen hu area of the South China Sea, and analyzed the influence of effective consolidation stress and hydrate saturation on the drained shear strength.

In summary, the articles published in this Special Issue encompass a wide range of research topics related to the engineering characteristics of marine soils and offshore foundations. All articles are open access, aiming to provide readers with a comprehensive understanding of the advancements in marine engineering research. Furthermore, it is hoped that this foundation will inspire readers to pursue more in-depth research and thereby advance the development of this field.

Acknowledgments: As a Guest Editor of the Special Issue "Engineering Properties of Marine Soils and Offshore Foundations", I would like to express my deep appreciation to all the authors whose valuable work was published under this Special Issue and thus contributed to the success of the edition.

Conflicts of Interest: The authors declare no conflicts of interest.

References

1. Peng, D.; Cheng, F.; She, X.; Zheng, Y.; Tang, Y.; Fan, Z. Three-Dimensional Ultrasonic Reverse-Time Migration Imaging of Submarine Pipeline Nondestructive Testing in Cylindrical Coordinates. *J. Mar. Sci. Eng.* **2023**, *11*, 1459. [CrossRef]
2. Peng, D.; Cheng, F.; Xu, H.; Zong, Y. An Application of 3D Cross-Well Elastic Reverse Time Migration Imaging Based on the Multi-Wave and Multi-Component Technique in Coastal Engineering Exploration. *J. Mar. Sci. Eng.* **2024**, *12*, 522. [CrossRef]
3. Gao, Y.; Guo, Z.; Yuan, Q. Pile Driving and the Setup Effect and Underlying Mechanism for Different Pile Types in Calcareous Sand Foundations. *J. Mar. Sci. Eng.* **2024**, *12*, 133. [CrossRef]
4. Chen, R.; Zhao, T.; Wu, Z.; Hao, D.; Xue, N.; Yuan, C. Experimental Investigation on Particle Breakage Behavior of Marine Silica Sand under High-Stress Triaxial Shear. *J. Mar. Sci. Eng.* **2023**, *11*, 1825. [CrossRef]
5. Shan, W.; Yuan, X.; Chen, H.-e.; Li, X.; Li, J. Permeability of High Clay Content Dredger Fill Treated by Step Vacuum Preloading: Pore Distribution Analysis. *J. Mar. Sci. Eng.* **2023**, *11*, 1714. [CrossRef]
6. Han, C.; Xie, H.; Bai, B.; Zhang, R.; Gao, Y.; Zhao, Z. Experimental Study on Mechanical Properties of Marine Mud Slurry Treated by Flocculation-Solidification-High Pressure Filtration Combined Method. *J. Mar. Sci. Eng.* **2023**, *11*, 2270. [CrossRef]
7. Yao, T.; Li, W. An Evaluation of Treatment Effectiveness for Reclaimed Coral Sand Foundation in the South China Sea. *J. Mar. Sci. Eng.* **2023**, *11*, 2288. [CrossRef]
8. Yu, C.; Yuan, J.; Cui, C.; Zhao, J.; Liu, F.; Li, G. Ontology Framework for Sustainability Evaluation of Cement–Steel-Slag-Stabilized Soft Soil Based on Life Cycle Assessment Approach. *J. Mar. Sci. Eng.* **2023**, *11*, 1418. [CrossRef]
9. Yuan, C.; Hao, D.; Chen, R.; Zhang, N. Numerical Investigation of Uplift Failure Mode and Capacity Estimation for Deep Helical Anchors in Sand. *J. Mar. Sci. Eng.* **2023**, *11*, 1547. [CrossRef]
10. Li, G.; Zhang, J.; Liu, J.; Xi, Y.; Kou, H. Pullout Bearing Capacity of End-Bearing Torpedo Anchors in Cohesive Soil Seabed. *J. Mar. Sci. Eng.* **2023**, *11*, 1548. [CrossRef]
11. Jiang, J.; Wang, D.; Fu, D. Vertical Monotonic and Cyclic Responses of a Bucket in Over-Consolidated Clay. *J. Mar. Sci. Eng.* **2023**, *11*, 2044. [CrossRef]
12. Wang, X.; Zhao, X.-L.; Sui, S.-H.; Hu, Z.-B.; Deng, W.-N.; Song, Q.-M. Bearing Characteristics of Tripod Bucket Foundation under Horizontal and Moment Load in Sand. *J. Mar. Sci. Eng.* **2023**, *11*, 1631. [CrossRef]
13. Lian, J.; Zhou, H.; Dong, X. A Theoretical Methodology and Measurement of Dynamic Characteristics of Wind Turbines with Composite Bucket Foundations. *J. Mar. Sci. Eng.* **2024**, *12*, 106. [CrossRef]
14. Dong, Y.; Zhao, E.; Cui, L.; Li, Y.; Wang, Y. Dynamic Performance of Suspended Pipelines with Permeable Wrappers under Solitary Waves. *J. Mar. Sci. Eng.* **2023**, *11*, 1872. [CrossRef]
15. Wei, R.; Liu, L.; Jia, C.; Dong, X.; Bu, Q.; Zhang, Y.; Liu, C.; Wu, N. Undrained Triaxial Shear Tests on Hydrate-Bearing Fine-Grained Sediments from the Shenhu Area of South China Sea. *J. Mar. Sci. Eng.* **2023**, *11*, 1604. [CrossRef]

Journal of
Marine Science and Engineering

MDPI

Article

An Application of 3D Cross-Well Elastic Reverse Time Migration Imaging Based on the Multi-Wave and Multi-Component Technique in Coastal Engineering Exploration

Daicheng Peng [1], Fei Cheng [2], Hao Xu [3,*] and Yuquan Zong [4]

1 Key Laboratory of Exploration Technologies for Oil and Gas Resource, Ministry of Education, Yangtze University, Wuhan 430100, China; pengdc_geo@126.com
2 Hubei Key Laboratory of Marine Geological Resources, China University of Geosciences, Wuhan 430074, China; chengfly03@126.com
3 School of Civil Engineering and Architecture, Wuhan Polytechnic University, 68 Xuefu South Road, Wuhan 430023, China
4 Central Southern China Electric Power Design Institute Co., Ltd. of China Power Engineering Consulting Group, Wuhan 430071, China; zyq5821@csepdi.com
* Correspondence: xuhao33011@163.com; Tel.: +86-18771099927

Abstract: Precise surveys are indispensable in coastal engineering projects. The extensive presence of sand in the coastal area leads to significant attenuation of seismic waves within unsaturated loose sediments. As a result, it becomes challenging for seismic waves to penetrate the weathered zone and reach the desired depth with significant amount of energy. In this study, the application of three-dimensional (3D) cross-well elastic reverse time migration (RTM) imaging based on multi-wave and multi-component techniques in coastal engineering exploration is explored. Accurate decomposition of vector compressional (P) and shear (S) waves is achieved through two wavefield decoupling algorithms without any amplitude and phase distortion. Additionally, compressional wave pressure components are obtained, which facilitates subsequent independent imaging. This study discusses and analyzes the imaging results of four imaging strategies under cross-correlation imaging conditions in RTM imaging. The analysis leads to the conclusion that scalarizing vector wavefields imaging yields superior imaging of P- and S-waves. Furthermore, the imaging results obtained through this approach are of great physical significance. In order to validate the efficacy of this method in 3D geological structure imaging in coastal areas, RTM imaging experiments were performed on two representative models. The results indicate that the proposed 3D elastic wave imaging method effectively generates accurate 3D cross-well imaging of P- and S-waves. This method utilizes the multi-wave and multi-component elastic wave RTM imaging technique to effectively leverage the Earth's elastic medium without increasing costs. It provides valuable information about the distribution of subsurface rock layers, interfaces, and other structures in coastal engineering projects. Importantly, this can be achieved without resorting to extensive excavation or drilling operations. This method addresses the limitations of current cross-well imaging techniques, thereby providing abundant and accurate geological and geophysical information for the analysis and interpretation of 3D geological structures in coastal engineering projects. It has important theoretical and practical significance in real-world production, as well as for the study of geological structures in coastal engineering.

Keywords: coastal engineering exploration; multi-wave and multi-component; cross-well seismic exploration; 3D reverse time migration imaging; elastic wave decomposition

check for updates

Citation: Peng, D.; Cheng, F.; Xu, H.; Zong, Y. An Application of 3D Cross-Well Elastic Reverse Time Migration Imaging Based on the Multi-Wave and Multi-Component Technique in Coastal Engineering Exploration. *J. Mar. Sci. Eng.* **2024**, *12*, 522. https://doi.org/10.3390/jmse12030522

Academic Editor: Merrick C. Haller

Received: 1 February 2024
Revised: 19 March 2024
Accepted: 20 March 2024
Published: 21 March 2024

1. Introduction

During coastal engineering projects, extensive surveys and investigations are crucial for gathering information on geological structures, sediment distribution, and soil properties in coastal areas. These investigations involve analyses by drilling, sampling, and geophysical exploration to evaluate the geological formations, soil types, lithology, and groundwater levels [1–3]. Among the numerous geophysical exploration methods, seismic exploration is widely employed due to its ability to provide superior subsurface resolution [4]. However, the extensive presence of sand in the coastal area leads to a significant attenuation of seismic waves within unsaturated loose sediments. Surface seismic methods are highly susceptible to variations in near-surface conditions and encountering difficulties in transmitting a significant amount of energy through the weathered zone to reach the targeted depths [5–7]. Compared to surface seismic reflection or refraction surveys, cross-well seismic acquisition methods emerge as an optimal choice for effectively targeting specific coastal areas [8,9].

Cross-well seismic can be applied to various aspects of oil and gas exploration, including detailed imaging of structures and precise characterization of reservoirs [10–12]. Furthermore, cross-well seismic has also been rapidly developed in the field of engineering [13,14] and has been applied in diverse areas, such as geological engineering [15–17], hydrogeological surveys [18–20], and quality inspections in civil engineering projects [21–23]. Currently, most cross-well seismic studies focus on two-dimensional (2D) tomographic imaging between adjacent wells [24,25]. It requires data collection and travel-time inversion between two wells to acquire subsurface structural profiles. Nevertheless, this imaging approach has certain limitations. For example, in 2D computed tomography (CT) imaging, only adjacent wells can be imaged, and information from multiple wells distributed in 3D space cannot be effectively utilized [26]. As a result, the utilization rate of spatial well locations remain relatively low. Furthermore, the 2D cross-well CT imaging technique can only provide information on the geological structure of the profile between the two wells, failing to capture the lateral structure information of the profile [27]. As a result, the imaging provides poor lateral continuity, hindering the evaluation of 3D geological structures.

Additionally, there are issues associated with imaging steep-dip interfaces (interfaces with dip angles exceeding 45°) [28]. In recent years, the rapidly developing RTM method follows the full-wave wave equation during wavefield extrapolation and is not limited by angles [29–31]. Among various pre-stack migration methods, RTM is considered the most accurate imaging technique. However, current studies on cross-well RTM imaging are limited to 2D space or 3D P-waves RTMs [32,33]. With the advancement of seismic exploration technology, the challenges have become increasingly complex. Conventional compressional wave exploration can no longer satisfy the needs of engineering projects [34]. Improvements need to be made in two key aspects: First, it is necessary to image areas with weak compressional wave reflection energy or no compressional wave reflection signals to provide reliable structural imaging information. Second, it is necessary to provide imaging information for elastic multi-wave analysis, thus establishing a reliable foundation for subsequent elastic wave amplitude analysis and rock physics parameter inversion. Conventional seismic exploration techniques commonly employ compressional wave sources. To address the limitations of compressional wave reflection energy, the shear waves generated through compressional wave conversion need to be imaged [35,36]. This approach aims to fully utilize the Earth's elastic media to construct comprehensive imaging information of subsurface structures without increasing costs. Simultaneously, the imaging results must include multi-wave imaging results for subsequent amplitude variation with angle (AVA) analysis and the inversion of rock physics parameters [37,38]. Compared with the conventional compressional wave exploration techniques, the theoretical assumptions of multi-wave and multi-component seismic exploration technology are more consistent with the actual characteristics of subsurface media, offering significant advantages in addressing practical challenges.

To effectively address the above issues, it is essential to fully utilize the distinctive features of multi-wave and multi-component seismic data acquired between wells. Therefore, this paper investigates the application of 3D cross-well elastic wavefield RTM imaging in coastal engineering exploration based on multi-wave and multi-component data. By applying multi-wave and multi-component elastic wave RTM imaging techniques, precise 3D geological structures between wells can be obtained, laying the foundation for subsequent lithological analysis and inversion. This approach addresses the limitations of current cross-well imaging techniques, thus providing rich geological and geophysical information for analyzing and interpreting 3D geological structures in coastal regions. It has important theoretical and practical significance in actual production, as well as for the investigation of geological structures in coastal engineering.

2. Methodology

To study the application of 3D cross-well seismic exploration in coastal engineering, it is essential to first conduct research on the numerical simulation of elastic waves. Numerical simulations encompass the simulation of the propagation characteristics of elastic waves in various media through mathematical methods based on the known physical parameters of the detection model, combined with elastic wave propagation theory [39]. Numerical simulations can be divided into three categories: wave equation numerical methods, integral methods, and ray tracing methods. Among these, wave equation numerical methods primarily include the finite difference method (FDM) [40], pseudo spectral method (PSM) [41], finite element method (FEM) [42], spectral element method (SEM) [43], and boundary element method (BEM) [44]. FDM is extensively employed in numerical simulations of elastic waves due to its advantageous features, including fast computation speed and high accuracy.

In this section, we begin by discussing the 3D elastic wave equations of motion and grid discretization in cylindrical coordinates. Then, we utilize two different methods for elastic wave decomposition. Following this, we introduce the principles of RTM imaging and provide detailed explanations of four different cross-correlation imaging conditions. Lastly, we propose the Poynting vector and Laplace filtering to attenuate RTM imaging artifacts.

2.1. 3D Elastic Wave Equations of Motion and Grid Discretization

The basic laws governing the propagation of elastic waves are described by the wave equation of elastic waves. Based on the equations of motion, Cauchy equations, and physical equations in a three-dimensional Cartesian coordinate system, the first-order velocity-stress equation in an isotropic medium in cylindrical coordinates can be expressed as follows [45]:

$$\begin{cases} \rho\frac{\partial v_x}{\partial t} = \frac{\partial \tau_{xx}}{\partial x} + \frac{\partial \tau_{xy}}{\partial y} + \frac{\partial \tau_{xz}}{\partial z} + f_x \\ \rho\frac{\partial v_y}{\partial t} = \frac{\partial \tau_{xy}}{\partial x} + \frac{\partial \tau_{yy}}{\partial y} + \frac{\partial \tau_{yz}}{\partial z} + f_y \\ \rho\frac{\partial v_z}{\partial t} = \frac{\partial \tau_{xz}}{\partial x} + \frac{\partial \tau_{yz}}{\partial y} + \frac{\partial \tau_{zz}}{\partial z} + f_z \end{cases} \tag{1}$$

$$\begin{cases} \frac{\partial \tau_{xx}}{\partial t} = (\lambda+2\mu)\frac{\partial v_x}{\partial x} + \lambda\left(\frac{\partial v_y}{\partial y} + \frac{\partial v_z}{\partial z}\right) + g_{xx} \\ \frac{\partial \tau_{yy}}{\partial t} = (\lambda+2\mu)\frac{\partial v_y}{\partial y} + \lambda\left(\frac{\partial v_x}{\partial x} + \frac{\partial v_z}{\partial z}\right) + g_{yy} \\ \frac{\partial \tau_{zz}}{\partial t} = (\lambda+2\mu)\frac{\partial v_z}{\partial z} + \lambda\left(\frac{\partial v_y}{\partial y} + \frac{\partial v_x}{\partial x}\right) + g_{zz} \\ \frac{\partial \tau_{xy}}{\partial t} = \mu(\frac{\partial v_x}{\partial y} + \frac{\partial v_y}{\partial x}) + g_{xy} \\ \frac{\partial \tau_{yz}}{\partial t} = \mu(\frac{\partial v_y}{\partial z} + \frac{\partial v_z}{\partial y}) + g_{xz} \\ \frac{\partial \tau_{xz}}{\partial t} = \mu(\frac{\partial v_z}{\partial x} + \frac{\partial v_x}{\partial z}) + g_{yz} \end{cases} \tag{2}$$

where $v = \left[v_x, v_y, v_z\right]^T$ denotes the velocity; $\tau = \left[\tau_{xx}, \tau_{yy}, \tau_{zz}, \tau_{xy}, \tau_{yz}, \tau_{xz}\right]^T$ denotes the stress vectors; $f = \left[f_x, f_y, f_z\right]^T$ represents the point force source; $g = \left[g_{xx}, g_{yy}, g_{zz}, g_{xy}, , g_{xz}, g_{yz}\right]^T$ represents the coupling; ρ is the density; λ and μ refer to the Lamé constants.

2.2. Elastic Wave Decomposition

For elastic wave RTM, the key to obtaining multi-wave and multi-component migration results is the decoupling of P- and S-waves. The elastic wavefields are separated during the reconstruction of the source and receiver wavefields to obtain pure P-wave and S-wave wavefields [46]. The following two methods are employed in this study.

2.2.1. Wave Equation Decoupling Method

Based on the homogeneous isotropic media and the first-order velocity-stress equations, the velocities of separated P- and S-waves in 3D space can be expressed as follows [47]:

$$\begin{cases} v_x = v_{px} + v_{sx} \\ v_y = v_{py} + v_{sy} \\ v_z = v_{pz} + v_{sz} \\ \frac{\partial v_{px}}{\partial t} = \frac{\lambda+2\mu}{\rho(3\lambda+2\mu)}\left(\frac{\sigma_{xx}}{\partial x} + \frac{\sigma_{yy}}{\partial x} + \frac{\sigma_{zz}}{\partial x}\right) \\ \frac{\partial v_{py}}{\partial t} = \frac{\lambda+2\mu}{\rho(3\lambda+2\mu)}\left(\frac{\sigma_{xx}}{\partial y} + \frac{\sigma_{yy}}{\partial y} + \frac{\sigma_{zz}}{\partial y}\right) \\ \frac{\partial v_{pz}}{\partial t} = \frac{\lambda+2\mu}{\rho(3\lambda+2\mu)}\left(\frac{\sigma_{xx}}{\partial y} + \frac{\sigma_{yy}}{\partial y} + \frac{\sigma_{zz}}{\partial y}\right) \\ \frac{\partial v_{sx}}{\partial t} = \frac{1}{\rho}\left(\frac{\sigma_{xy}}{\partial y} + \frac{\sigma_{xz}}{\partial z}\right) + \frac{1}{\rho(3\lambda+2\mu)}\left[(\lambda+2\mu)\left(\frac{\sigma_{yy}}{\partial x} + \frac{\sigma_{zz}}{\partial x}\right) - 2\lambda\frac{\sigma_{xx}}{\partial x}\right] \\ \frac{\partial v_{sy}}{\partial t} = \frac{1}{\rho}\left(\frac{\sigma_{xy}}{\partial x} + \frac{\sigma_{yz}}{\partial z}\right) + \frac{1}{\rho(3\lambda+2\mu)}\left[(\lambda+2\mu)\left(\frac{\sigma_{xx}}{\partial y} + \frac{\sigma_{zz}}{\partial y}\right) - 2\lambda\frac{\sigma_{yy}}{\partial y}\right] \\ \frac{\partial v_{sz}}{\partial t} = \frac{1}{\rho}\left(\frac{\sigma_{xz}}{\partial x} + \frac{\sigma_{yz}}{\partial z}\right) + \frac{1}{\rho(3\lambda+2\mu)}\left[(\lambda+2\mu)\left(\frac{\sigma_{xx}}{\partial z} + \frac{\sigma_{yy}}{\partial z}\right) - 2\lambda\frac{\sigma_{zz}}{\partial z}\right] \end{cases} \quad (3)$$

In Equation (3), v_{px} and v_{sx} represent the P- and S-wave components in the x-direction, respectively. Similarly, v_{py} and v_{sy} represent the P- and S-wave components in the y-direction, and v_{pz} and v_{sz} represent the P- and S-wave components in the z-direction, respectively. By solving Equation (3), the wavefields of pure P- and S-waves can be obtained. With this method, the amplitude and phase information of the P- and S-waves can be better retained.

2.2.2. Auxiliary Variables Method

In addition to the six velocity component equations incorporated in the equation for conventional elastic wave, the method of auxiliary variables introduces an additional component to achieve the vector decomposition of P- and S-waves, which is the compressional wave pressure component equation [48,49]. The expression of this equation is as follows:

$$\begin{cases} \frac{\partial \tau_p}{\partial t} = (\lambda+2\mu)\left(\frac{\partial v_x}{\partial x} + \frac{\partial v_y}{\partial y} + \frac{\partial v_z}{\partial z}\right) \\ \frac{\partial v_{px}}{\partial t} = \frac{1}{\rho}\frac{\partial \tau_p}{\partial x} \quad v_{sx} = v_x - v_{px} \\ \frac{\partial v_{py}}{\partial t} = \frac{1}{\rho}\frac{\partial \tau_p}{\partial y} \quad v_{sy} = v_y - v_{py} \\ \frac{\partial v_{pz}}{\partial t} = \frac{1}{\rho}\frac{\partial \tau_p}{\partial z} \quad v_{sz} = v_z - v_{pz} \end{cases} \quad (4)$$

In the method of auxiliary variables, the introduction of the τ_p component allows for additional wavefield information to be obtained. The τ_p component is analogous to the acoustic pressure component in the acoustic wave equation. In the context of elastic waves, it represents the pressure component of the P-wave [50].

2.3. Cross-Correlation Imaging Condition

The RTM algorithm, with its significant impact on imaging accuracy, relies heavily on the imaging condition [51,52]. Different imaging conditions employed in the imaging process can lead to different imaging results. The following sections introduce four cross-correlation imaging conditions used in elastic RTM.

2.3.1. Imaging by Vector Velocity Fields

In conventional elastic RTM imaging, the complete wavefield is employed for the imaging process. For multi-component seismic data, the velocity wavefields of the source ($v^S(x,t)$) and the receiver ($v^R(x,t)$) in the subsurface at each position x can be obtained through forward modeling and wavefield extrapolation. The imaging condition utilizing the three-component velocity wavefields can be expressed as follows:

$$I_{ij}(\mathbf{x}) = \int_0^{T_{\max}} v_i^S(\mathbf{x},t) \cdot v_j^R(\mathbf{x},t)dt \tag{5}$$

where S and R represent the velocity wavefields of the source and receiver, respectively; i and j correspond to the three Cartesian components: x, y, and z; and I_{ij} represents the imaging structure generated by cross-correlating the i-component of the source wavefield with the j-component of the receiver wavefield. $T_{\max}$ is the maximum time of the seismic records. This method can introduce crosstalk between P- and S-waves in both the source wavefield and the receiver wavefield.

2.3.2. Imaging by Scalar and Vector Potentials

Under the elastic RTM imaging condition, the second imaging method is implemented. It utilizes the Helmholtz decomposition during the extrapolation of the wavefield, effectively separating the source wavefield and receiver wavefield into pure P- and S-wave wavefields. The expression of the imaging condition is as follows:

$$I_{ij}(\mathbf{x}) = \int_0^{T_{\max}} m_i^S(\mathbf{x},t) \cdot m_j^R(\mathbf{x},t)dt \tag{6}$$

where S and R represent the velocity wavefields of the source and receiver, respectively; i and j correspond to different wave modes, specifically the P- and S-waves. $T_{\max}$ is the maximum time of the seismic records. This method modifies the phase and amplitude characteristics of the original wavefield when employing the Helmholtz decomposition.

2.3.3. Imaging by Pure Vector P- and S-Waves

Using the separation method described in Equation (3), pure P- and S-wave wavefields in various directions can be obtained. Afterward, during the process of wavefield extrapolation, these separated wavefields are utilized as boundary conditions. The imaging results of the P- and S-wave vector components can be acquired through correlating the source wavefield with the receiver wavefield. Figure 1 illustrates the imaging process for this method.

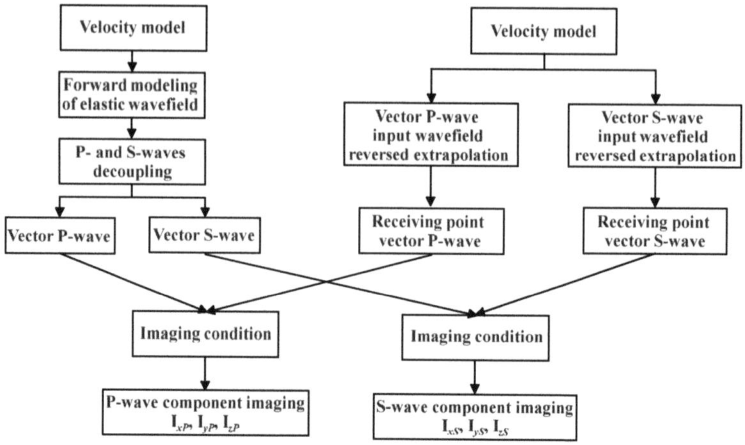

Figure 1. The flowchart illustrating the imaging process using pure vector P- and S-waves.

The following expression represents the imaging condition for elastic wave RTM using vector P- and S-wave velocity wavefields:

$$I_{ij}(\mathbf{x}) = \int_0^{T_{\max}} m_i^S(\mathbf{x},t) \cdot m_j^R(\mathbf{x},t)dt \tag{7}$$

where i represents the direction of the source wavefield and receiver wavefield (x, y, z); j represents the wavefield type of the source wavefield and receiver wavefield (P for P-wave, S for S-wave). $T_{\max}$ is the maximum time of the seismic records. Since the decoupled velocity–stress equations are used in the calculation of P- and S-waves and the input wavefields are either pure P-wave or pure S-wave, the utilization of the imaging condition described in Equation (7) leads to minimized interference in the obtained imaging results.

2.3.4. Imaging by Scalarizing Vector Wavefields

Both the pure P- and S-wave wavefields obtained through decoupling are vector fields. However, the results obtained through RTM imaging are not align with the expected reflection coefficients depicted in the imaging profile, lacking clear physical interpretation. To overcome this issue, a scalarization imaging condition is proposed. It converts the vector wavefields obtained from wavefield separation into scalar quantities before cross-correlation imaging. Figure 2 describes the detailed imaging process of this method.

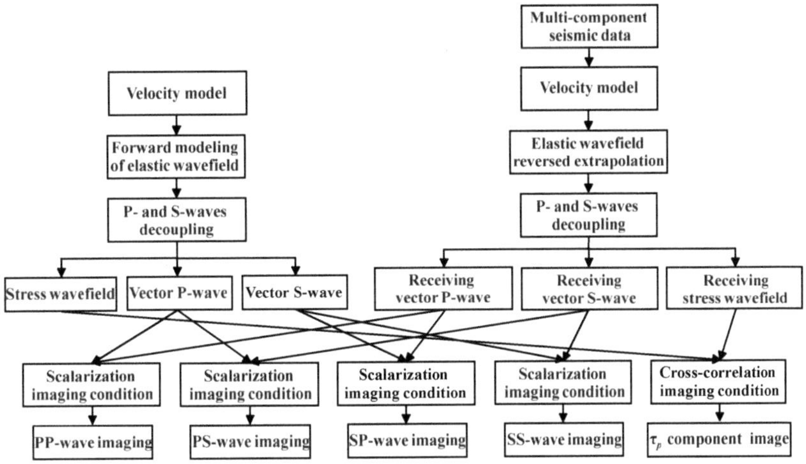

Figure 2. The flowchart illustrating the imaging process using scalarizing vector wavefields.

The following expression represents the imaging condition:

$$\begin{cases} I^{PP}(\xi) = \int_0^T \mathbf{S}^P(\xi, T-t) \cdot \mathbf{R}^P(\xi, T-t)dt \\ I^{PS}(\xi) = \int_0^T \mathbf{S}^P(\xi, T-t) \cdot \mathbf{R}^S(\xi, T-t)dt \\ I^{SP}(\xi) = \int_0^T \mathbf{S}^S(\xi, T-t) \cdot \mathbf{R}^P(\xi, T-t)dt \\ I^{SS}(\xi) = \int_0^T \mathbf{S}^S(\xi, T-t) \cdot \mathbf{R}^S(\xi, T-t)dt \end{cases} \quad (8)$$

where I^{PP}, I^{PS}, I^{SP}, and I^{SS} denote PP, PS, SP, and SS images, respectively; $\mathbf{S}^P$ and $\mathbf{S}^S$ denote P- and S-wave vector velocity fields at the source side, respectively; and $\mathbf{R}^P$ and $\mathbf{R}^S$ denote P- and S-wave vector velocity fields at the receiver side, respectively.

2.4. Denoising of the RTM Imaging Results

Although RTM is effective in imaging complex structures, low-frequency artifacts can occur when applying imaging conditions [53,54]. On this basis, the Poynting vector and Laplace filtering are proposed to attenuate RTM imaging artifacts. In elastic wavefields, the Poynting vector can be calculated by considering the stress and velocity components of the particles, as follows:

$$\begin{cases} E_x = -(\tau_{xx}v_x + \tau_{xy}v_y + \tau_{xz}v_z) \\ E_y = -(\tau_{yx}v_x + \tau_{yy}v_y + \tau_{yz}v_z) \\ E_z = -(\tau_{zx}v_x + \tau_{zy}v_y + \tau_{zz}v_z) \end{cases} \quad (9)$$

In the equations, τ_{ij} ($i, j = x, y, z$) and v_i ($i = x, y, z$) represent the stress and velocity components, respectively. E_x, E_y, and E_z represent the vectors of energy flux density in the x, y, and z directions, respectively. A positive value of E_i ($i = x, y, z$) indicates that the propagation of the wavefield in the positive direction along the corresponding axis. Conversely, a negative value of E_i ($i = x, y, z$) indicates that the propagation of the wavefield in the negative direction along the corresponding axis. In the 3D case, the Laplace operator can be expressed as follows:

$$\nabla^2 \xrightarrow{FFT} = -(k_x^2 + k_y^2 + k_z^2) = -\frac{4\omega^2 \cos^2\theta}{V^2} \quad (10)$$

where ∇^2 denotes the Laplace operator; k_x, k_y, and k_z represent the wave numbers along the coordinate axes; θ represents the incident angle; V denotes the medium's velocity; and ω is the angular frequency. Following Laplace filtering, the imaging noise is entirely eliminated within the imaging profile when the incident angle is $\theta = 90°$. When $\theta < 90°$, the imaging noise is partially suppressed to a certain extent.

3. Verifications and Discussion

In the following section, we start by discussing the vector decoupling for elastic wave separation. We then validate and discuss the imaging results under four different imaging conditions. Lastly, we verify and discuss the noise suppression methods for RTM imaging.

3.1. Discussion on Vector Decoupling for Elastic Wave Separation

A cross-well seismic model was designed to evaluate the effectiveness of the elastic wave separation algorithm in cross-well seismic exploration, as depicted in Figure 3. The effective computational domain of the model is 100.0 m × 100.0 m × 150.0 m. The physical properties of the three-layered medium, arranged from top to bottom, are as follows: the P-wave velocities are 2200.0 m/s, 2600.0 m/s, and 3300.0 m/s, respectively; the S-wave velocities are 1200.0 m/s, 1450.0 m/s, and 1850.0 m/s, respectively; the densities of the medium are 1800.0 kg/m^3, 2000.0 kg/m^3, and 2500.0 kg/m^3, respectively; and the layer thicknesses are 60.0 m, 60.0 m, and 30.0 m, respectively. A sixth-order finite difference algorithm with a spatial grid size of 1.0 m in each dimension was used for the forward modeling simulations. The seismic source employed in the study utilized a Gaussian first derivative wavelet with a dominant frequency of 120.0 Hz. The seismic source type was an explosive source located at coordinates (50.0 m, 50.0 m, 3.0 m). The horizontal position of the receiving well was (20.0 m, 0.0 m), covering the entire well range. The interline spacing between receivers was set at 2.0 m, resulting in a total of 75 receiver points. The recording length was set to 200.0 ms.

The three-component wavefield recordings are depicted in Figure 4. Figure 4a–c represent the v_x component, while Figure 4d–f represent the v_y component, and Figure 4g–i represent the v_z component. Figure 4a illustrates the undecomposed v_x component recording, Figure 4d illustrates the undecomposed v_y component recording, and Figure 4g illustrates the undecomposed v_z component recording. In these wavefields, all three components contain information about both P- and S-waves. However, Figure 4b,e,h display the wavefield recordings of the P-wave in the three components, exclusively capturing P-wave information. Similarly, Figure 4c,f,i show the wavefield recordings of the S-wave in the three components, exclusively containing shear wave information.

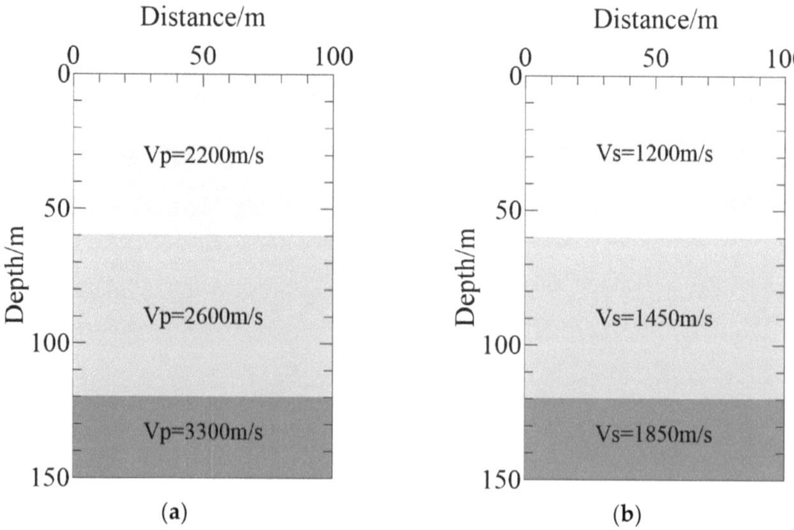

Figure 3. The layered model. (**a**) XOZ plane; (**b**) YOZ plane.

Due to the use of an explosive seismic source, the recorded direct wave includes a compressional P-wave. Records obtained within the well are more complex than surface records. They encompass transmitted compressional waves TP1 and TP2 through interfaces and reflected compressional waves RPP1 and RPP2 originating from two interfaces. Furthermore, transmitted converted shear waves TPS1 and TPS2 and reflected converted shear waves RPS1 and RPS2 are also obtained, accompanied by interbed multiple conversions. In the wavefield recordings of the P-wave (Figure 4b,e,h), it is evident that the main energy is occupied by P-waves (TP, RPP). However, in the wavefield recordings of the S-wave (Figure 4c,f,i), the energy of P-waves is greatly attenuated, and the S-waves (TPS, RPS) are the predominant component in the wavefield. The wavefield recordings are comprehensive, offering valuable information for understanding the wave propagation process. This observation strongly illustrate that the method employed in this study effectively on vector decoupling for elastic wave separation in the three-component cross-well recorded wavefield. It demonstrates the effectiveness and feasibility of the proposed method.

A single seismic trace was extracted from the inter-well Vz component, Vpz component, and Vsz component at a depth of 30.0 m for comparison. The comparison results are presented in Figure 5. It can be seen that for cross-well seismic recordings, the intermediate variable method successfully decomposes the data into the horizontal P-wave component Vpz, which exclusively contains all the P-wave information of the Vz component. The amplitudes and phase characteristics of the P-wavefields in the Vpz recordings precisely match those of the Vz component. Similarly, the decomposed S-wave component Vsz closely matches the S-wave information in the Vz component. The above results suggest that the method is effective in separating P- and S-waves from the elastic wavefield in a 3D cross-well environment. Notably, amplitude preservation is realized during decomposition, allowing for the isolation of P- and S- waves without interference. This characteristic enables the calculation of cross-well seismic source wavefield propagation directions and improves the accuracy of imaging results during subsequent RTM processes.

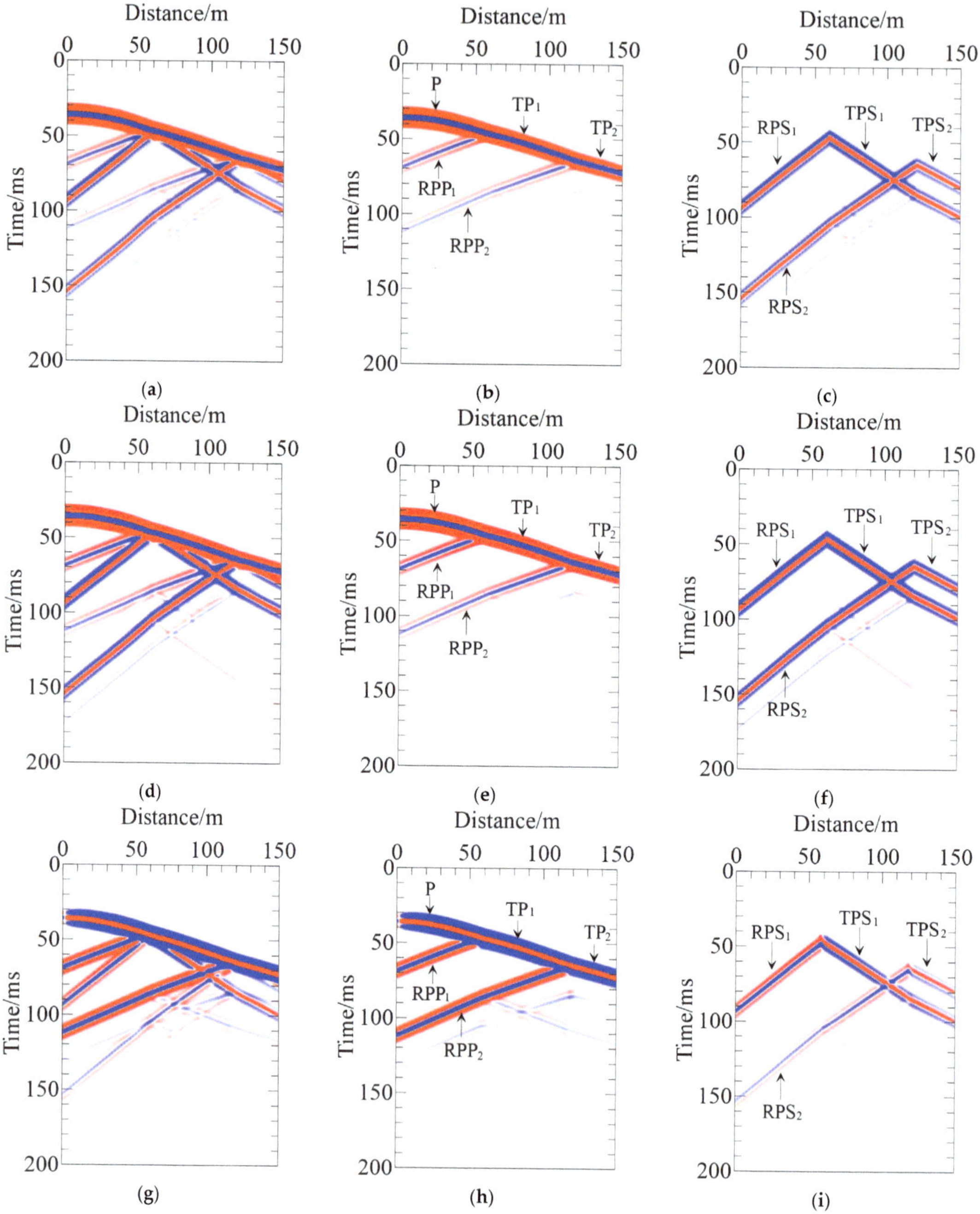

Figure 4. Cross-well three-component recordings and decomposed P- and S-wave recordings. (**a–c**) v_x component; (**d–f**) v_y component; (**g–i**) v_z component.

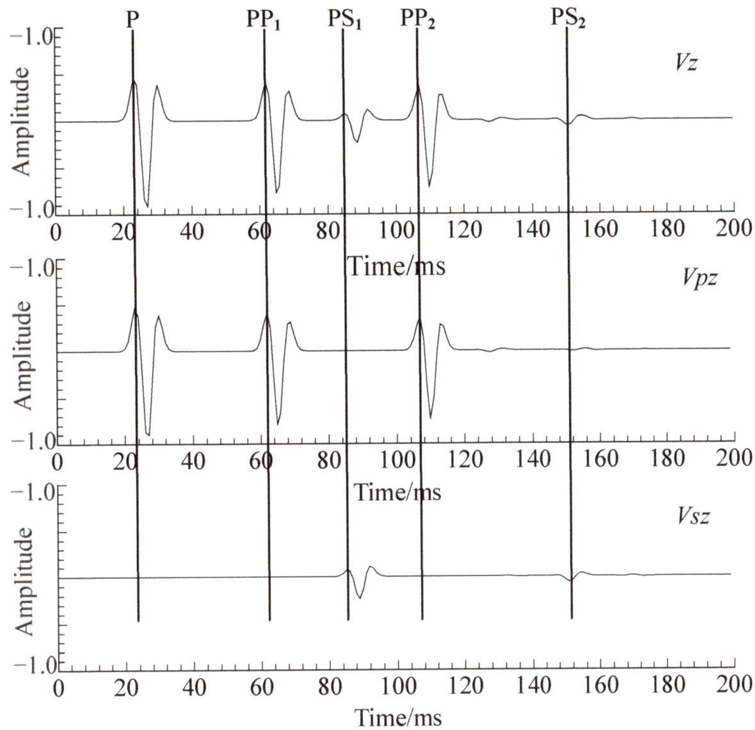

Figure 5. The seismic trace comparison between the cross-well *Vz* component and the *Vpz* and *Vsz* components.

In order to quantitively analyze the two wavefield separation methods, single-trace records extracted from the *Vz* component were normalized and compared. The comparison of the single-trace records is illustrated in Figure 6a. It can be seen that the characteristics of both methods are consistent after normalizing the amplitudes. However, the key difference is that the intermediate variable method generates an independent record of the auxiliary variables τ_p, as illustrated in Figure 6b.

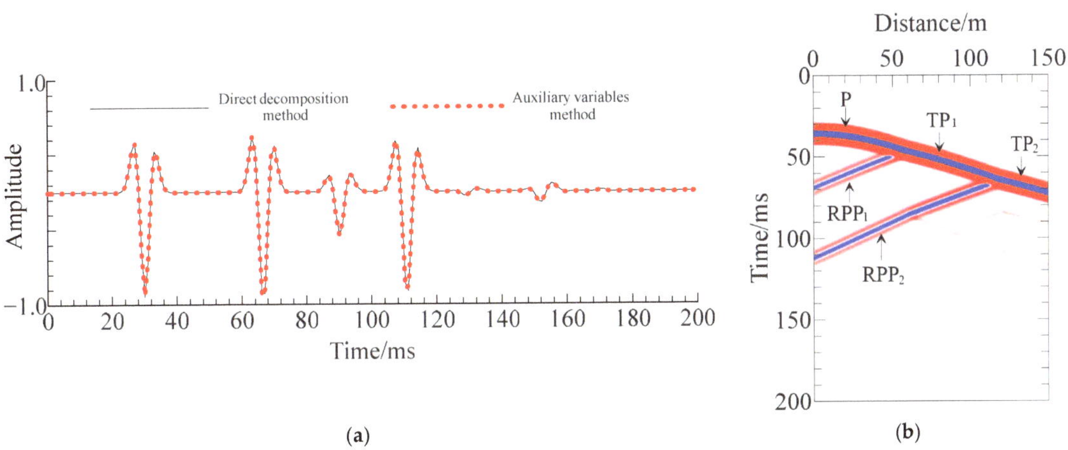

Figure 6. (**a**) The single-trace comparison plot; (**b**) the record of the τp component within the well.

3.2. Discussion on the Imaging Results under Various Imaging Conditions

The model depicted in Figure 7 was established to analyze the validity of each imaging result. It comprises a two-layer model with a computational domain of 100.0 m × 100.0 m × 80.0 m. The physical properties of the two layers from top to bottom are P-wave velocities of 2200.0 m/s and 2600.0 m/s, S-wave velocities of 1200.0 m/s and 1450.0 m/s, densities of 1800.0 kg/m^3 and 2000.0 kg/m^3, respectively. The interface depth is 60.0 m. The parameters used for the forward modeling and RTM image are consistent with those described in Section 3.1.

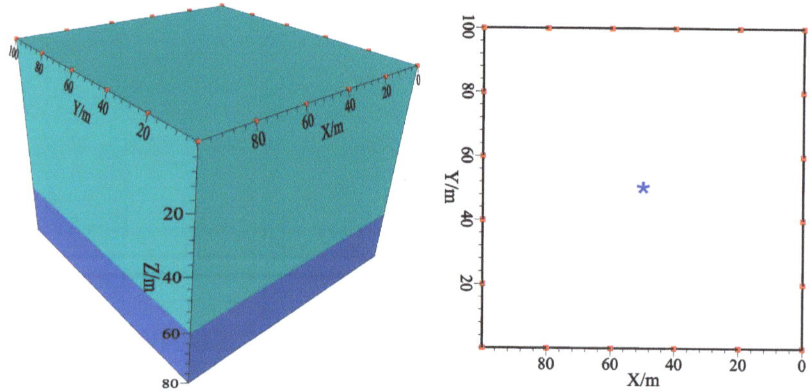

Figure 7. Schematic representation of a layered model and an observational system. The small red rectangle represents the position of the wells. The blue asterisk represents the location of the source.

3.2.1. Imaging Results Obtained Utilizing Vector Velocity Fields

Numerical simulation and imaging were conducted using the observational system described above, and the results obtained with the imaging strategy in Section 2.3.1 are shown in Figure 8. Due to the implementation of single-shot recordings, the completeness of the in-well 3D imaging results and the visual depiction of the structure may not be as good as conventional surface-based imaging. However, an interface can be observed at a depth of 60.0 m despite relatively low continuity. This limitation is due to the constraints imposed by the cross-well observation system. I_{xx} and I_{yy} exhibit a significant similarity, differing only by 90 degrees in orientation, which aligns with the expected theoretical results. These two components rely heavily on the v_x and v_y components, highlighting the imaging results of PS-waves. However, since the seismic source is centered at XOY plane and the receiving well is located at the outer boundary, the central region of the model lacks crucial wavefield information, making it unsuitable for imaging. As a result, imaging can only be performed in the boundary region. The imaging results for I_{zz} are mainly dependent on the v_z component, thus primarily representing the imaging results of PP-waves. Similarly, imaging in the central region of the model is not feasible due to the absence of wavefield information. Furthermore, high levels of noise can be observed in the imaging results. This noise originates from the direct correlation imaging of velocity components, leading to interference among different wave modes and a large number of noise artifacts. As a result, the quality of the imaging results is compromised. The lack of a clear physical interpretation of the offset imaging results, represented by the horizontal and vertical components, hinders subsequent tasks such as interpretation and inversion.

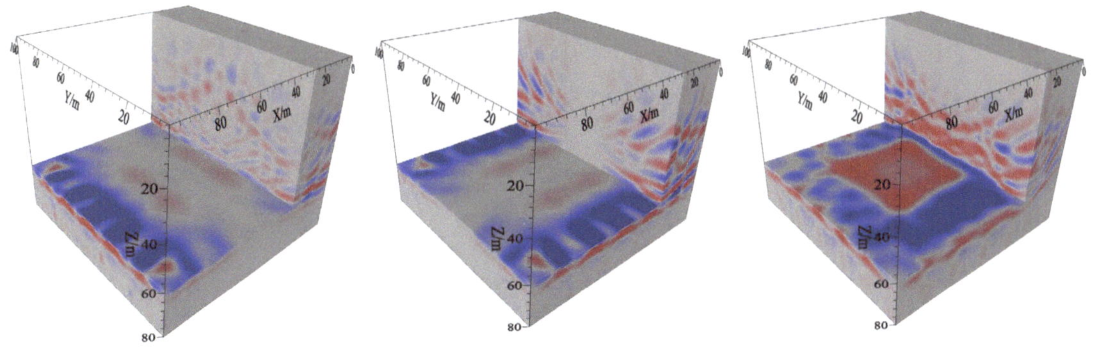

(**a**) (**b**) (**c**)

Figure 8. Imaging results obtained using vector velocity fields for all three components. (**a**) I_{xx} component imaging results; (**b**) I_{yy} component imaging results; (**c**) I_{zz} component imaging results.

3.2.2. Imaging Results Obtained Utilizing Scalar and Vector Potentials

Figure 9 illustrates the imaging results obtained using the imaging strategy described in Section 2.3.2. It can be seen that due to the implementation of single-shot recordings and the constraints of the cross-well observation system, the 3D imaging results within the well have relatively poor interface continuity. The imaging results of the PP-waves are close to the source well area, while those of the PS-waves are close to the receiver well area, consistent with the actual situation. Due to the interferences among different mode waves, the imaging results of PS-waves exhibit significantly higher noise levels compared to those of PP-waves. In 3D space, both the imaging interfaces of PP- and PS-waves exhibit polarity reversals. This phenomenon poses challenges for the stacking of multiple shots.

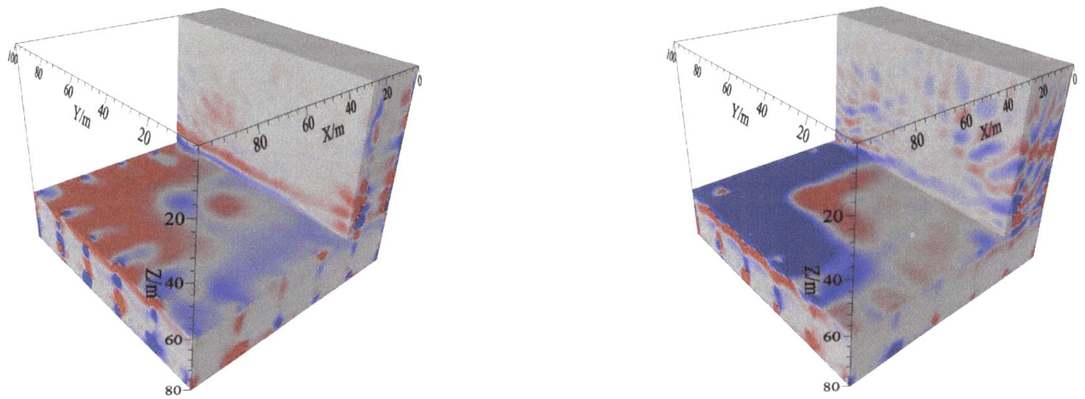

(**a**) (**b**)

Figure 9. Imaging results obtained utilizing scalar and vector potentials. (**a**) PP-wave imaging results; (**b**) PS-wave imaging results.

3.2.3. Imaging Results Obtained Using Pure Vector P- and S-Wave

The P-wave components (I_{xP}, I_{yP}, and I_{zP}) were imaged in various directions and modes using the imaging strategy outlined in Section 2.3.3. The results are presented in Figure 10a–c. Additionally, the imaging results for the S-wave components (I_{xS}, I_{yS}, and I_{zS}) are depicted in Figure 10d–e. It can be observed that the imaging results of the six components in Figure 10 are relatively similar to those in Figure 8. The imaging results of the two horizontal components also exhibit a 90° rotation, and the central region of the model still lacks wavefield information. It can be clearly observed that employing the separated wave equations and utilizing either pure compressional or shear waves as boundary conditions results in images with significantly reduced noise levels compared to Figure 8. The separated wave equations and the input of single-mode waves effectively reduce interferences in imaging, significantly increasing the signal-to-noise ratio of the imaging results. However, different directions and modes produce more images, and the physical significance of the offset imaging results expressed by the two horizontal components and the vertical component is unclear, which is inconducive to the superimposition of the imaging results.

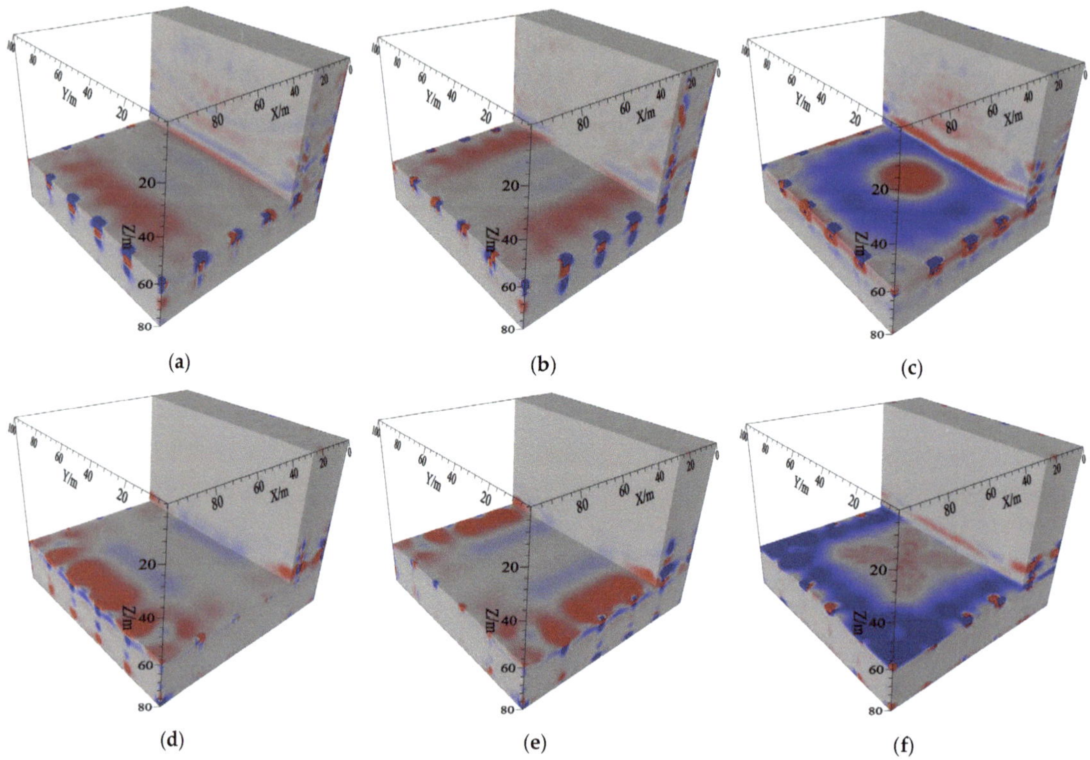

Figure 10. Imaging results obtained using pure vector P- and S-waves. (**a**) I_{xP} component imaging results; (**b**) I_{yP} component imaging results; (**c**) I_{zP} component imaging results; (**d**) I_{xS} component imaging results; (**e**) I_{yS} component imaging results; (**f**) I_{zS} component imaging results.

3.2.4. Imaging Results Obtained Using Scalarizing Vector Wavefields

In order to achieve easier stacking of multiple shots and clearer physical interpretation of the imaging results, the imaging strategy described in Section 2.3.4 was implemented. This strategy involved converting the vectorial P-waves and vectorial S-waves into scalar quantities. The scalarized imaging results for PP-waves and PS-waves are illustrated in Figure 11. It is evident from Figure 11a,b that the scalarized imaging results of PP-waves and PS-waves are relatively continuous and have no directionality. This characteristic facilitates the stacking of multiple shots after imaging. In addition, the τ_p component can be extracted from the wavefield using the auxiliary variables method, which is also applicable for imaging purposes. The imaging results obtained using this component are illustrated in Figure 11c. The imaging results for the τ_p component are similar to the imaging results of PP-waves, with an improved signal-to-noise ratio and continuity. These results can be effectively employed in subsequent applications.

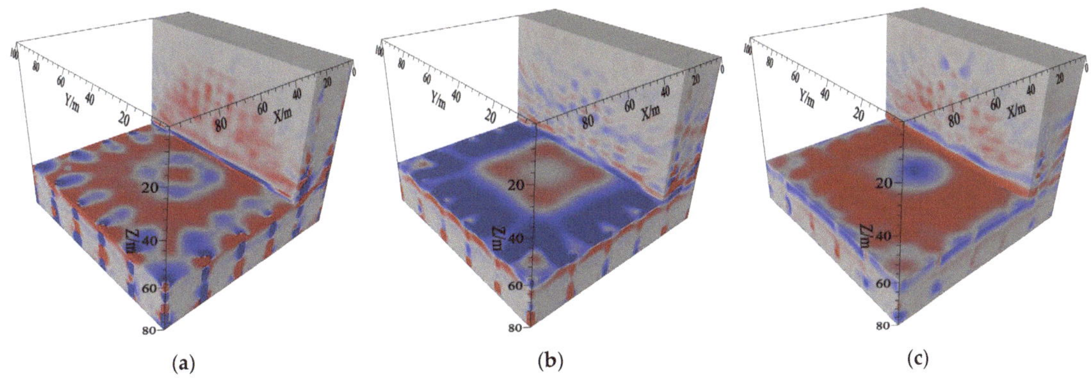

Figure 11. Imaging results obtained utilizing pure vector P- and S-waves. (**a**) PP-wave imaging results; (**b**) PS-wave imaging results; (**c**) τp component imaging results.

According to the four different imaging results mentioned above, firstly, the imaging results by vector velocity fields exhibit significant noise, resulting in low-quality images. Furthermore, the imaging results lack clear physical meaning, which hinders subsequent interpretation and inversion tasks. Secondly, the imaging by scalar and vector potentials show relatively poor interface continuity, with polarity reversals observed on both the imaging interfaces of PP waves and PS waves. Moreover, the imaging results by pure vector P- and S-waves exhibit a high-quality representation with minimal noise interference. However, the physical significance of the imaging results is also unclear. Finally, the imaging results by scalarizing vector wavefields exhibit a high degree of continuity and lack directional bias. This characteristic makes them more suitable for superposition, enabling easier overlaying of multiple images and facilitating a clearer physical interpretation of the imaging results.

3.3. RTM Imaging Artifacts Attenuation

Figure 12 compares the pre- and post-denoised imaging profiles of single-shot PP-waves. It can be seen that without noise suppression, the imaging profiles exhibit a large amount of low-frequency noise with high energy. This noise is mainly distributed above the interfaces, significantly interfering with the imaging quality of the interfaces. By employing the Poynting vector to suppress the reflected waves and applying Laplace filtering, the low-frequency noise is effectively eliminated. Consequently, the imaging effect is improved, and the actual position and tilt angle of the interface are more accurately reflected.

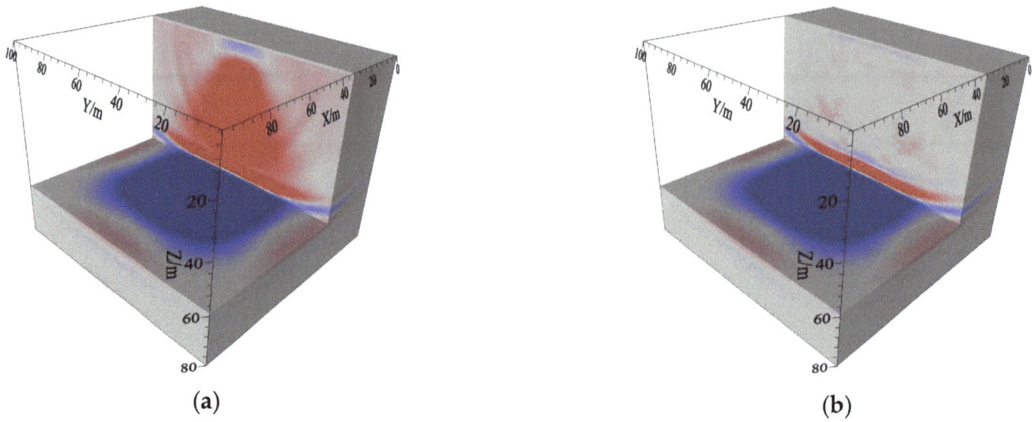

Figure 12. Pre- and post-denoised imaging profiles of single-shot PP-waves: (**a**) image results before denoising; (**b**) image results after denoising.

4. Multiple Sensors Cross-Well 3D RTM Image Results and Analysis

In this section, we first discuss the sensor settings and design of the observation system for 3D cross-well seismic exploration. Subsequently, using the observation system, we conduct 3D imaging of layered media model and high-velocity ellipsoidal boulder models. Meanwhile, a comprehensive and detailed analysis of the imaging results is provided.

4.1. Sensor Settings and Observation System

The cross-well seismic observation system distinguishes itself from the surface-based 3D seismic methods by enabling the large-scale and high-density deployment of seismic acquisition lines. However, the cross-well seismic observation system is influenced by the actual well layout. Based on practical engineering considerations, this paper discusses the layout of the 3D multi-sensors cross-well observation system in commonly encountered square exploration areas, as depicted in Figure 13. Within a 50.0 m $\times$ 50.0 m square plane, a total of 20 well locations were arranged for the analysis. The specific coordinates of each well in the XOY plane can be found in Table 1. The data acquisition was performed in the following steps: Initially, the source was subjected to an explosion in Well-1, while the other 19 were used to receive seismic signals. Then, the source was sequentially moved to the next well, and the remaining 19 wells continued to receive the seismic signals. This process lasted until Well-20 was reached.

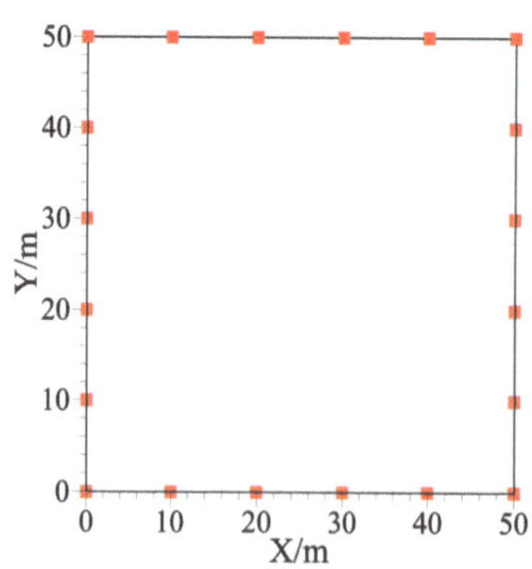

Figure 13. Three-dimensional multi-cross-well observation system diagram. The small red rectangle represents the position of the wells.

Table 1. The well locations within the square area in the XOY plane.

No.	Well-1	Well-2	Well-3	Well-4	Well-5	Well-6	Well-7	Well-8	Well-9	Well-10
X	0.0	10.0	20.0	30.0	40.0	50.0	0.0	10.0	20.0	30.0
Y	0.0	0.0	0.0	0.0	0.0	0.0	50.0	50.0	50.0	50.0
No.	**Well-11**	**Well-12**	**Well-13**	**Well-14**	**Well-15**	**Well-16**	**Well-17**	**Well-18**	**Well-19**	**Well-20**
X	40.0	50.0	0.0	0.0	0.0	0.0	50.0	50.0	50.0	50.0
Y	50.0	50.0	10.0	20.0	30.0	40.0	10.0	20.0	30.0	40.0

4.2. Layered Medium Model

Figure 14a illustrates a 3D layered medium model. The physical parameters of the two layers from top to bottom are as follows: P-wave velocities of 2200.0 m/s and 2600.0 m/s, S-wave velocities of 1200.0 m/s and 1450.0 m/s, densities of 1800.0 kg/m^3 and 2000.0 kg/m^3, and an interface depth of 60.0 m. After removing the direct waves from the recordings, the decoupled 3D elastic wave equation RTM can be used to image the layered medium in 3D across walls. The corresponding imaging results are presented in Figure 14b–d. Among them, Figure 14b depicts the imaging results for PP-waves, Figure 14c illustrates the imaging results for PS-waves, and Figure 14d shows the imaging results for τ_p. The wavefield energy is precisely concentrated at the actual interface, which is very similar to the real model. This observation result indicates that using cross-well elastic wave data and sensors can ensure accurate imaging of the 3D cross-well layered medium, encompassing both P- and S-waves. The accurate imaging results demonstrate the feasibility of the proposed method. Notably, the source locations are all positioned above the interface, and the imaging results successfully reveal the structural characteristics of the underlying interface. Compared to the traditional first-arrival travel-time imaging method, the RTM imaging method based on reflected waves has a wider exploration range. It eliminates the depth limitation of traditional observing systems for exploration.

4.3. High-Velocity Ellipsoid Boulder Model

Figure 15 displays the high-velocity ellipsoid boulder model. The physical parameters of the two layers from top to bottom are as follows: P-wave velocities of 2200.0 m/s and 2600.0 m/s, S-wave velocities of 1200.0 m/s and 1450.0 m/s, densities of 1800.0 kg/m^3 and 2000.0 kg/m^3, and an interface depth of 60.0 m. The high-speed ellipsoidal boulder has three axis radii: a = b = 5.0 m, c = 10.0 m. The ellipsoidal body is centered at (25.0 m, 25.0 m, 30.0 m). The physical properties of the ellipsoidal body match those of the second-layer medium. After eliminating the direct waves from the recorded data, the cross-well 3D elastic wave imaging results for the high-velocity ellipsoid boulder can be obtained using the decoupled 3D elastic wave RTM equation. These results are illustrated in Figure 15b–d. Specifically, Figure 15b depicts the imaging result of the PP-wave, Figure 15c illustrates the imaging result of the PS-wave, and Figure 15d shows the imaging result of τ_p. It can be seen that the wavefield energy is concentrated at the interfaces and the position of the ellipsoid boulder in the center. The locations of the interfaces align with those of the coherent events, demonstrating a high consistency with the actual model. Based on this result, it can be concluded that the 3D cross-well elastic wave RTM imaging method with multiple wells and sensors can capture the P- and S-wave structures of localized heterogeneous bodies and perform 3D imaging. In addition to horizontal interfaces, the geological structures of lateral media are effectively revealed, and variations in lateral structures are accurately characterized. Moreover, the limitations of traditional 2D imaging methods in imaging lateral media are successfully addressed. In conclusion, the proposed method provides a fundamental basis for extracting dynamic parameters and facilitates the acquisition of comprehensive geological and geophysical information about 3D geological structures. Therefore, this method can be used as a theoretical reference and a solid foundation for detecting the geological formations of high-velocity ellipsoid boulders, especially in the field of engineering exploration.

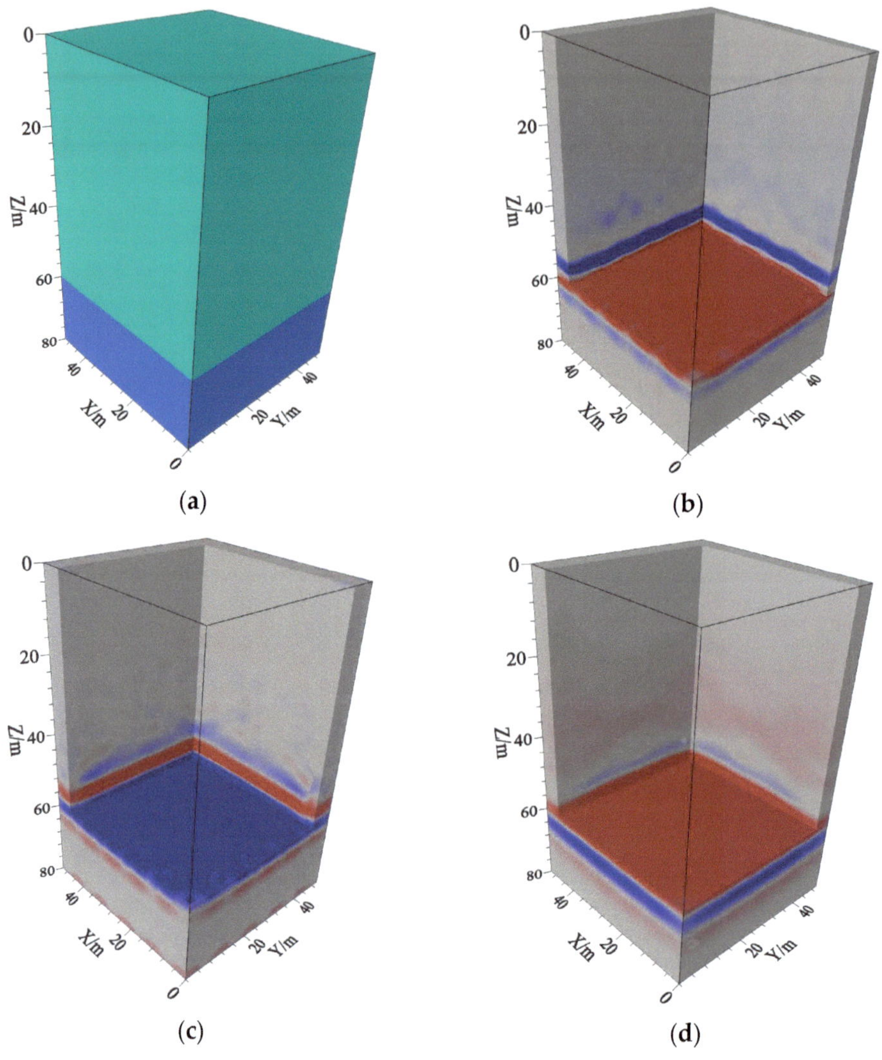

Figure 14. Three-dimensional cross-well layered medium model and the elastic waves RTM imaging results. (**a**) Layered medium model; (**b**) PP-waves imaging results; (**c**) PS-waves imaging results; (**d**) τp component imaging results.

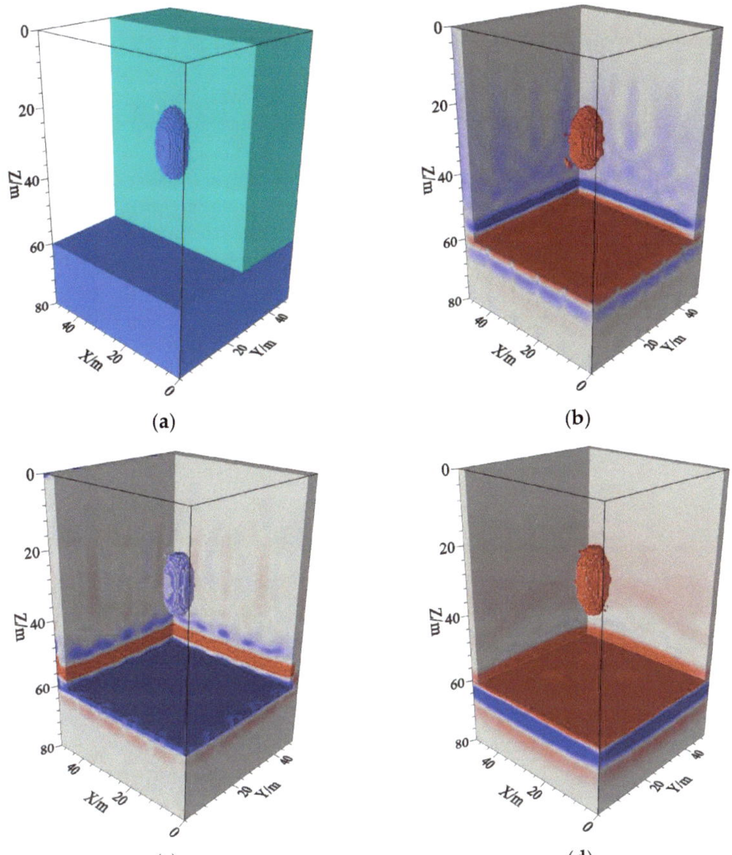

Figure 15. Three-dimensional cross-well high-velocity ellipsoid boulder model and the elastic waves RTM imaging results. (**a**) High-velocity ellipsoid boulder model; (**b**) PP-waves imaging results; (**c**) PS-waves imaging results; (**d**) τp component imaging results.

5. Conclusions

In coastal engineering, a large amount of unsaturated loose sediments (plentiful sand) in the coastal zone leads to a strong attenuation of seismic waves. It poses significant challenges for ground seismic methods to transmit sufficient energy through the weathered zone to reach the desired depth. Compared to surface seismic reflection or refraction surveys, cross-well seismic acquisition methods emerge as an optimal choice for effectively targeting specific coastal areas. However, current studies on cross-well RTM imaging are limited to 2D space or 3D P-wave RTMs. To address the limitations of current studies, this study proposed the 3D cross-well elastic RTM imaging based on a multi-wave and multi-component technique in coastal engineering exploration. The practical benefit of this method is the utilization of the Earth's elastic medium without increasing costs, which is used to obtain information about the distribution of subsurface rock layers, interfaces, and other structures in coastal engineering projects. Importantly, this can be achieved without resorting to extensive excavation or drilling operations.

Firstly, based on the vector decoupled elastic wave equation, this study compares and analyzes the amplitude-preserving separation algorithms for P- and S-waves implemented using the direct decomposition method and the auxiliary variable method. The wavefield decoupling algorithm achieves accurate separation of vector P- and S-waves without introducing any distortions in amplitude and phase. This feature facilitates accurate imaging of P- and S-waves in subsequent stages. Moreover, the algorithm incorporating intermediate variables integrates the P-wave pressure equation into the traditional elastic wave equations, offering significant advantages for independent imaging in the later stages. Then, this study focuses on analyzing four imaging methods under cross-correlation imaging conditions in RTM. The imaging performances of these four types of imaging strategies are compared and analyzed using experimental models. The results indicate that well-defined P- and S-wave imaging profiles can be obtained using scalarizing vector wavefield imaging conditions, and the imaging results have unique physical significance. Finally, the layer model and high-velocity ellipsoid boulder model were subjected to an RTM imaging experiment. The results indicate that the proposed 3D elastic wave imaging method can effectively generate accurate 3D cross-well profiles of P- and S-waves, thus accurately describing the geological structure. The implementation of multi-wave and multi-component RTM imaging significantly improves the utilization of wavefield information during the imaging process, thereby providing novel insights for cross-well seismic exploration. Moreover, abundant geological and geophysical information can be obtained for analyzing and interpreting 3D geological structures in coastal areas. The findings of this study have crucial theoretical significance and practical implications for exploration and development in real-world production, as well as for the investigation of geological structures in coastal engineering projects.

Author Contributions: Conceptualization, D.P. and H.X.; Methodology, D.P., F.C. and H.X.; Software, D.P., F.C. and Y.Z.; Validation, D.P., H.X. and Y.Z.; Formal analysis, D.P. and F.C.; Investigation, F.C. and Y.Z.; Resources, D.P. and H.X.; Data curation, D.P., F.C. and H.X.; Writing—original draft, D.P.; Writing—review & editing, F.C., H.X. and Y.Z.; Visualization, D.P.; Supervision, H.X.; Project administration, D.P. and H.X.; Funding acquisition, D.P. and F.C. All authors have read and agreed to the published version of the manuscript.

Funding: This research was funded by the Open Funds for Hubei Key Laboratory of Marine Geological Resources, China University of Geosciences (No. MGR202201); the Open Fund of the Key Laboratory of Exploration Technologies for Oil and Gas Resources (Yangtze University), Ministry of Education, No. K2023-01; and the National Natural Science Foundation of China (No. 41704146).

Institutional Review Board Statement: Not applicable.

Informed Consent Statement: Not applicable.

Data Availability Statement: The data presented in this study are available on request from the corresponding author.

Conflicts of Interest: Yuquan Zong was employed by Central Southern China Electric Power Design Institute Co., Ltd. of China Power Engineering Consulting Group. The remaining authors declare that the research was conducted in the absence of any commercial or financial relationships that could be construed as a potential conflict of interest. The funders had no role in the design of the study; in the collection, analyses, or interpretation of data; in the writing of the manuscript; or in the decision to publish the results.

Glossary

Seismic waves	Seismic waves refer to the propagating vibrations generated during an earthquake. Seismic waves can be classified into two main types: body waves and surface waves. Body waves are the waves that propagate through the Earth's interior and include compressional waves (P-waves) and shear waves (S-waves).
P-waves	Compressional waves, also known as P-waves or primary waves, are longitudinal waves that cause particles in the material to move in the same direction as the wave propagation.
S-waves	Shear waves, also known as S-waves or secondary waves, are transverse waves that cause particles to move perpendicular to the direction of wave propagation.
Reverse time migration (RTM)	RTM is a seismic imaging technique used to generate high-resolution subsurface images in geophysics and exploration geology.
Cross-well seismic exploration	Cross-well seismic exploration is a geophysical technique used to obtain detailed subsurface information between two or more wells.
Imaging conditions	In seismic imaging, imaging conditions refer to the mathematical relationships and criteria used to convert seismic data into subsurface images.
Elastic wave decomposition	Elastic wave decomposition is a technique used in seismic data processing to separate the recorded seismic data into its constituent wave modes. It aims to isolate and analyze the individual components of the seismic wavefield, such as compressional (P) waves and shear (S) waves.
Amplitude variation with angle (AVA)	AVA is a phenomenon observed in seismic data where the amplitude of reflected seismic waves changes as a function of the angle of incidence and reflection at interfaces within the subsurface. It is an important attribute used in seismic analysis to infer properties of subsurface rock formations and fluid content.
Sensor settings	In seismic exploration Sensor settings refer to the settings of the geophone. A geophone is a type of sensor used in seismic exploration and monitoring to detect and measure ground vibrations caused by seismic waves. It is a critical component in seismic data acquisition systems and plays a fundamental role in studying the Earth's subsurface.

References

1. Esteban, M.D.; López-Gutiérrez, J.-S.; Negro, V.; Neves, M.G. Coastal Engineering: Sustainability and New Technologies. *J. Mar. Sci. Eng.* **2023**, *11*, 1562. [CrossRef]
2. Sun, Q.; Wang, Q.; Shi, F.; Alves, T.; Gao, S.; Xie, X.; Wu, S.; Li, J. Runup of landslide-generated tsunamis controlled by paleogeography and sea-level change. *Commun. Earth Environ.* **2022**, *3*, 244. [CrossRef]
3. Dong, Y. Reseeding of particles in the material point method for soil–structure interactions. *Comput. Geotech.* **2020**, *127*, 103716. [CrossRef]
4. Wang, H.; Lin, J.; Dong, X.; Lu, S.; Li, Y.; Yang, B. Seismic velocity inversion transformer. *Geophysics* **2023**, *88*, R513–R533. [CrossRef]
5. Nakata, R.; Jang, U.; Lumley, D.; Mouri, T.; Nakatsukasa, M.; Takanashi, M.; Kato, A. Seismic Time-Lapse Monitoring of Near-Surface Microbubble Water Injection by Full Waveform Inversion. *Geophys. Res. Lett.* **2022**, *49*, e2022GL098734. [CrossRef]
6. Wang, W.; Ma, J. Velocity model building in a crosswell acquisition geometry with image-trained artificial neural networks. *Geophysics* **2020**, *85*, U31–U46. [CrossRef]
7. Habib, M.A.; Zarillo, G.A. Construction of a Real-Time Forecast Model for Coastal Engineering and Processes Nested in a Basin Scale Model. *J. Mar. Sci. Eng.* **2023**, *11*, 1263. [CrossRef]
8. Liu, J.P.; Wang, Y.Y.; Liu, Z.; Pan, X.K.; Zong, Y.Q. Progress and application of near-surface reflection and refraction method. *Chin. J. Geophys.* **2015**, *58*, 3286–3305.

9. Lu, A.; Asli, B.H.S. Seismic Image Identification and Detection Based on Tchebichef Moment Invariant. *Electronics* **2023**, *12*, 3692. [CrossRef]
10. Huang, G.; Chen, X.; Luo, C.; Bai, M.; Chen, Y. Time-lapse seismic difference-and-joint prestack AVA inversion. *IEEE Trans. Geosci. Remote Sens.* **2020**, *59*, 9132–9143. [CrossRef]
11. Li, G.; Cai, H.; Li, C.F. Alternating joint inversion of controlled-source electromagnetic and seismic data using the joint total variation constraint. *IEEE Trans. Geosci. Remote Sens.* **2019**, *57*, 5914–5922. [CrossRef]
12. da Silva, J.A., Jr.; Poliannikov, O.V.; Fehler, M.; Turpening, R. Modeling scattering and intrinsic attenuation of crosswell seismic data in the Michigan Basin. *Geophysics* **2018**, *83*, WC15–WC27. [CrossRef]
13. Zhang, X.; Zheng, Z.; Wang, L.; Cui, H.; Xie, X.; Wu, H.; Liu, X.; Gao, B.; Wang, H.; Xiang, P. A Quasi-Distributed optic fiber sensing approach for interlayer performance analysis of ballastless Track-Type II plate. *Opt. Laser Technol.* **2024**, *170*, 110237. [CrossRef]
14. Angioni, T.; Rechtien, R.D.; Cardimona, S.J.; Luna, R. Crosshole seismic tomography and borehole logging for engineering site characterization in Sikeston, MO, USA. *Tectonophysics* **2003**, *368*, 119–137. [CrossRef]
15. Chen, H.; Hu, Y.; Jin, J.; Ran, Q.; Yan, L. Fine stratigraphic division of volcanic reservoir by uniting of well data and seismic data—Taking volcanic reservoir of member one of Yingcheng Formation in Xudong area of Songliao Basin for an example. *J. Earth Sci.* **2014**, *25*, 337–347. [CrossRef]
16. Sagong, M.; Park, C.S.; Lee, B.; Chun, B.-S. Cross-hole seismic technique for assessing in situ rock mass conditions around a tunnel. *Int. J. Rock Mech. Min. Sci.* **2012**, *53*, 86–93. [CrossRef]
17. Marelli, S.; Manukyan, E.; Maurer, H.; Greenhalgh, S.A.; Green, A.G. Appraisal of waveform repeatability for crosshole and hole-to-tunnel seismic monitoring of radioactive waste repositories. *Geophysics* **2010**, *75*, Q21–Q34. [CrossRef]
18. Lin, J.; Ma, R.; Sun, Z.; Tang, L. Assessing the connectivity of a regional fractured aquifer based on a hydraulic conductivity field reversed by multi-well pumping tests and numerical groundwater flow modeling. *J. Earth Sci.* **2023**, *34*, 1926–1939. [CrossRef]
19. Pongrac, B.; Gleich, D.; Malajner, M.; Sarjaš, A. Cross-Hole GPR for Soil Moisture Estimation Using Deep Learning. *Remote Sens.* **2023**, *15*, 2397. [CrossRef]
20. Binley, A.; Hubbard, S.S.; Huisman, J.A.; Revil, A.; Robinson, D.A.; Singha, K.; Slater, L.D. The emergence of hydrogeophysics for improved understanding of subsurface processes over multiple scales. *Water Resour. Res.* **2015**, *51*, 3837–3866. [CrossRef]
21. Liu, Q.; Hu, L.; Bayer, P.; Xing, Y.; Qiu, P.; Ptak, T.; Hu, R. A numerical study of slug tests in a three-dimensional heterogeneous porous aquifer considering well inertial effects. *Water Resour. Res.* **2020**, *56*, e2020WR027155. [CrossRef]
22. Zhai, J.; Wang, Q.; Xie, X.; Qin, H.; Zhu, T.; Jiang, Y.; Ding, H. A New Method for 3D Detection of Defects in Diaphragm Walls during Deep Excavations Using Cross-Hole Sonic Logging and Ground-Penetrating Radar. *J. Perform. Constr. Facil.* **2023**, *37*, 04022065. [CrossRef]
23. Luo, N.; Zhao, Z.; Illman, W.A.; Zha, Y.; Mok, C.M.W.; Yeh, T.-C.J. Three-Dimensional Steady-State Hydraulic Tomography Analysis with Integration of Cross-Hole Flowmeter Data at a Highly Heterogeneous Site. *Water Resour. Res.* **2023**, *59*, e2022WR034034. [CrossRef]
24. White, D.; Daley, T.M.; Paulsson, B.; Harbert, W. Borehole seismic methods for geologic CO_2 storage monitoring. *Lead. Edge* **2021**, *40*, 434–441. [CrossRef]
25. Liu, X.; Liu, S.; Tian, S.; Zhao, Q.; Lu, Q.; Wang, K. Slowness High-resolution Tomography of Cross-hole Radar Based on Deep Learning. *IEEE Geosci. Remote Sens. Lett.* **2024**, *21*, 3000505. [CrossRef]
26. Becht, A.; Bürger, C.; Kostic, B.; Appel, E.; Dietrich, P. High-resolution aquifer characterization using seismic cross-hole tomography: An evaluation experiment in a gravel delta. *J. Hydrol.* **2007**, *336*, 171–185. [CrossRef]
27. Cheng, F.; Peng, D.; Yang, S. 3D Reverse-Time Migration Imaging for Multiple Cross-Hole Research and Multiple Sensor Settings of Cross-Hole Seismic Exploration. *Sensors* **2024**, *24*, 815. [CrossRef]
28. Cheng, F.; Liu, J.; Wang, J.; Zong, Y.; Yu, M. Multi-hole seismic modeling in 3-D space and cross-hole seismic tomography analysis for boulder detection. *J. Appl. Geophys.* **2016**, *134*, 246–252. [CrossRef]
29. Wang, H.; Lin, C.P.; Liu, H.C. Pitfalls and refinement of 2D cross-hole electrical resistivity tomography. *J. Appl. Geophys.* **2020**, *181*, 104143. [CrossRef]
30. Bae, H.S.; Kim, W.-K.; Son, S.-U.; Park, J.-S. Imaging of Artificial Bubble Distribution Using a Multi-Sonar Array System. *J. Mar. Sci. Eng.* **2022**, *10*, 1822. [CrossRef]
31. Zhong, Y.; Gu, H.; Liu, Y.; Luo, X.; Mao, Q. Elastic reverse time migration method in vertical transversely isotropic media including surface topography. *Geophys. Prospect.* **2022**, *70*, 1528–1555. [CrossRef]
32. Shin, Y.; Ji, H.-G.; Park, S.-E.; Oh, J.-W. 4D Seismic Monitoring with Diffraction-Angle-Filtering for Carbon Capture and Storage (CCS). *J. Mar. Sci. Eng.* **2023**, *11*, 57. [CrossRef]
33. Wang, Y.; Hu, X.; Harris, J.M.; Zhou, H. Crosswell seismic imaging using Q-compensated viscoelastic reverse time migration with explicit stabilization. *IEEE Trans. Geosci. Remote Sens.* **2022**, *60*, 1–11. [CrossRef]
34. Cao, L.; Li, X.; Cao, H.; Liu, L.; Wei, T.; Yang, X. 3-D Crosswell electromagnetic inversion based on IRLS norm sparse optimization algorithms. *J. Appl. Geophys.* **2023**, *214*, 105072. [CrossRef]
35. Li, X.; Lei, X.; Li, Q.; Chen, D. Influence of bedding structure on stress-induced elastic wave anisotropy in tight sandstones. *J. Rock Mech. Geotech. Eng.* **2021**, *13*, 98–113. [CrossRef]
36. Sun, R.; McMechan, G.A.; Hsiao, H.; Chow, J. Separating P- and S-waves in prestack 3D elastic seismograms using divergence and curl. *Geophysics* **2004**, *69*, 286–297. [CrossRef]

37. Sun, R.; McMechan, G.A.; Lee, C.-S.; Chow, J.; Chen, C.-H. Prestack scalar reverse-time depth migration of 3D elastic seismic data. *Geophysics* **2006**, *71*, S199–S207. [CrossRef]
38. Zong, Z.; Ji, L. Model parameterization and amplitude variation with angle and azimuthal inversion in orthotropic media. *Geophysics* **2021**, *86*, R1–R14. [CrossRef]
39. Asli, B.H.S.; Flusser, J.; Zhao, Y. 2-D Generating Function of the Zernike Polynomials and their Application for Image Classification. In Proceedings of the 2019 Ninth International Conference on Image Processing Theory, Tools and Applications (IPTA), Istanbul, Turkey, 6–9 November 2019; pp. 1–6.
40. Alterman, Z.; Karal, F.C. Propagation of elastic waves in layered media by finite difference methods. *Bull. Seismol. Soc. Am.* **1968**, *58*, 367–398.
41. Furumura, T.; Takenaka, H. A wraparound elimination technique for the pseudospectral wave synthesis using an antiperiodic extension of the wavefield. *Geophysics* **1995**, *60*, 302–307. [CrossRef]
42. Tan, C.J.; Aslian, A.; Honarvar, B.; Puborlaksono, J.; Yau, Y.H.; Chong, W.T. Estimating surface hardening profile of blank for obtaining high drawing ratio in deep drawing process using FE analysis. *IOP Conf. Ser. Mater. Sci. Eng.* **2015**, *103*, 012047. [CrossRef]
43. Seriani, G.; Priolo, E. Spectral element method for acoustic wave simulation in heterogeneous media. *Finite Elem. Anal. Des.* **2012**, *16*, 337–348. [CrossRef]
44. Bouchon, M.; Campillo, M.; Gaffet, S. A boundary integral equation-discrete wavenumber representation method to study wave propagation in multilayered media having irregular interfaces. *Geophysics* **1989**, *54*, 1134–1140. [CrossRef]
45. Pan, X.; Lu, C.; Zhang, G.; Wang, P.; Liu, J. Seismic characterization of naturally fractured reservoirs with monoclinic symmetry induced by horizontal and tilted fractures from amplitude variation with offset and azimuth. *Surv. Geophys.* **2022**, *43*, 815–851. [CrossRef]
46. Chen, J.B.; Cao, J. An average-derivative optimal scheme for modeling of the frequency-domain 3D elastic wave equation. *Geophysics* **2018**, *83*, T209–T234. [CrossRef]
47. Yang, J.; He, X.; Chen, H. Crosswell frequency-domain reverse time migration imaging with wavefield decomposition. *J. Geophys. Eng.* **2023**, *20*, 1279–1290. [CrossRef]
48. Gu, B.; Li, Z.; Ma, X.; Liang, G. Multi-component elastic reverse time migration based on the P-and S-wave separated velocity-stress equations. *J. Appl. Geophys.* **2015**, *112*, 62–78. [CrossRef]
49. Xiao, X.; Leaney, W.S. Local vertical seismic profiling (VSP) elastic reverse-time migration and migration resolution: Salt-flank imaging with transmitted P-to-S waves. *Geophysics* **2010**, *75*, S35–S49. [CrossRef]
50. Du, Q.; Guo, C.; Zhao, Q.; Gong, X.; Wang, C.; Li, X.-Y. Vector-based elastic reverse time migration based on scalar imaging condition. *Geophysics* **2017**, *82*, S111–S127. [CrossRef]
51. Wang, W.; McMechan, G.A.; Zhang, Q. Comparison of two algorithms for isotropic elastic P and S vector decomposition. *Geophysics* **2015**, *80*, T147–T160. [CrossRef]
52. Peng, D.; Cheng, F.; She, X.; Zheng, Y.; Tang, Y.; Fan, Z. Three-Dimensional Ultrasonic Reverse-Time Migration Imaging of Submarine Pipeline Nondestructive Testing in Cylindrical Coordinates. *J. Mar. Sci. Eng.* **2023**, *11*, 1459. [CrossRef]
53. Claerbout, J.F. Toward a unified theory of reflector mapping. *Geophysics* **1971**, *36*, 467–481. [CrossRef]
54. Moradpouri, F.; Moradzadeh, A.; Pestana, R.; Ghaedrahmati, R.; Monfared, M.S. An improvement in wavefield extrapolation and imaging condition to suppress reverse time migration artifacts. *Geophysics* **2017**, *82*, S403–S409. [CrossRef]

Journal of
Marine Science and Engineering

MDPI

Article

Pile Driving and the Setup Effect and Underlying Mechanism for Different Pile Types in Calcareous Sand Foundations

Yan Gao [1,2,*], Zixin Guo [3] and Quan Yuan [4]

1 School of Earth Sciences and Engineering, Sun Yat-sen University, Zhuhai 519082, China
2 Southern Marine Science and Engineering Guangdong Laboratory (Zhuhai), Zhuhai 519082, China
3 Research Center of Coastal and Urban Geotechnical Engineering, College of Civil Engineering and Architecture, Zhejiang University, Zijingang Compus, Hangzhou 310058, China; guozx9@mail3.sysu.edu.cn
4 Guangzhou Metro Design and Research Institute Co., Ltd., Guangzhou 510010, China; yuanquan@dtsjy.com
* Correspondence: gaoyan25@mail.sysu.edu.cn; Tel.: +86-159-1451-8020

Abstract: The mechanical response and deformation characteristics in calcareous sand foundations during pile driving and setup were studied using model tests combined with the technical methods of tactile pressure sensors and close-range photogrammetry. Different types of piles were considered, including a pipe pile, square pile and semi-closed steel pipe pile. The test results show that during pile driving, the pile tip resistance of different piles increases with an increase in the pile insertion depth, and an obvious fluctuation is also obtained due to the particle breakage of the calcareous sand and energy dissipation. Different degrees of particle breakage generated by different type piles make the internal stress variations different, as with the pile tip resistance. The pile tip resistance of model pile A, which simulates a pipe pile, is the highest, followed by model pile B, the simulated square pile. Model pile C, which simulates a semi-closed steel pipe pile, has the smallest pile tip resistance because its particle breakage is the most obvious and the pile tip energy cannot be continuously accumulated. The induced deformation such as sag or uplift on the surface and the associated influence range for the calcareous sand foundation are the smallest for model pile C, followed by model pile B and then model pile A. Model pile A has the most obvious pile driving effect. During the pile setup process after piling, the increase in the total internal stress of model pile B is the largest, and the improvement of the potential bearing capacity is the most obvious, followed by model pile A and model pile C. During the pile setup, the induced uplift deformation in pile driving is recovered and the potential bearing capacity increases due the redistribution and uniformity of the vertical and radial stress distributions in the calcareous sand foundation. Considering the potential bearing capacity of different model piles, the influence range of pile driving, foundation deformation and the pile setup effect, it is suggested to use a pointed square pile corresponding to model pile B in pile engineering in calcareous sand foundations.

Keywords: pile foundation; pile tip resistance; stress distributions; particle breakage

Citation: Gao, Y.; Guo, Z.; Yuan, Q. Pile Driving and the Setup Effect and Underlying Mechanism for Different Pile Types in Calcareous Sand Foundations. *J. Mar. Sci. Eng.* **2024**, *12*, 133. https://doi.org/10.3390/jmse12010133

Academic Editors: Dmitry A. Ruban and George Kontakiotis

Received: 16 October 2023
Revised: 14 December 2023
Accepted: 5 January 2024
Published: 9 January 2024

1. Introduction

Calcareous sand is widely distributed between 30 degrees south latitude and 30 degrees north latitude, covering many countries and regions [1], especially the South China Sea Islands, most of which are coral reefs covered with thick layers of calcareous sand [2,3]. These islands and their surrounding areas are rich in marine mineral resources and also have an important strategic position in the social economy, national defense and scientific research. With an increase in the scale and quantity of island and reef engineering, engineering problems are becoming more and more prominent [4].

Calcareous sand is different from siliceous sand and characterized by a high porosity ratio [5,6], prominent edges [7,8] and easy breakage [9,10]. Its engineering mechanical properties are so complex and special that the research results of and engineering experience with traditional soil materials cannot be directly applied to calcareous sand [11–19].

As one of the most common foundation forms, it is of great significance to study the application of pile foundations to calcareous sand, especially to explore the effect of pile driving and engineering mechanical properties in calcareous sand foundations. Up to now, previous studies have mostly focused on the macroscopic mechanical properties of pile foundations in calcareous sand [20–24], and have mainly considered the effects of the calcareous sand particle size distribution [25], cementation degree, pile length [26,27] and load conditions [28–32] on the bearing characteristics of piles. McDowell et al. [33], using small-size model tests on the bearing characteristics of a single pile, pointed out that the peek resistance of a single pile is a strong function of the initial particle size distribution, and the peak resistance in well-graded calcareous sand is greater than that in uniformly graded calcareous sand. In terms of the influence of the cementation degree, Houlsby et al. [34] proposed a functional relationship between the ratio of the cemented layer thickness to the pile diameter and the ultimate bearing capacity. Wang et al. [35], using a model test of jacked piles, pointed out that the jacking pile pressure increases nonlinearly with the increase in pile length, in which the radial stress of the piles gradually converges with passive Earth pressure with the increase in the pile length. By analyzing the datasets of over 37 real projects on super-large and long piles, Thien et al. [36] pointed out that pile length has a certain influence on the pile tip resistance, and they are positively correlated.

As for the influence of the load conditions, Al-Douri et al. [37] pointed out that under cyclic axial loads with the same amplitude, the total displacement increased with an increase in load level and the number of cycles. Among them, Roozbeh et al. [38] believe that asymmetric two-way loading is the most destructive type of cyclic loading. The research by Wang et al. [39] showed that the shaft resistance gradually decreased during dynamic loading, and the ultimate pile tip resistance decreased with an increase in the dynamic load ratio; furthermore, the strength of the pile–sand interface was weakened under dynamic loading, and the calcareous sand particles were broken seriously, resulting in shear shrinkage.

Considering the distribution of the pile foundation and whether the pile ends are closed or not, Jiang [40,41] and Yang et al. [42] conducted experimental studies on the single pile and pile group effects of open-ended and closed-ended piles in calcareous sand, and the results showed that the pile group effect in calcareous sand was significantly different from that in quartz sand. The bearing capacity of these piles was mainly provided by the pile tip resistance, and the mechanism of pile tip resistance and shaft resistance was also analyzed; however, they did not take a deep dive into the bearing characteristics of different pile types in calcareous sand foundations. Wan et al. [43,44] constructed experimental studies on the axial and horizontal bearing capacity of post-grouting piles in calcareous sand, pointing out that the bearing capacity of a single pile after grouting is improved, and, due to the influence of particle breakage in calcareous sand, the increase ratio of the bearing capacity after grouting in calcareous sand is higher than that of siliceous sand. Yu et al. [45], using a model test of cast-in-place concrete X-section and circular section pile groups, pointed out that under the same section area, the bearing capacity of X-section piles is much greater than that of circular section piles.

Some researchers have also analyzed the breakage of calcareous sand particles in the process of pile driving using simulations based on the discrete element method (DEM), so as to explore the mechanical response of calcareous sand and its mechanism [46,47]. However, the existing experimental studies have not revealed the mechanical characteristics and mechanism of different pile types in calcareous sand foundations. Therefore, this paper carried out pile model tests of different pile types in calcareous sand, combined with the use of the technical methods of tactile pressure sensors and close-range photogrammetry, to explore the pile driving and setup of different pile types in calcareous sand foundations, and reveal the mechanical characteristics of the whole process of pile driving.

2. Experimental Details

2.1. Model Piles

According to three traditional pile end shapes of precast piles, a prestressed pipe pile with a side length of Φ600 mm, a precast reinforced concrete square pile with a side length of 600 mm and a semi-closed-end steel pipe pile with a side length Φ600 mm were selected as prototypes, and reduced to small-size models with a ratio of 40:1. In the process of pile driving using a static load, whether the pile is hollow or not and whether the pile is open-ended or close-ended, there is little influence on the bearing characteristics [48]. Therefore, model piles with a solid structure are adopted to ensure the small-size model piles are rigid enough. The geometric shape and size of different model piles are shown in Figure 1. Among them, model pile A simulates the prestressed pipe pile, model pile B simulates the precast reinforced concrete square pile and model C simulates the semi-closed-end steel pipe pile.

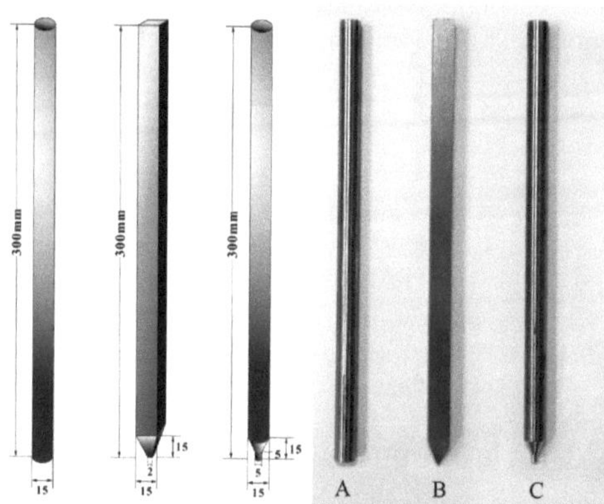

Figure 1. Schematic diagram of the model piles with different shapes and sizes.

2.2. Model Box and Test Material

The model box is cubic and made of acrylic plates with a thickness of 10 mm, which have the advantage of transparency and visibility under the premise of meeting the rigidity requirements. The refraction effect of acrylic plates in different model tests is relatively consistent, which makes sure there are only small measurement errors in the comparison of particle breakage under different tests. Considering the requirements of the particle size effect, boundary effect and pile distance from the bottom of the box [49,50], the interior dimensions of the model box are designed to be 200 mm long, 120 mm wide and 220 mm high. A reduced model would affect the magnitude of the displacement and stress, which may be different from large-scale prototypes, while the underlying mechanics will not be changed.

The calcareous sand used in this paper is from an island in the South China Sea and comprises unconsolidated and loose particles. The main component of calcareous sand is calcium carbonate with a content of more than 90%. Due to its sedimentary origin and limited geological transportation, the shape of calcareous sand particles exhibits a high degree of irregularity, characterized by various shapes including flakes, blocks, strips, dendrites and others. According to the similarity theory, the influence of the particle size effect on the test results should be reduced as much as possible. Therefore, dry calcareous sand with a particle size of 2–5 mm is used as the foundation soil material for the model test, and its maximum and minimum dry densities are 1290 kg/m^3 and 1180 kg/m^3, respectively. The calcareous sand was divided into eight layers and laid layer by layer to the designed height of 200 mm in the model box, in which the relative density of the calcareous sand was controlled to be 0.60, and the corresponding dry density was 1240 kg/m^3.

2.3. Testing Design

In order to explore the characteristics of calcareous sand foundations during pile driving and setup, tactile pressure sensors (mode PFS5051, Tekscan Inc., Norwood, MA, USA) with a measuring range of 334.74 kPa are utilized, as shown in Figure 2a. The sensing area of the sensors is 56 mm × 56 mm, and 1936 measuring points are distributed in a matrix of 44 rows and 44 columns, among which the area of a single measuring point is only 1.62 mm^2. To obtain the vertical and radial stress distribution during pile driving in calcareous sand foundations, two tactile pressure sensors are deployed in the vertical and horizontal directions, as shown in Figure 2b. The horizontal tactile pressure sensor is buried 10 mm below the pile end at the maximum driving depth, that is, 110 mm away from the bottom of the model box. The circumferential tactile pressure sensor is buried around the center of the pile, about 15 mm away from the pile shaft, and the bottom of the sensor is buried 15 mm above the horizontal tactile pressure sensor, that is, 125 mm away from the bottom of the model box.

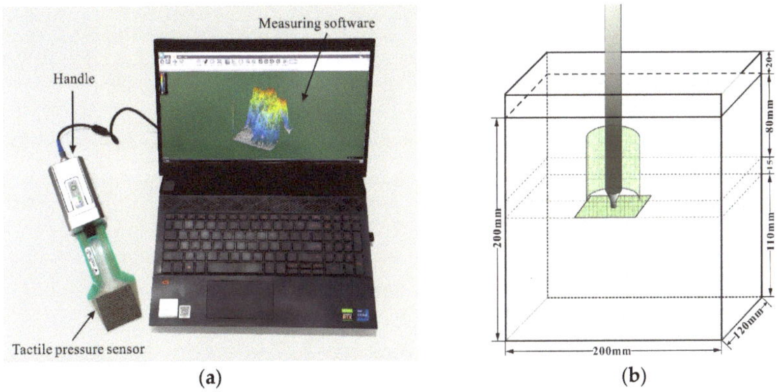

Figure 2. Measurement system of tactile pressure sensors and arrangement: (**a**) measurement system of tactile pressure sensors; (**b**) the arrangement of tactile pressure sensors.

In order to simulate pile driving using a static load, the MTS electromechanical testing system (Exceed, model E45) is used to apply loading at a constant speed of 0.02 mm/s until the pile driving depth reaches 80 mm, and the stress distribution calcareous sand foundation is measured every 5 mm using the tactile pressure sensor. After the pile driving is completed, the external force on the model pile is removed, and the depth of the pile is kept at 80 mm to enter the static process, i.e., the pile setup process. The pile setup is retained for 24 h, and the stress distribution calcareous sand foundation is measured at 5 min intervals within the first hour, and then measured and photographed at time points of 90 min, 120 min, 240 min, 420 min, 540 min, 1020 min, 1260 min and 1440 min, respectively. Fixed-point photogrammetry is also carried out in front of the model box to observe the deformation and particle crushing of the calcareous sand foundation during the pile driving and setup processes.

3. Results

3.1. Pile Tip Resistance during Pile Driving

In the process of pile driving, the pile tip resistance and pile tip displacement are measured and recorded using the MTS system, and the results of different model piles are shown in Figure 3. It is readily seen that with the increase in pile depth, the pile tip resistance of different model piles continuously increases, but it also fluctuates obviously. The reason for this is that the calcareous sand particles break during the process of pile driving and dissipate energy, which makes the pile tip resistance decrease slightly; with the deepening of pile driving, the pile tip resistance increases, which induces a higher degree of particle breakage and the more obvious fluctuation in pile tip resistance. The pile tip resistance of different pile types is different. In general, the pile tip resistance of model pile A, which simulates the shape of a pipe pile, is the largest during pile driving, and the influence of particle breakage on it is the least significant, which also leads to the small fluctuation in pile tip resistance. The pile tip resistance of model pile B, which simulates the shape of a pointed square pile, is the second greatest. Model pile C, which simulates a semi-closed steel pipe pile, has the least pile tip resistance, experiencing the greatest influence of particle breakage and the biggest fluctuation in pile tip resistance. Even after a certain pile depth (>40 mm), the increase in the pile tip resistance is small and tends to be stable. These macroscopic mechanical changes can be explained by the particle breakage and the contact force variation between particles corresponding to the micro-evidence.

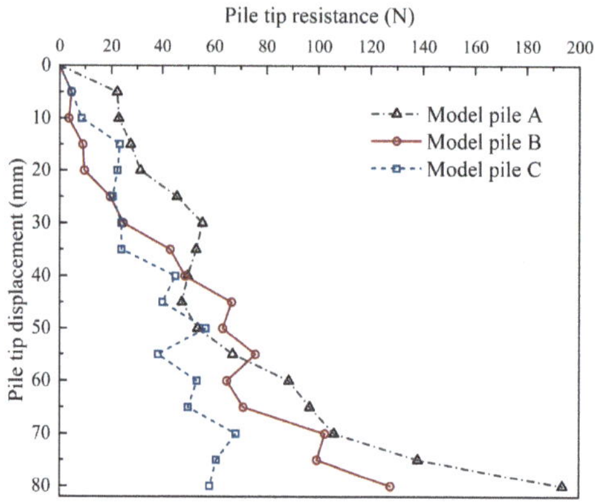

Figure 3. Pile tip resistance–displacement curves of different pile models during pile driving.

3.2. Particle Breakage during Pile Driving

In terms of the particle breakage, using fixed-point photogrammetry, the changes in the calcareous sand foundation with different model piles under different pile depths are obtained, as shown in Figure 4. Representative in clearly showing the particle variation during pile driving, the changes in the calcareous sand particles near model pile A and on the surface at different pile depths are extracted, as shown in Figure 5. It can be found that with the increase in pile depth, the calcareous sand particles at the bottom of the pile tip undergo a cyclic process from thinning to densification and from a large particle size to a small particle size. When the pile depth is small (0~20 mm), the calcareous sand particles under the pile tip are continuously compacted, and the bearing capacity increases. After reaching a certain stress limit (pile depth at 20–40 mm), the calcareous sand particles under the pile tip are broken, resulting in more fine particles and a sparse arrangement. The energy dissipation and foundation bearing capacity decrease, which provides micro-evidence supporting the slight decline in the pile tip resistance of model pile A when the pile depth is 30–45 mm. When the pile depth is around 40–50 mm, with the re-compaction, the calcareous sand at the bottom of the pile tip reaches a much denser state, and the final bearing capacity increases. The position of the calcareous sand particles at the bottom of the pile tip changed significantly with a pile depth at 50~80 mm, and the flat calcareous sand particles on both sides of the pile gradually filled the gap between the pile shaft and the wall of the model box, resulting in the particle breakage not being obvious.

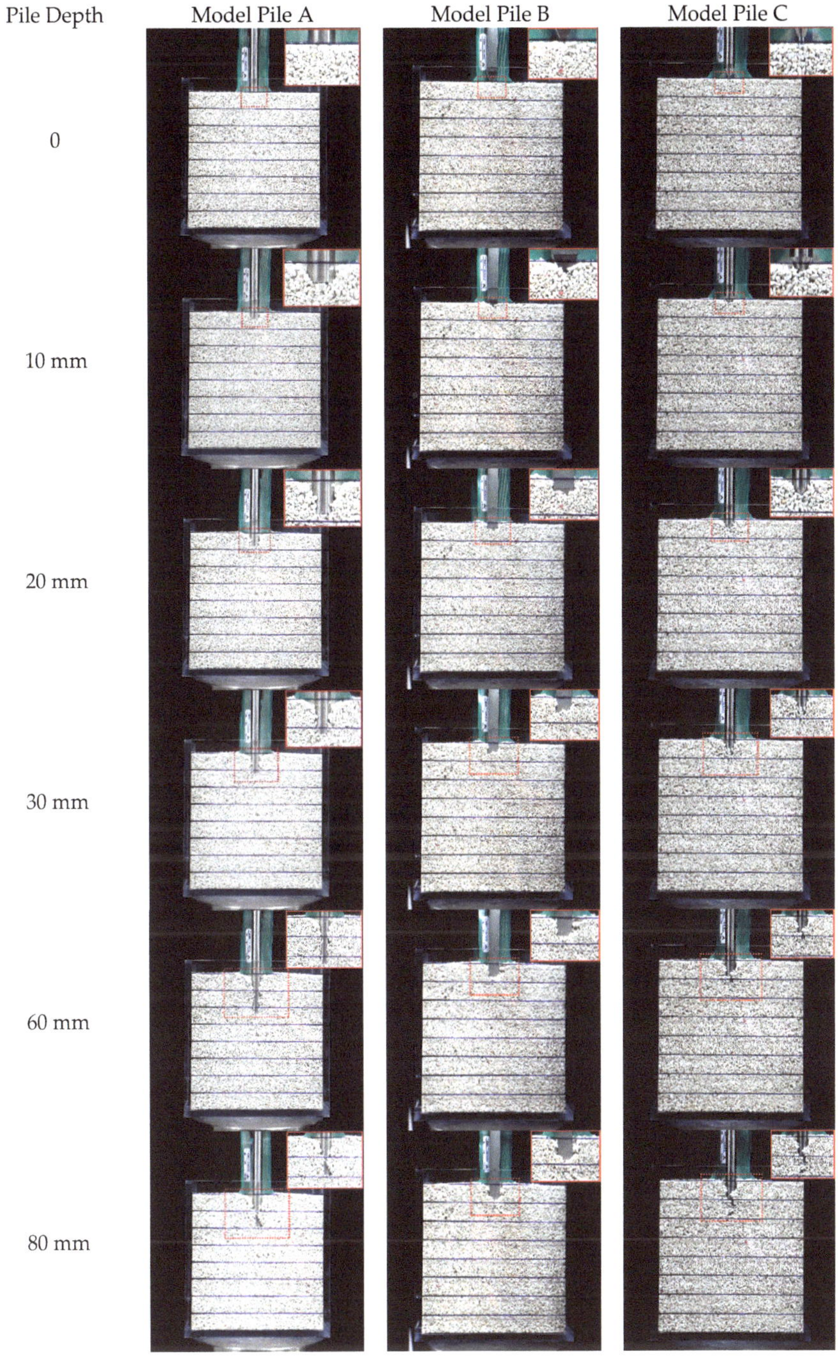

Figure 4. Changes in calcareous sand foundation during pile driving process for different types of piles. The soil around the pile is also enlarged as show in the marked box.

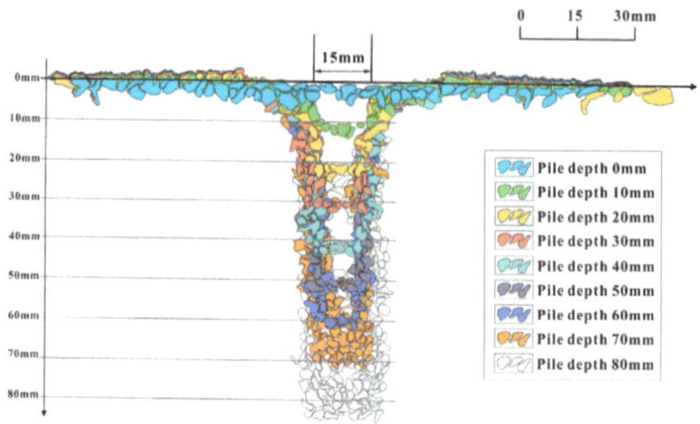

Figure 5. Changes in calcareous sand particles of model pile A under different pile depths.

Likewise, similar change rules can also be found in model pile B and model pile C, but the corresponding particle breakage and energy accumulation of the different model piles are different, resulting in differences in the value and change points of pile tip resistance.

3.3. Deformation Characteristics during Pile Driving

Based on the image obtained using fixed-point photography in front of the model box, Figure 6 shows the surface deformation of the calcareous sand foundation obtained using tracing and extraction in the process of pile driving with different model piles. With the continuous driving of model pile A (Figure 6a), a funnel-shaped groove is gradually formed on the surface of the calcareous sand foundation within the range of one pile diameter from the pile shaft, while the calcareous sand outside the funnel-shaped groove produces a slight extrusion uplift. Until a distance of five times the pile diameter from the pile shaft, the driving of the model pile hardly caused deformation on the surface of the calcareous sand foundation. However, model pile B is different from model pile A (Figure 6b). When the pile depth is relatively small (0~40 mm), the surface of the calcareous sand foundation near the pile produces an extrusion uplift; meanwhile, when the pile depth is relatively large (50–80 mm), the uplifts close to both sides of the pile body gradually move outward, and a small funnel-shaped groove is formed within one of the pile diameter from the pile shaft, but the groove depth does not change significantly with the pile depth. The surface displacement of model pile C (Figure 6c) is also similar to that of model pile A, but the depth of the funnel-shaped groove is the smallest among the different model piles.

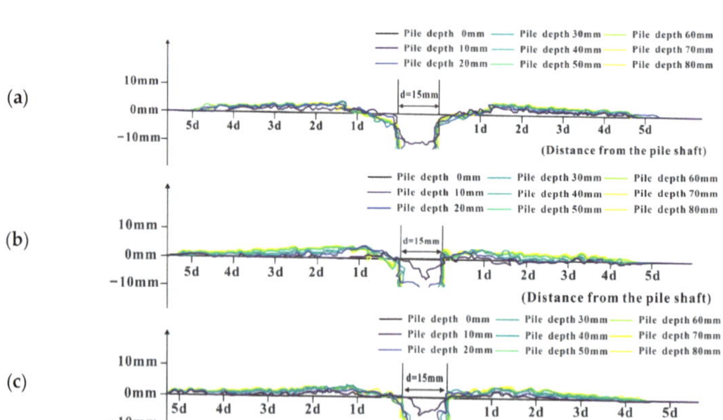

Figure 6. Surface deformation of calcareous sand foundation in the process of pile driving: (**a**) model pile A; (**b**) model pile B; (**c**) model pile C.

By comparing the changes in the surface deformation of the calcareous sand foundation with different model piles, it can be found that model pile C has the smallest induced vertical deformation and influence range on the surface of the calcareous sand foundation, followed by model pile B, while model pile A has the most obvious pile driving effect and the most significant effect on the deformation and destruction of the calcareous sand around the pile.

3.4. Internal Stress Evolution in Calcareous Sand Foundation during Pile Driving

3.4.1. Vertical Stress Distribution

In terms of the contact force between particles, the vertical stress distribution in the calcareous sand foundation at 10 mm below the pile tip at the maximum driving depth is measured using the tactile pressure sensor. The distributions of and changes in the vertical stress generated by different model piles in the calcareous sand foundation during the driving process are shown in Figure 7. It can be seen that during pile driving, the stress distribution in the calcareous sand foundation is not uniform, whereas the influence range of pile driving is limited, and, further, the stress concentration is mainly generated in the range of about double the pile diameter around the pile. When the pile depth reaches 60 mm, an obvious stress concentration zone is formed in the calcareous sand foundation under the center of the pile. The vertical stress distribution of model pile B and model pile C at different pile depths is more uniform than that of model pile A, and their stress concentration zone is relatively less obvious.

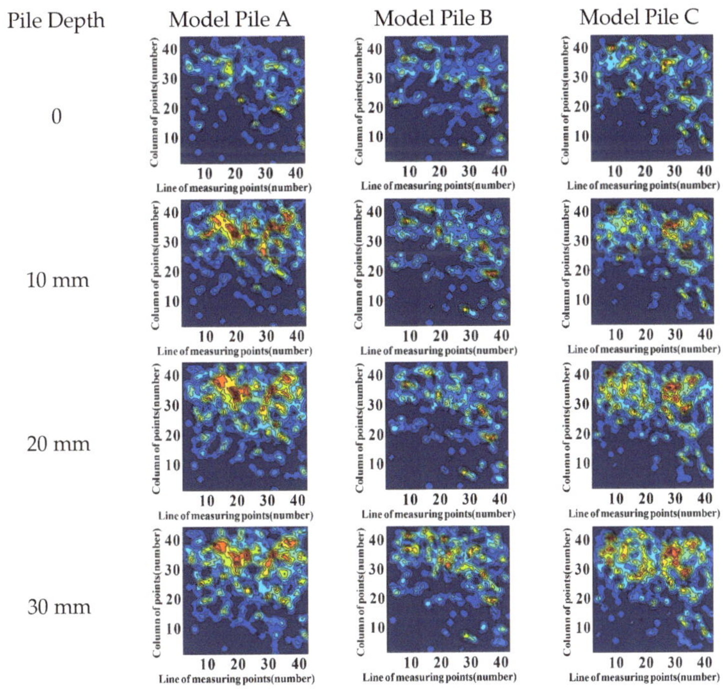

Figure 7. *Cont.*

60 mm

80 mm

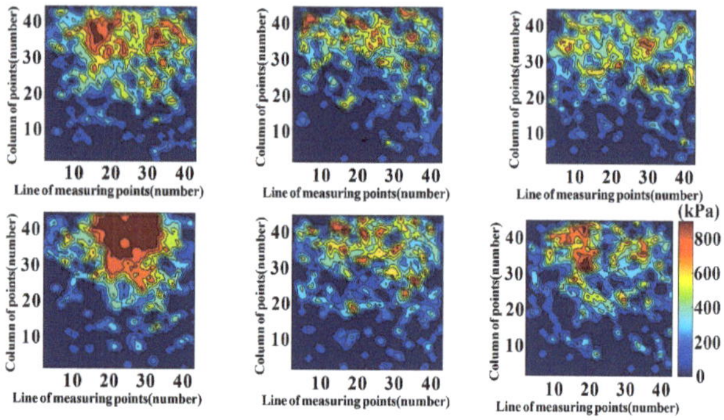

Figure 7. Vertical stress distributions of different types of model piles at different pile depths.

3.4.2. Radial Stress Distribution in a Calcareous Sand Foundation during Pile Driving

The radial stress distribution and the variation generated by different model piles in a calcareous sand foundation during pile driving are obtained. Among them, the original results of the radial stress distribution of model pile A measured using a tactile pressure are representatively shown in Table 1. It can be seen that with the increase in the pile depth, the radial stress around model pile A in the calcareous sand foundation first increases until it reaches a peak at a pile depth of 30 mm, and then decreases, which reflects the influence of the pile driving and the relative position in the foundation. As the pile depth increases, the pile tip gradually approaches, passes through and finally moves away from the position of the circumferential tactile pressure sensor. The radial stresses of model pile B and model pile C also have the same change trend during the process of pile driving in the calcareous sand foundation.

Table 1. Radial stress distribution of model pile A at different pile depths.

	Pile Depth 0 mm	Pile Depth 30 mm	Pile Depth 80 mm	Stress (kPa)
Two-dimensional stress contour map	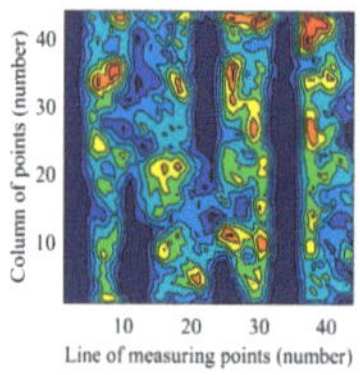	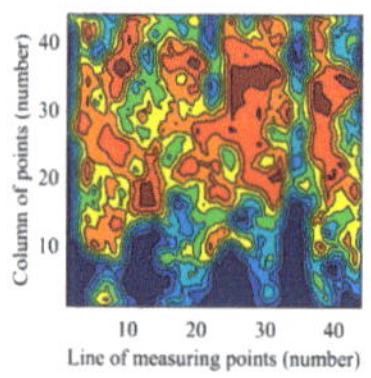	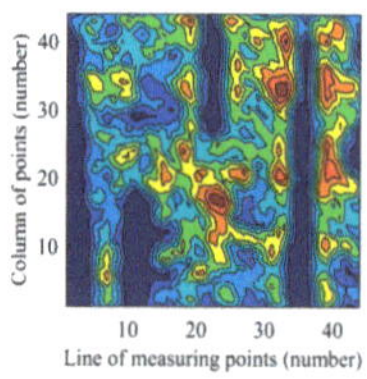	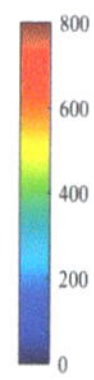
Three-dimensional stress distribution cloud image	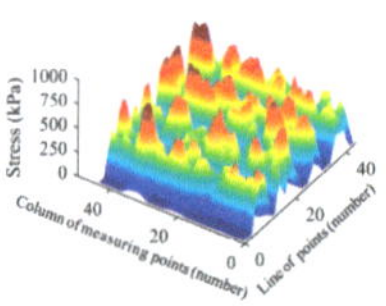	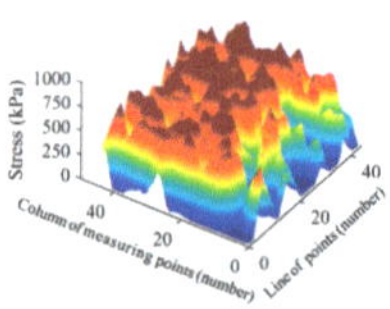	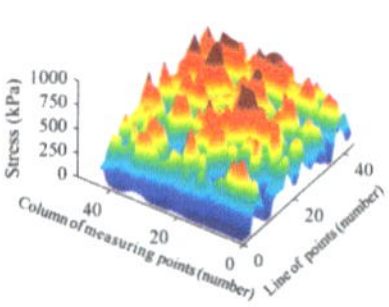	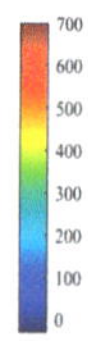

3.5. Deformation during Pile Setup

By comparing the surface deformation of the calcareous sand foundation with different model piles before and after the pile setup (Table 2), it can be seen that after 24 h, the uplift on the surface of the calcareous sand foundation basically is recovered and becomes flat, and the calcareous sand particles near the piles are consolidated under the self-weight. However, the range and degree of compaction for different model piles are different, which corresponds to the change in vertical and radial stress. Given the above, during the pile setup, the deformation of the foundation and the settlement of the piles themselves is small and negligible, but with the extension in time, the internal forces in the calcareous sand foundation is redistributed, and both the vertical and radial stresses in the foundation increase. In other words, when there is no obvious particle breakage in the calcareous sand foundation, the pile tip resistance and shaft resistance increase.

Table 2. Changes in calcareous sand foundation for different model piles under different conditions. The soil around the pile is also enlarged as show in the marked box.

Model Pile	After Pile Driving	Before Pile Setup	After Pile Setup
Model plie A	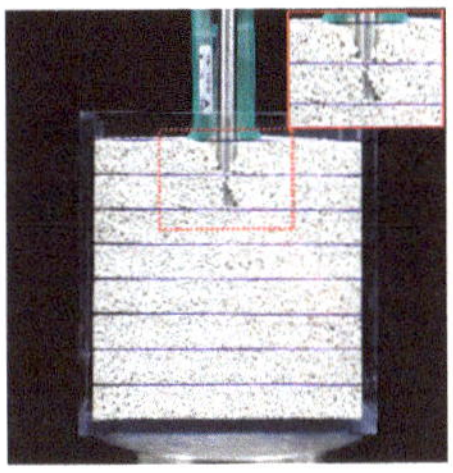	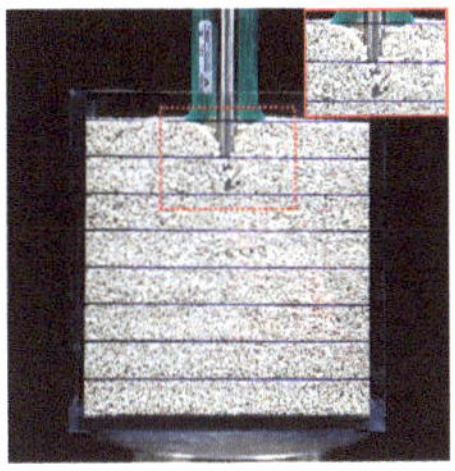	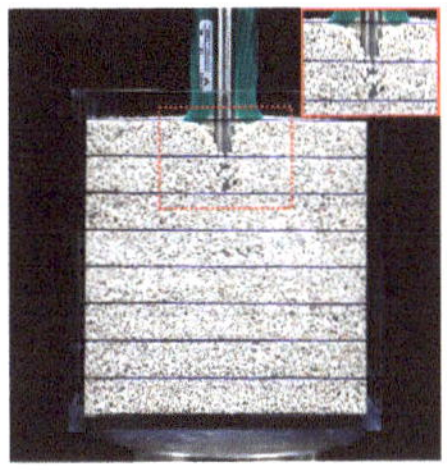
Model plie B	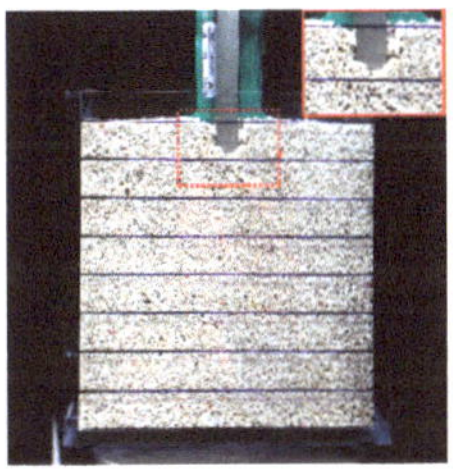	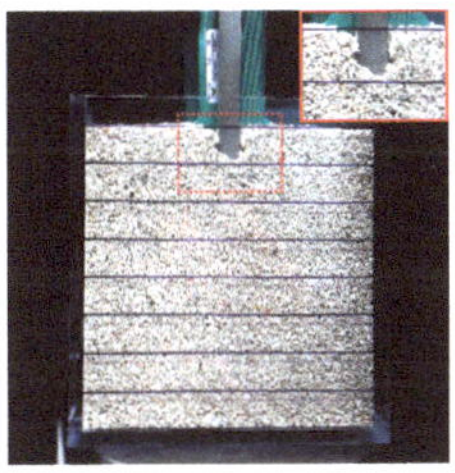	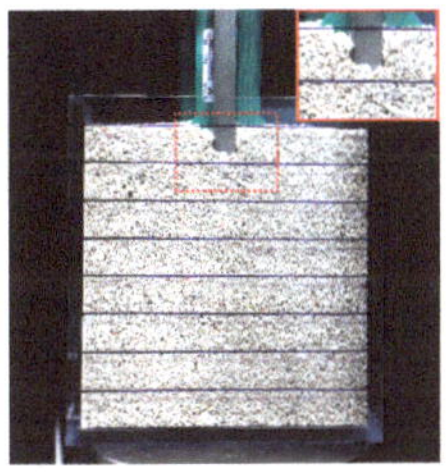
Model plie C	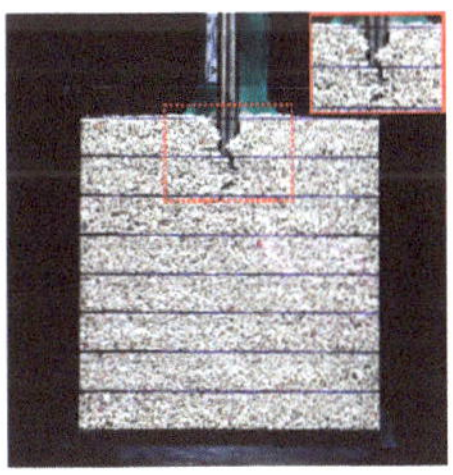	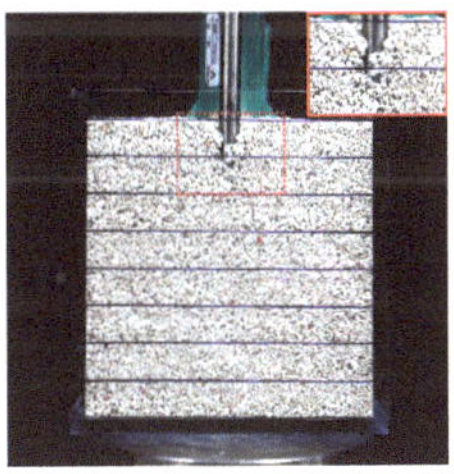	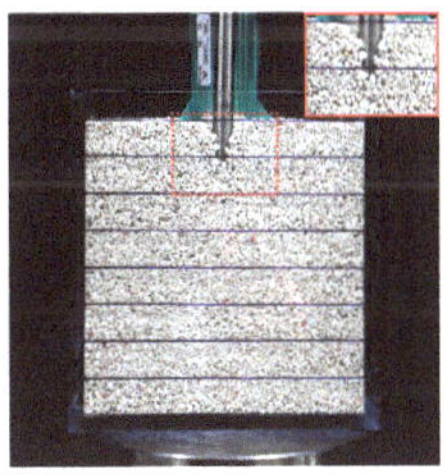

4. Discussion

The macro-behavior is highly related to the micro-information. The underlying mechanism of different pile type effects in the calcareous sand foundation can be explored according to the internal stress obtained using the tactile pressure sensors. The mean stress and coefficient of variation in stress are used to characterize the stress distribution, in which the coefficient of variation is the ratio of standard deviation to the mean value, ranging from 0 to 1. The smaller the coefficient of variation, the smaller the dispersion degree of the data and the more concentrated it is in a smaller range. And the data processing method of normalization is used: the data before the test (i.e., the pile depth is 0 mm) are taken as the initial value, then the normalized data are the ratio between the data and the initial value.

4.1. During Pile Driving

By comparing the vertical stress at the bottom of the pile tip in the calcareous sand foundation across different model piles (Figure 8), the results show that the vertical stress increases with the increase in the pile depth, among which model pile A has the largest increment, suggesting the highest potential vertical bearing capacity. Since the pile tip area of model pile B and model pile C is small, the interparticle stress can be easily highly concentrated, which induces more particle breakage and energy dissipation and, in turn, decreases the interparticle stress. This is why their corresponding vertical stress increases are relatively small, and the related bearing capacity provided by the pile tip is also limited. The vertical stress of model pile C tends to be constant at a certain depth with the largest relative particle breakage, which is consistent with the change in the pile tip resistance in Figure 3 and the conclusion reached by Gao et al. [46] using DEM numerical simulation. The coefficient of variation in vertical stress decreases with the increase in the pile depth, that is, the stress distribution tends to be uniform.

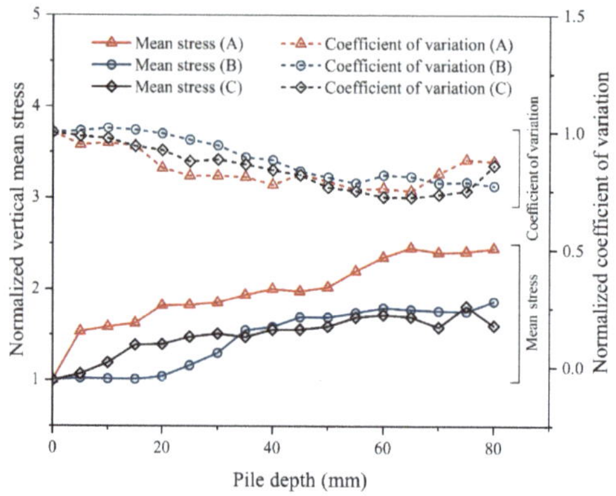

Figure 8. Variations in vertical stresses for different model piles during pile driving process.

The changes in the normalized radial mean stress and coefficient of variation of different model piles with pile depth are shown in Figure 9. It can be found that with an increasing depth of pile driving, the coefficient of variation is slightly changed compared with that for vertical stress. It decreases first (indicating that the stress tends to be uniform) and then increases (indicating that the stress is more concentrated). Model pile C needs a larger pile depth to reach the maximum radial stress, reflecting a higher particle breakage degree, and it is difficult for model pile C to accumulate energy compared with other model piles, which is consistent with the above analysis. According to the increment in the mean radial stresses of different model piles, it is considered that model pile A has the largest influence range on the lateral force in the calcareous sand foundation, while model pile B has the smallest influence, which is due to the difference in the pile end shape.

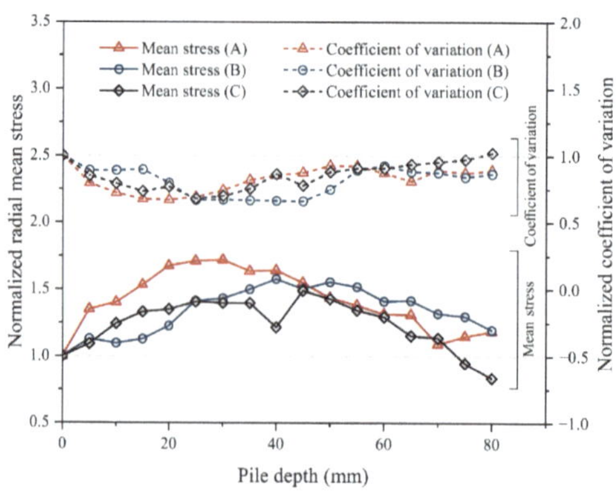

Figure 9. Changes in radial stresses for different model piles during pile driving process.

4.2. During Pile Setup

In the actual pile foundation, after pile driving is completed, the piles are left in place without external forces until monitoring is deemed satisfactory, or the next phase of construction commences. Consequently, compared to previous studies [43,45,51], this paper presents a more comprehensive analysis of the pile setup in a calcareous sand foundation, thereby revealing a deeper understanding of the underlying mechanism within real-world pile foundation engineering.

After the pile was driven to the target depth, the external force on the model pile was removed. That is, the buried depth of the model pile was kept at the designed pile depth of 80 mm for 24 h to develop the setup process, and the change in the stress distribution in the calcareous sand foundation was monitored. It is found that the global stress distribution changes a little with some local changes. The clear variations in the normalized vertical mean stress and coefficient of variation for different model piles along with increasing time are shown in Figure 10. It can be seen that the coefficient of variation in vertical stress decreases and the vertical mean stress increases after 24 h only under gravity. In other words, stress redistribution occurs in the calcareous sand foundation over time, and it tends to be more uniform, which induces a slight increase in its bearing capacity.

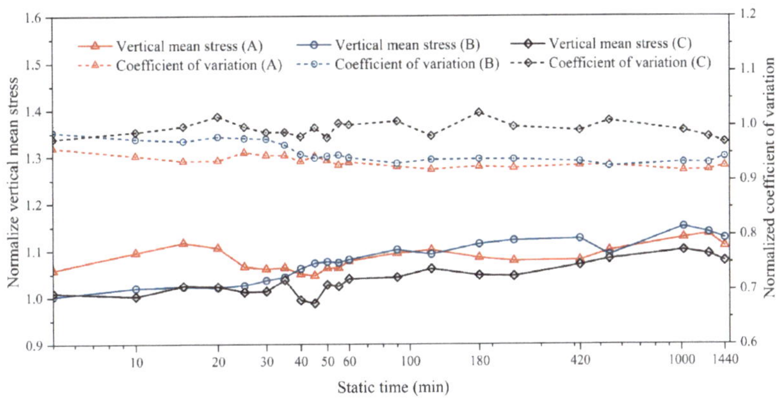

Figure 10. The variations in vertical mean stress and coefficient of variation over time.

The radial stress distribution of different model piles before and after the pile setup shows little global change but some local changes. The much clearer changes in the normalized radial mean stress and the coefficient of variation in radial stress for different model piles over time are shown in Figure 11. It is found that the radial mean stress in the calcareous sand foundation with model pile A and model pile B increases with time, and the coefficient of variation decreases, although the decrease in the coefficient of variation is not as large as that of the vertical stress. The decreasing coefficient of variation indicates a more uniform stress distribution, which can give a more stable internal structure. But for model pile C, the corresponding radial stress has no obvious change after a certain time. Based on the changes in vertical and radial stress, it is concluded that model pile B has the largest increase in total stress and the most obvious increase in potential bearing capacity, followed by model pile A, and model pile C has the smallest increase.

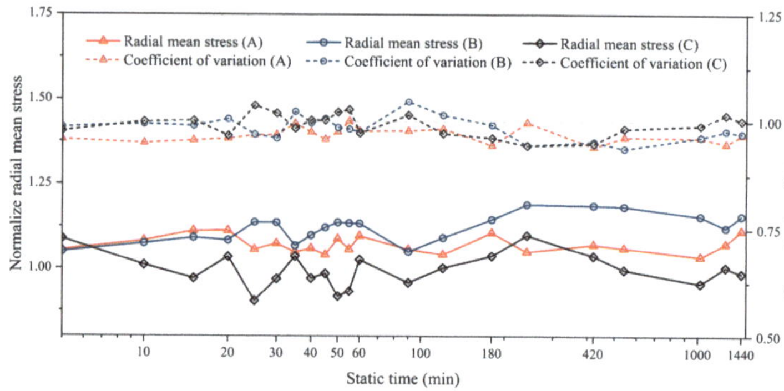

Figure 11. The variations in radial mean stress and coefficient of variation with time.

5. Conclusions

In this study, three different model piles were designed based on three common pile shapes, which were pipe piles, pointed square piles and semi-closed steel pipe piles. Using the model tests of pile driving and pile setup in a calcareous sand foundation, the pile driving and time effect and mechanical properties in calcareous sand foundations for different pile types were explored, and the main conclusions are as follows.

(1) In the process of pile driving, with an increase in pile depth, the pile tip resistance of different pile models increases in general, while there are fluctuations caused by the particle breakage of the calcareous sand and energy dissipation. As the pile is driven deeper, the fluctuation in the pile tip resistance becomes more pronounced. The pile tip resistance for model pile A simulating a pipe pile is the highest, followed by model pile B simulating a pointed square pile and model pile C simulating a semi-closed steel pipe pile has the least. It is considered that, at the completion of pile driving, the potential bearing capacity of model pile A is the highest, followed by model pile B, and model pile C has the smallest potential bearing capacity.

(2) The results on the interparticle stress distributions in the calcareous sand foundation during pile driving are well in agreement with the global pile tip resistance behavior. The vertical stress concentration is mainly generated in the range of double the pile diameter around the pile in calcareous sand foundations. The radial stress is increased by the pile passing through. With an increasing pile depth, the stress distribution tends to be uniform, which is characterized by the reducing coefficient of variation in interparticle stresses; the mean vertical stress for model pile A has the largest increase, followed by model pile B and model pile C, which can explain the change in pile tip resistance.

(3) During pile driving, the calcareous sand foundation surface gradually forms a funnel-shaped groove, and a slight extrusion uplift occurs outside the groove. Model pile C has the smallest deformation and impact range on the surface of the calcareous sand foundation, followed by model pile B. Model pile A has the most obvious pile driving effect with the most significant deformation and influence range on the calcareous sand foundation around the pile.

(4) With a certain duration of pile setup, the vertical and radial stresses in the calcareous sand foundation with different pile models also tend to be redistributed and become uniform, and the average vertical and radial stresses increase. This indicates that the total potential bearing capacity increases after a certain time. The increased magnitude in reverse order from large to small spans model pile B, A, then C. The uplift deformation of the ground surface is also recovered.

(5) Considering the potential bearing capacity of different model piles, the influence range of pile driving, foundation deformation and pile setup, model pile B has a good bearing capacity potential and a small deformation and damage effect on calcareous sand foundations. It is suggested to use a pointed square pile corresponding to model pile B in

pile foundation engineering for calcareous sand foundations. Large-scale prototype tests are also recommended for further identification in the future.

Author Contributions: Conceptualization, Y.G. and Z.G.; methodology, Y.G.; validation, Y.G., Z.G. and Q.Y.; formal analysis, Y.G. and Z.G.; investigation, Y.G. and Q.Y.; resources, Y.G.; data curation, Y.G.; writing—original draft preparation, Z.G.; writing—review and editing, Y.G. and Q.Y.; visualization, Z.G.; supervision, Y.G.; project administration, Y.G.; funding acquisition, Y.G. All authors have read and agreed to the published version of the manuscript.

Funding: This research was funded by National Natural Science Foundation of China, grant number 42072295, Guangdong Project, grant number 2017ZT07Z066, National Key R&D Program of China, grant number 2022YFC3005203, and Guangdong Provincial Key Laboratory of New Construction Technology for Urban Rail Transit Engineering (2017B030302009).

Institutional Review Board Statement: Not applicable.

Informed Consent Statement: Not applicable.

Data Availability Statement: The raw data supporting the conclusions of this article will be made available by the authors on request.

Conflicts of Interest: Author Quan Yuan was employed by the company Guangzhou Metro Design and Research Institute Co., Ltd. The remaining authors declare that the research was conducted in the absence of any commercial or financial relationships that could be construed as a potential conflict of interest.

References

1. Gao, Y.; Shi, T.; Yuan, Q.; Sun, K. The creep characteristics and related evolution of particle morphology for calcareous sand. *Powder Technol.* **2024**, *431*, 119077. [CrossRef]
2. Wang, X.; Jiao, Y.; Wang, R.; Hu, M.; Meng, Q.; Tan, F. Engineering characteristics of the calcareous sand in Nansha Islands, South China Sea. *Eng. Geol.* **2011**, *120*, 40–47. [CrossRef]
3. Ye, J.; Haiyilati, Y.; Cao, M.; Zuo, D.; Chai, X. Creep characteristics of calcareous coral sand in the South China Sea. *Acta Geotech.* **2022**, *17*, 5133–5155. [CrossRef]
4. Murff, J.D. Pile Capacity in Calcareous Sands: State if the Art. *J. Geotech. Eng.* **1987**, *113*, 490–507. [CrossRef]
5. Fan, Z.; Hu, C.; Zhu, Q.; Jia, Y.; Zuo, D.; Duan, Z. Three-dimensional pore characteristics and permeability properties of calcareous sand with different particle sizes. *Bull. Eng. Geol. Environ.* **2021**, *80*, 2659–2670. [CrossRef]
6. Zhu, C.; Wang, X.; Wang, R.; Chen, H.; Meng, Q. Experimental microscopic study of inner pores of calcareous sand. *Mater. Res. Innov.* **2014**, *18*, S2-207–S2-214. [CrossRef]
7. Shen, Y.; Zhu, Y.; Liu, H.; Li, A.; Ge, H. Macro-meso effects of gradation and particle morphology on the compressibility characteristics of calcareous sand. *Bull. Eng. Geol. Environ.* **2018**, *77*, 1047–1055. [CrossRef]
8. Touiti, L.; Kim, T.; Jung, Y.H. Analysis of calcareous sand particle shape using fourier descriptor analysis. *Int. J. Geo-Eng.* **2020**, *11*, 15. [CrossRef]
9. Kuang, D.; Long, Z.; Guo, R.; Zhao, T.; Wu, K. Experimental and numerical study on the fragmentation mechanism of a single calcareous sand particle under normal compression. *Bull. Eng. Geol. Environ.* **2021**, *80*, 2875–2888. [CrossRef]
10. Lyu, H.; Gu, J.; Zhou, J.; Li, B. Mechanical behavior and particle breakage of calcareous sand in triaxial test. *Mar. Geophys. Res.* **2023**, *44*, 18. [CrossRef]
11. Ma, C.; Qu, R.; Zhu, C.; Liu, H.; Wang, T. Microscopic study of the impact of particle morphology on the compressibility of calcareous sands. *Powder Technol.* **2024**, *433*, 119192. [CrossRef]
12. Shen, Y.; Rui, X.; Ma, Y.; Shen, J.; Xu, J. Anisotropic behaviors of calcareous sand dependent on loading direction and initial shear stress. *Appl. Ocean Res.* **2023**, *141*, 103775. [CrossRef]
13. Jin, H.; Zhou, L.; Guo, L.; Tong, J. Comparative Study on Liquefaction Behavior of Calcareous Sand and Siliceous Sand Under Simple Shear Loading. *J. Earthq. Eng.* **2023**, *27*, 3471–3489. [CrossRef]
14. Kurniawan, R.; Nghia-Nguyen, T.; Kikumoto, M. Mechanical behavior of porous crushable soils: Effect of intragranular porosity. *IOP Conf. Ser. Earth Environ. Sci.* **2023**, *1249*, 012024. [CrossRef]
15. Luo, M.; Zhang, J.; Liu, X.; Wu, C. Effect of Particle Breakage and Interlocking on Strength and Dilatancy Characteristics of Calcareous Sands. *KSCE J. Civ. Eng.* **2023**, *27*, 3270–3284. [CrossRef]
16. Xiang, C.; Xu, D.; Shen, J.; Wei, H.; Wang, R. Effect of particle size and particle distribution pattern on dynamic behavior of cemented calcareous sand. *Mar. Georesour. Geotechnol.* **2023**, *41*, 412–424. [CrossRef]
17. Gilchrist, J.M. Load Tests on Tubular Piles in Coralline Strata. *J. Geotech. Eng.* **1985**, *111*, 641–655. [CrossRef]
18. Neely, K.L.; Lewis, C.L.; Macaulay, K.A. Disparities in Spawning Times between in situ and ex situ Pillar Corals. *Front. Mar. Sci.* **2020**, *7*, 643. [CrossRef]

19. Zhu, C.Q.; Liu, H.F.; Wang, X.; Meng, Q.S.; Wang, R. Engineering geotechnical investigation for coral reef site of the cross-sea bridge between Malé and Airport Island. *Ocean Eng.* **2017**, *146*, 298–310. [CrossRef]
20. Wang, C.; Ding, X. Effect of Vertical Load on Lateral Response of Single Piles in Coral Sand. In Proceedings of the 2022 International Conference on Green Building, Civil Engineering and Smart City, Guilin, China, 8–10 April 2022. [CrossRef]
21. Lehane, B.M.; Schneider, J.A.; Lim, J.K.; Mortara, G. Shaft Friction from Instrumented Displacement Piles in an Uncemented Calcareous Sand. *J. Geotech. Geoenviron. Eng.* **2012**, *138*, 1357–1368. [CrossRef]
22. Hu, H.; Luo, L.; Lei, G.; Guo, J.; He, S.; Hu, X.; Guo, P.; Gong, X. The Transverse Bearing Characteristics of the Pile Foundation in a Calcareous Sand Area. *Materials* **2022**, *15*, 6176. [CrossRef] [PubMed]
23. Qin, Y.; Xu, D.; Fan, X. Effect of pile inclination on the lateral deformation behaviour of pipe piles in calcareous sand. *Mar. Georesour. Geotechnol.* **2022**, *40*, 589–599. [CrossRef]
24. Spagnoli, G.; Scheller, P. Doherty, In situ and laboratory tests on a novel offshore mixed-in-place pile for oil and gas platforms. *J. Pet. Sci. Eng.* **2016**, *145*, 502–509. [CrossRef]
25. Ouyang, H.; Dai, G.; Qin, W.; Zhang, C.; Gong, W. Dynamic behaviors of calcareous sand under repeated one-dimensional impacts. *Soil Dyn. Earthq. Eng.* **2021**, *150*, 106891. [CrossRef]
26. Salem, T.N.; Elkhawas, N.M.; Elnady, A.M. Behavior of Offshore Pile in Calcareous Sand—Case Study. *J. Mar. Sci. Eng.* **2021**, *9*, 839. [CrossRef]
27. Liu, H.; Wang, C.; Ding, X.; Zhang, Y. Comparative numerical analysis of the response of laterally loaded pile in coral and silica sands. *Acta Geotech.* **2023**, *18*, 4767–4787. [CrossRef]
28. Ding, X.; Deng, X.; Ou, Q.; Deng, W. Experimental study on the behavior of single pile foundation under vertical cyclic load in coral sand. *Ocean Eng.* **2023**, *280*, 114672. [CrossRef]
29. Xu, D.; Huang, F.; Rui, R. Investigation of Single Pipe Pile Behavior Under Combined Vertical and Lateral Loadings in Standard and Coral Sands. *Int. J. Geosynth. Ground Eng.* **2020**, *6*, 20. [CrossRef]
30. Wei, H.; Xin, L.; Li, W.; Li, X.; Yang, M.; Chen, Y. Oblique tensile tests on model anchor piles in calcareous sand deposits. *Ocean Eng.* **2024**, *291*, 116285. [CrossRef]
31. Wu, Q.; Ding, X.; Zhang, Y. Dynamic interaction of coral sand-pile-superstructure during earthquakes: 3D numerical simulations. *Mar. Georesour. Geotechnol.* **2023**, *41*, 774–790. [CrossRef]
32. Wu, Q.; Ding, X.; Chen, Z.; Zhang, Y. Shaking Table Tests on Seismic Responses of Pile-soil-superstructure in Coral Sand. *J. Earthq. Eng.* **2020**, *26*, 3461–3487. [CrossRef]
33. McDowell, G.R.; Bolton, M.D. Effect of Particle Size Distribution on Pile Tip Resistance in Calcareous Sand in the Geotechnical Centrifuge. *Granul. Matter* **2000**, *2*, 179–187. [CrossRef]
34. Houlsby, G.T.; Nutt, N.R.F.; Sweeney, M. End Bearing Capacity of Piles in Carbonate Soils. In Proceedings of the 13th International Conference on Soil Mechanics and Foundation Engineering, New Delhi, India, 5–10 January 1994.
35. Wang, Q.; Xiao, Z.; Zhao, X.; Feng, D. The Effects and Vertical Bearing Capacity of Two Jacked Model Piles in Sand. *Sustainability* **2022**, *14*, 14493. [CrossRef]
36. Huynh, T.Q.; Nguyen, T.T.; Nguyen, H. Base resistance of super-large and long piles in soft soil: Performance of artificial neural network model and field implications. *Acta Geotech.* **2023**, *18*, 2755–2775. [CrossRef]
37. Al-Douri, R.H.; Poulos, H.G. Predicted and Observed Cyclic Performance of Piles in Calcareous Sand. *J. Geotech. Eng.* **1995**, *121*, 1–16. [CrossRef]
38. Malakshah, R.R.; Moradi, M.; Ghalandarzadeh, A. Centrifuge Modelling of Monopiles in Calcareous Sand Subjected to Cyclic Lateral Loading. *Int. J. Civ. Eng.* **2022**, *20*, 195–206. [CrossRef]
39. Wang, S.; Lei, X.; Meng, Q.; Xu, J.; Wang, M.; Guo, W. Model Tests of Single Pile Vertical Cyclic Loading in Calcareous Sand. *Mar. Georesour. Geotechnol.* **2021**, *39*, 670–681. [CrossRef]
40. Jiang, H.; Wang, R.; Lv, Y.; Meng, Q. Test Study of Model Pile in Calcareous Sands. *Rock Soil Mech.* **2010**, *31*, 780–784. (In Chinese) [CrossRef]
41. Jiang, H.; Wang, R.; Lv, Y.; Meng, Q. Model Tests of Pile Groups in Calcareous Sands. *Chin. J. Rock Mech. Eng.* **2010**, *29*, 3023–3028. (In Chinese)
42. Yang, C.; Jiang, H. The Comparative Study of Single Pile Bearing Characteristics in Calcareous Sands Foundation and Quartz Sands Foundation Based on Model Experiment. *IOP Conf. Ser. Earth Environ. Sci.* **2020**, *455*, 012086. [CrossRef]
43. Wan, Z.; Liu, H.; Zhou, F.; Dai, G. Axial Bearing Mechanism of Post-Grouted Piles in Calcareous Sand. *Appl. Sci.* **2022**, *12*, 2731. [CrossRef]
44. Wan, Z.; Dai, G.; Gong, W. Study on the response of postside-grouted piles subjected to lateral loading in calcareous sand. *Acta Geotech.* **2022**, *17*, 3099–3115. [CrossRef]
45. Peng, Y.; Liu, J.; Ding, X.; Fang, H.; Jiang, C. Performance of X-Section Concrete Pile Group in Coral Sand Under Vertical Loading. *China Ocean Eng.* **2020**, *34*, 621–630. [CrossRef]
46. Gao, Y.; Yu, J.; Wang, Y.; Li, W.; Shi, T. Breakage Effect of Calcareous Sand on Pile Tip Resistance and the Surrounding Soil Stress. *Energy Rep.* **2022**, *8*, 183–190. [CrossRef]
47. Peng, Y.; Ding, X.; Yin, Z.; Wang, P. Micromechanical analysis of the particle corner breakage effect on pile penetration resistance and formation of breakage zones in coral sand. *Ocean Eng.* **2022**, *259*, 111859. [CrossRef]

48. Jiang, H. Research on Bearing Behavior of Pile Foundation in Calcareous Sands. Ph.D. Dissertation, The Chinese Academy of Sciences (Institute of Rock and Soil Mechanics), Wuhan, China, 2009. (In Chinese).
49. Ovensen, N.K. The Use of Physical Models in Design: The Scaling Law Relationships. In Proceedings of the 7th European Conference on Soil Mechanics and Foundation Engineering, Brighton, UK, 10–13 September 1979.
50. Dyson, G.J.; Randolph, M.F. Monotonic Lateral Loading of Piles in Calcareous Sand. *J. Geotech. Geoenviron. Eng.* **2001**, *127*, 346–352. [CrossRef]
51. Peng, Y.; Liu, H.; Li, C.; Ding, X.; Deng, X.; Wang, C. The detailed particle breakage around the pile in coral sand. *Acta Geotech.* **2021**, *16*, 1971–1981. [CrossRef]

Journal of
Marine Science and Engineering

MDPI

Article

A Theoretical Methodology and Measurement of Dynamic Characteristics of Wind Turbines with Composite Bucket Foundations

Jijian Lian [1,2], Huan Zhou [1,2] and Xiaofeng Dong [1,2,*]

1 State Key Laboratory of Hydraulic Engineering Simulation and Safety, Tianjin University, Tianjin 300350, China; jjlian@tju.edu.cn (J.L.); zhouhuan@tju.edu.cn (H.Z.)
2 School of Civil Engineering, Tianjin University, Tianjin 300350, China
* Correspondence: tju_skl@126.com

Abstract: A composite bucket foundation (CBF) is a new type of supporting structure in offshore wind engineering. Its huge transition part is the key difference compared to other offshore foundations. Firstly, the vibration measurement system of a wind turbine with the CBF is introduced. A finite element method (FEM) was developed, and the rigid deformation performance of the transition part was characterized. Then, to clarify the influence of the transition part brings to wind turbines with CBFs, a three-DOF theoretical model was established by simplifying the transition part as a rigid body. Horizontal and rotational foundation stiffness were considered to present the constraint effect below the mudline. Sensitivity studies were conducted on the parameters (including mass, moment of inertia and mass center height) of the transition part. Further, the vibration properties of the CBF structures under different operation load conditions were compared through the theoretical model and the in situ data. The results show that the relative errors between the theoretical model and FEM model are 3.78% to 5.03%, satisfying the accuracy requirements. The parameters of the transition part have varying degrees of influence on the natural frequency, foundation stiffness and vibration response of the wind turbines with CBFs. Compared to wind and 1P loads, the 3P load has a greater influence if the 3P frequency is close to the natural frequency of the wind turbine.

Keywords: composite bucket foundation; transition part; theoretical model; measurement; operation load

Citation: Lian, J.; Zhou, H.; Dong, X. A Theoretical Methodology and Measurement of Dynamic Characteristics of Wind Turbines with Composite Bucket Foundations. *J. Mar. Sci. Eng.* **2024**, *12*, 106. https://doi.org/10.3390/jmse12010106

Academic Editor: José António Correia

Received: 1 December 2023
Revised: 30 December 2023
Accepted: 3 January 2024
Published: 5 January 2024

1. Introduction

The current global energy structure still primarily relies on non-renewable fossil fuels, and the issues of the energy crisis and environmental pollution are becoming increasingly prominent. Developing renewable energy is an important way to ensure sustainable development for humanity. Wind energy is one of the most widely used and rapidly developing renewable energy sources, including onshore and offshore wind energy. Offshore wind energy has become a global focus due to its abundant reserves, minimal land use, proximity to load centers and suitability for large-scale development. According to the Global Wind Energy Council (GWEC) statistics, the global offshore market grew on average by 21% each year in the past decade, bringing total installations to 64.3 GW at the end of 2022, and GWEC Market Intelligence expects more than 380 GW of new offshore wind capacity to be added over the next decade [1].

The foundation types of offshore wind structures have shown a diversified development trend in the efficient development and utilization of offshore wind energy. Several types of offshore foundations have been applied and developed to different extents, such as the mono-pile foundation [2], multi-pile foundation [3], jacket foundation [4] and bucket foundation [5]. The bucket foundation is mainly composed of a bucket inserted into the soil and a transition part above the mudline. Compared to other foundations, it has several advantages, such as easy construction and installation, good resistance to overturning and steel material savings. In recent years, the bucket foundations have occupied a certain market share as some designs have successfully been applied in engineering, including the mono-bucket foundation (MBF) [6], bucket jacket foundation (BJF) [7] and composite bucket foundation (CBF) [8], as shown in Figure 1. The MBF consists of a single suction bucket and a single steel column transition part. The BJF is composed of the suction bucket(s) and jacket-type transition part. The CBF is proposed by China's Tianjin University. It consists of a single suction bucket and a curved reinforced concrete transition part, making it suitable for the widely distributed weak coastal soils in China.

(a) (b) (c)

Figure 1. Bucket foundation applications: (a) MBF; (b) BJF; (c) CBF.

Studying the dynamic characteristics of the bucket foundation structure is of great significance for its design and the operation safety of wind turbines. Houlsby et al. [9,10] conducted model tests of the MBF and tetrapod foundation under cyclic load conditions and observed the variation in the foundation stiffness. Nielsen SD et al. [11] studied the behavior of an MBF in situ after a half year of measurement. In [12], Wang B et al. compared the dynamic performance of a monopile and an MBF in conjunction with the geological conditions of the East China Sea; it is recommended to use bucket foundations in the deep sea. Zhang PY et al. [13] conducted shaking table tests for the MBF and CBF and compared their responses under seismic load conditions. The results show that CBFs are safer than MBFs during earthquakes. Yu TS et al. [14,15] studied the dynamic response characteristics of the CBF, considering the combination of wave and current loads. Liu GW et al. [16] conducted a three-dimensional numerical simulation to calculate the wave forces on the CBF and suggest the selection of the Morison equation or diffraction theory using the relative diameter of the CBF. Ding HY et al. [17,18] carried out extensive tests on three- and four-bucket BJFs and compared their dynamic performance under seismic load conditions; the results prove the inhibition effects of the three-bucket BJFs on the seismic responses of soils. Jalbi et al. [19] developed analytical solutions to predict the natural frequencies of the BJF wind turbines, which may impact the choice of foundations for jackets.

It is clear that many studies have been conducted on the dynamic characteristics of the bucket foundations, but there is still a lack of studies in some respects. On the one hand, the studies are based on the whole bucket foundation structure, there are few studies focusing on the transition part and there is a lack of suitable model to describe the CBF and its transition part. On the other hand, the dynamic analyses in the tests and numerical simulations are mainly conducted under seismic, wind and wave load conditions, and there are few studies focusing on operation loads. Hence, the main scope of this work was to study the dynamic characteristics of the CBF transition part at a theoretical level and to compare the vibration properties of different operation loads. For calculation convenience, the transition part should be simplified, and the finite element method (FEM) is necessary to demonstrate the simplification. To build a rational FEM model of the CBF structure, a practical CBF structure is needed. Therefore, this paper mainly consists of four parts. Firstly, Section 2 introduces the main information of a CBF wind turbine and the structural vibration monitoring system. The FEM was used to study the deformation of the CBF transition part, as outlined in Section 3. The establishment and verification of the theoretical model of the CBF structure is presented in Section 4. Thirdly, Section 5 contains the detailed analyses of the influence of the transition part parameters (mass, moment of inertia and mass center height) on the CBF structure (natural frequency, foundation stiffness and vibration properties). Finally, the vibration performance of the operation loads (1P/3P/wind load) is compared in Section 6.

2. In Situ Measurement of a Composite Bucket Foundation Structure

2.1. Structure and Measurement System

The study site is a wind farm in Xiangshui, Jiangsu, China. Its center is about 10 km offshore, and the water depth is from 8 m to 12 m [20], as shown in Figure 2a. The farm has a total capacity of 202 MW, consisting of 55 wind turbines. The world's first commercially applied CBF was installed in this wind farm to support a 3.0 MW turbine, as shown in Figure 2b. The CBF was towed from factory to the installation location for 112 h and 290 nautical miles, and the installation work took 8 h, with a horizontal accuracy of 0.03%. The main parameters of the whole CBF structure are listed in Table 1.

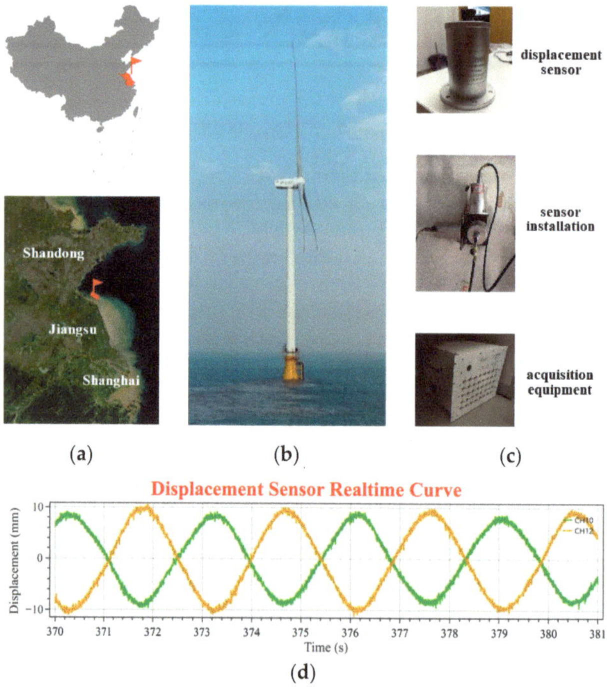

Figure 2. Information of the CBF structure: (**a**) Xiangshui wind farm location; (**b**) the CBF structure; (**c**) vibration displacement sensor and acquisition equipment; (**d**) displacement sensor real-time curve.

Table 1. Main parameters of the CBF structure.

	Parameter	Value		Parameter	Value
Turbine	head mass (t)	190	Suction bucket	height (m)	10
	rotation frequency range (RPM [1])	7.5~13.5		diameter (m)	30
Tower	mass (t)	207	Transition part	mass (t)	1949
	height (m)	73.59		height (m)	23
	diameter (m)	3.2~4.3		mass center height (m)	5.5
	thickness (mm)	14~48		thickness (mm)	600
				diameter (m)	4.3~20.6

[1] RPM = revolutions per minute.

To determine the vibration properties of the structure, a monitoring system was equipped in the wind turbine. Low-frequency vibration displacement sensors have been installed at the tower top to obtain its dynamic movement with the lowest frequency of 0.1 Hz [21], and acquisition equipment is located at the entrance platform with the sampling frequency of 300 Hz, as shown in Figure 2c. Figure 2d shows the sensors' real-time curve of measured displacement.

Based on the modal analysis method SSI (stochastic subspace identification) [22], Table 2 lists the first three orders of natural frequency results of the in situ measurement of the CBF structure. Detailed analysis of the displacement may be found in [21].

Table 2. The first three orders of natural frequency results of the in situ measurement of the CBF structure.

First Order (Hz)	Second Order (Hz)	Third Order (Hz)
0.350	2.501	5.010

2.2. Operation and Environmental Measurement

The structure has been tested in situ for a long period after installation. During vibration testing, operation and environmental information (such as wind speed, rotation speed, power and pitch angle) were also recorded simultaneously by the supervisory control and data acquisition (SCADA) system of the wind turbine. Figure 3 shows the history of wind and rotation speed and their relationship from 2 November 2017 to 8 November 2017. The value of a single point is the mean value of every 100 s. The wind speed range during the test period is from 0 to 15.57 m·s^{-1}, and the rotation speed changes from 0 to 13.5 RPM, including all turbine working conditions from park to rated generation. From Figure 3c, the operation strategy is clear: With the increase in wind speed, the wind turbine changes from park condition to generation state, the cut-in rotation speed is 7.5 RPM. Then, when the wind speed grows from 4.5 m·s^{-1} to 8 m·s^{-1}, the rotation speed grows until it reaches the rated rotation speed of 13.5 RPM. Subsequently, the rotor works stably at the rated speed, though wind speed may continually grow.

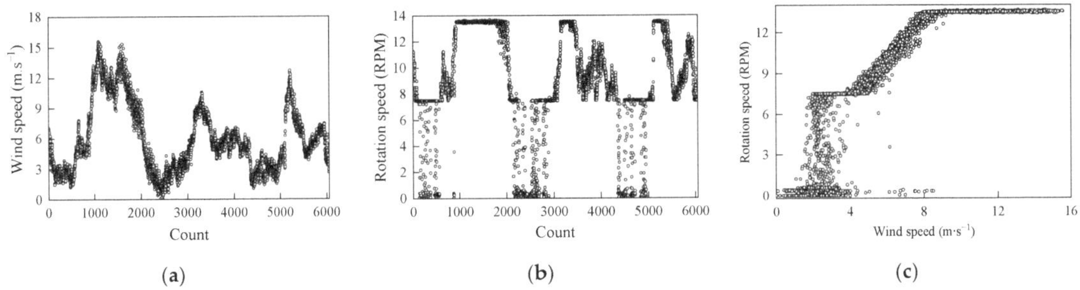

Figure 3. The history of wind and rotation speed and their relationship: (**a**) wind history; (**b**) rotation speed history; (**c**) relationship between wind and rotation speed.

3. Finite Element Modeling

Taking the above mentioned CBF structure as the research object, an ABAQUS model was established using the finite element method, as shown in Figure 4a. The head of the wind turbine was simplified as a mass point placed at the top of the tower. The tower material is Q345E steel, and the transition part is prestressed concrete structure, with C60 concrete and Q235 steel. The steel materials have a density of 7850 kg·m^{-3}, elastic modulus of 206 GPa and Poisson's ratio of 0.3, and the concrete has a density of 2500 kg·m^{-3}, elastic modulus of 36 GPa and Poisson's ratio of 0.25. Rayleigh damping was used for structural damping, with a damping ratio of 2%. An elastic constitutive model was adopted for the whole CBF structure, and the soil was simulated using the Mohr–Coulomb constitutive model. A tie connection was adopted between the tower and foundation, while surface-to-surface contact was used to simulate the interaction between the CBF and soil. Hard contact was applied in the normal direction, and the friction coefficient was set to 0.3 in the tangent direction [23]. The tower and CBF were simulated using 3D shell elements (S4R); the transition part and soil were simulated using 3D solid elements (C3D8), and the grid meshing was performed using the sweep technique. In order to ensure the computational efficiency and numerical accuracy, the local area involved in contact interaction of the model was refined, and mesh density sensitivity analysis was performed. The total number of meshes was about 100,000, and the minimum element size was $0.005D_s$ (D_s is the diameter of the foundation skirt). To eliminate boundary effects, the radius of the soil was taken as $4D_s$, and the depth was taken as $6H_s$ (H_s is the height of the foundation skirt). The bottom boundary of the foundation was fixed, and the lateral boundary only allowed vertical displacement. The soil parameters are listed in Table 3.

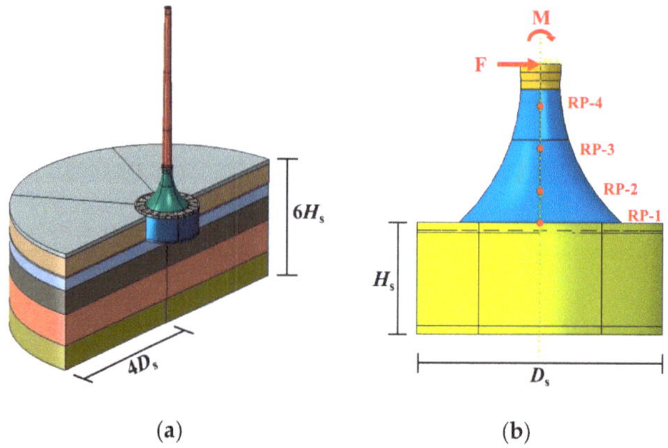

Figure 4. The finite model of the CBF structure: (**a**) whole structure; (**b**) CBF model.

Table 3. The soil parameters of the CBF structure.

Layer	Soil Type	Thickness (m)	Submerged Unit Weight (kN·m^{-3})	Compression Modulus (MPa)	Friction Angle (°)	Cohesion (kPa)
1	Sandy clay	2.1	9.7	4.0	33.8	5.0
2	Silty clay	9.5	8.7	4.8	11.9	22.0
3	Silt	6.6	8.9	6.4	29.3	8.4
4	Sandy clay	11.6	8.7	6.4	12.7	23.9
5	Silt	15.4	8.9	6.4	29.3	8.4
6	Fine sand	11.8	8.7	12.6	12.0	25.0

At the equivalent points on the top of the transition part, horizontal forces and bending moments were applied in the horizontal and rotational directions, respectively, as shown in Figure 4b. The displacements of four reference points (RPs) were observed. The four RPs are key nodes of the transition part, and the heights are 1.20 m, 7.34 m, 11.98 m and 17.00 m from the mudline, respectively, as shown in Figure 4b.

The displacements at four RPs under horizontal force conditions of 50 kN, 100 kN, 250 kN, 500 kN and 1000 kN are shown in Figure 5a. It can be observed that the displacements at different points under the same force conditions are linearly related. There is no relative distortion between the top and bottom of the transition part, indicating that the CBF transition part has linear deformation under horizontal force conditions.

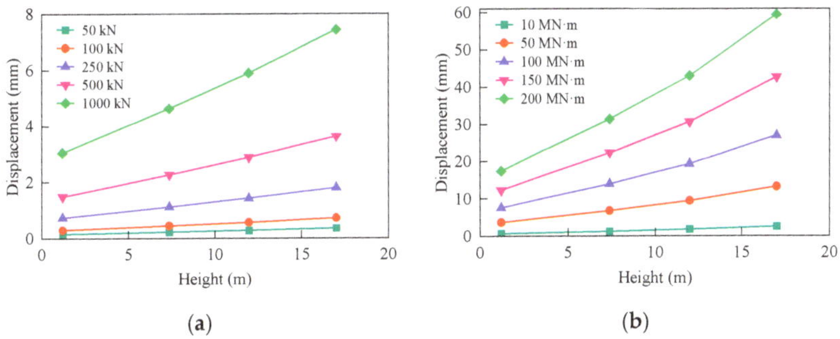

Figure 5. The deformation of the CBF transition part: (**a**) under horizontal force conditions; (**b**) under bending moment conditions.

When applying bending moments, with values of 10 MN·m, 50 MN·m, 100 MN·m, 150 MN·m and 200 MN·m, the displacements at four RPs are shown in Figure 5b. Similarly, when the bending moment does not exceed 100 MN·m, the displacements at the four RPs still show a linear relationship. Only when the moment reaches 150 MN·m does turning occur in the line, but the trend is not significant. So, it can still be approximately considered that the CBF transition part has linear deformation under bending moment conditions.

Therefore, the transition part of the CBF can be simplified as a rigid body and transformed into a mass point with mass m_f and moment of inertia I_f placed at the mass center height.

4. Establishing the Dynamic Model of the CBF Structure

The load of a practical wind structure is complicated; the FEM has a complex modeling process and a time-consuming calculation process, so neither method is suitable for structural dynamic analysis. To make clear how the transition part influences the CBF structure, it is better to establish a theoretical model so that the factors can be quantitatively analyzed one-by-one.

4.1. Motion Equation

The theoretical model of the CBF structure was established in the *xz* plane, as shown in Figure 6. The head of the wind turbine (including impeller, hub and nacelle, etc.) was simplified to a centralized mass point *m* with a rigid connection to the tower, ignoring the mass overhang. The tower was simplified to an elastic beam, with length *h* and bending stiffness k_t. Based on the analysis in Section 3, the transition part (height *H*) is regarded as a mass point, with mass m_f and moment of inertia I_f. The distance from the mass point to the mud surface is the mass center height h_f. The constraint effect of the CBF below the mudline to the wind turbine is equivalent to the horizontal stiffness k_L and rotational stiffness k_R.

As can be seen, the theoretical model vibrates in the *xz* plane and has three degrees of freedom (DOFs), q_1 represents the tower-top relative deformation with respect to the foundation and q_2 and q_3 donate the horizontal displacement and rotation angle of the foundation, respectively.

The Lagrange motion Equation (1) is used to derive the motion equation of the CBF structure [24]:

$$\frac{\mathrm{d}}{\mathrm{d}t}\left(\frac{\partial T}{\partial q'_k}\right) - \frac{\partial T}{\partial q_k} + \frac{\partial V}{\partial q_k} + \frac{\partial D}{\partial q'_k} = F_k,\ k = 1,\ 2,\ \cdots,\ n \tag{1}$$

where *T* is the kinetic energy of the entire vibration system; *V* is the potential energy of the entire vibration system; *D* is the system energy dissipation function, which is defined as the work carried out by the damping force of the system during the vibration process; q_k (k = 1, 2, $\cdots$, n) are generalized coordinates; F_k is the generalized force corresponding to the k-th generalized coordinate; and n is the number of DOFs of the system, n is 3 for this model.

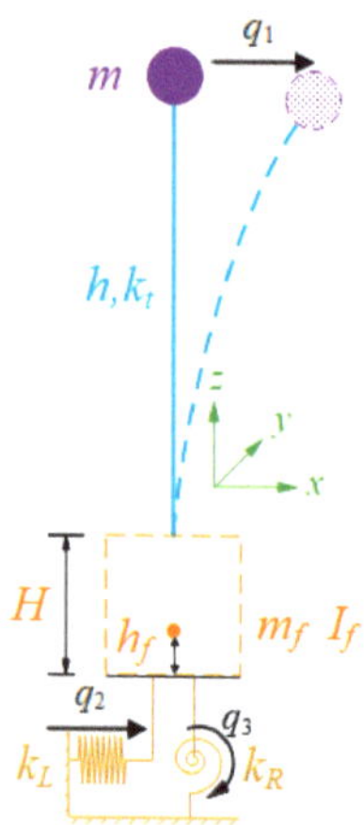

Figure 6. The theoretical model of a CBF structure.

From Figure 6, the absolute displacement of the head (u_h), tower (u_t) and foundation (u_f) of the structure can be expressed as:

$$u_h = q_1 + q_2 + L\tan(q_3) \approx q_1 + q_2 + Lq_3 \tag{2}$$

$$u_t = q_1\varphi_{1t} + q_2 + z\tan(q_3) \approx q_1\varphi_{1t} + q_2 + zq_3 \tag{3}$$

$$u_f = q_1\varphi_{1t,h_f} + q_2 + h_f\tan(q_3) \approx q_1\varphi_{1t,h_f} + q_2 + h_f q_3 \tag{4}$$

where $L = h + H$, φ_{1t} is the vibration mode of the tower, and φ_{1t,h_f} refers to the value of vibration mode at the height of transition part mass center ($z = h_f$).

The total kinetic energy T of the CBF structure can be obtained with the following equation:

$$T = \frac{1}{2}m{u'_h}^2 + \frac{1}{2}\int_H^L \widetilde{m}{u'_t}^2 dz + \frac{1}{2}m_f{u'_f}^2 + \frac{1}{2}I_f{q'_3}^2 \tag{5}$$

where $\widetilde{m}$ is the mass per unit length of the tower.

The total potential energy V of the whole system is:

$$\begin{aligned} V &= \tfrac{1}{2}k_t q_1^2 + \tfrac{1}{2}k_L q_2^2 + \tfrac{1}{2}k_R q_3^2 \\ k_t &= \int_H^L EI(z)(\varphi''_{1t})^2 dz \end{aligned} \tag{6}$$

where k_t is the stiffness of the tower [25].

System energy dissipation function D [26] is defined as:

$$D = \frac{1}{2}c_h{u'_h}^2 + \frac{1}{2}c_L{q'_2}^2 + \frac{1}{2}c_R{q'_3}^2 \tag{7}$$

where c_h is the damping at the head, and c_L and c_R refer to the horizontal and rotational damping of the foundation, respectively.

Introducing Formulas (5)–(7) into Formula (1), the motion equation of the CBF structure can be obtained as follows:

$$\boldsymbol{M}\boldsymbol{q}'' + \boldsymbol{C}\boldsymbol{q}' + \boldsymbol{K}\boldsymbol{q} = \boldsymbol{F}(t) \tag{8}$$

where

$$\boldsymbol{M} = \begin{bmatrix} m + M_{1t} & m + M_{2t} & mL + M_{3t} \\ m + M_{2t} & m + \widetilde{m}h + m_f & mL + \frac{1}{2}\widetilde{m}(L^2 - H^2) + m_f h_f \\ mL + M_{3t} & mL + \frac{1}{2}\widetilde{m}(L^2 - H^2) + m_f h_f & mL^2 + \frac{1}{3}\widetilde{m}(L^3 - H^3) + m_f {h_f}^2 + I_f \end{bmatrix} \tag{9}$$

$$M_{1t} = \widetilde{m}\int_H^L \varphi_{1t}^2 dz + m_f \varphi_{1t,h_f}^2,\ M_{2t} = \widetilde{m}\int_H^L \varphi_{1t} dz + m_f \varphi_{1t,h_f},\ M_{3t} = \widetilde{m}\int_H^L \varphi_{1t} z dz + m_f h_f \varphi_{1t,h_f} \tag{10}$$

$$\boldsymbol{C} = \begin{bmatrix} c_h & 0 & 0 \\ 0 & c_L & 0 \\ 0 & 0 & c_R \end{bmatrix} \tag{11}$$

$$\boldsymbol{K} = \begin{bmatrix} k_t & 0 & 0 \\ 0 & k_L & 0 \\ 0 & 0 & k_R \end{bmatrix} \tag{12}$$

For such a low damping structure like the wind turbine structure, the coupling effect caused by the non-diagonal elements in matrix $\boldsymbol{C}$ is negligible compared with that of diagonal elements, so only the diagonal elements are retained [27]. The damping ratios in this study were all set to 2% Rayleigh damping [28].

4.2. Foundation Stiffness

A genetic algorithm (GA) is a random search algorithm that borrows from the natural selection and genetic mechanisms in the biological world. Starting from any initial population, it continuously reproduces and evolves generation-by-generation through random selection, crossover and mutation operation and finally converges to a group of individuals which is the most fitted to the environment to find the optimal solution to problems [29]. The GA has high parallel computing capabilities and good scalability and has widely used in wind engineering to optimize wind farm layout, power production and structural design [30–32]. Before solving the motion Equation (9), the foundation stiffness k_L and k_R are obtained by the GA to ensure that the natural frequency results of the CBF structure meet the requirements. The target function is defined as the sum of percentage difference between the natural frequency results and the natural frequency targets, as shown in Equation (13). The smaller the sum, the more accurate the result is. The calculation process of GA to solve the foundation stiffness is shown in Figure 7, and the GA parameters used are shown in Table 4.

$$\min\ target = \sum_{i=1}^{N} \frac{\left| f - f_{np} \right|}{f_{np}} \times 100\% \tag{13}$$

where min represents minimum of the sum; N is the order of the natural frequencies, specifically, $N = 3$ in this study; f is the natural frequency results simulated by the GA; and f_{np} is the natural frequency target.

4.3. Loading

Compared with other types of tall structures, operation loads are additional loads for wind turbines due to rotation of the rotor. Operation loads mainly include the fluctuating wind load, 1P load and 3P load [33]. The 1P/3P loads are derived from the 1P/3P vibration generated by the rotation of the rotor. Their load frequencies are equal to the rotor rotation frequency (1P frequency) and the blade passing frequency (3P frequency), respectively. In recent studies, there is a lack of understanding of the influence of each operation load. Therefore, this study investigated the dynamic characteristics of the CBF structure by separately considering the effects of the 1P load, 3P load and wind load.

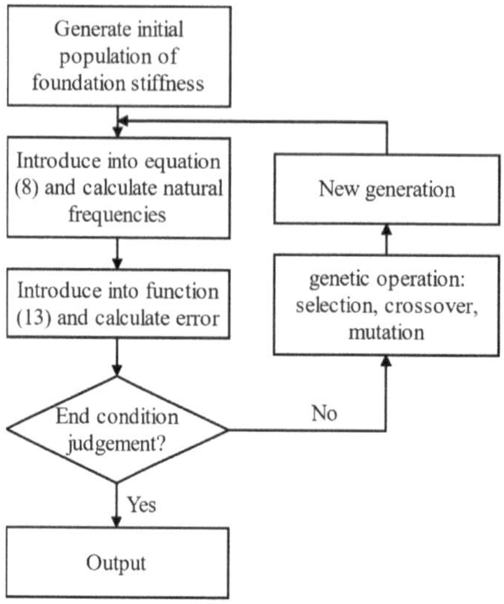

Figure 7. Flowchart of foundation stiffness calculation by the GA.

Table 4. The GA parameters for solving foundation stiffness.

Initial Population	Maximum Genetic Generation	Selection Type	Crossover Rate	Mutation Rate
300	100	roulette	0.6	0.05

In practical engineering, the 1P/3P load can be quite complex. For simplicity of calculation, sinusoidal functions were adopted in this study to simplify the 1P/3P load.

$$\begin{aligned} F_{1P} &= \sqrt{2} f_{1P} \sin(2\pi\omega t) \\ F_{3P} &= \sqrt{2} f_{3P} \sin(6\pi\omega t) \end{aligned} \tag{14}$$

where F_{1P} and F_{3P} are the 1P load and 3P load, respectively; f_{1P}, f_{3P} are the root mean square (RMS) value of the loads; and ω is the wind turbine rotation frequency.

The characteristic parameters reflecting the properties of fluctuating wind include turbulence intensity, turbulence integral length and pulsating wind speed power spectrum. The Det Norske Veritas (DNV) specification [34] recommends using the Kaimal spectrum to obtain the required wind speed, and the wind speed spectrum can be written as follows:

$$S_U(f) = \frac{4I^2 U_{10} L_k}{(1 + 6fL_k/U_{10})^{5/3}} \tag{15}$$

where f represents the frequency, U_{10} is the average wind speed in 10 min, I is the turbulence intensity, and L_k is the turbulence integral length, which can be determined by the following formula:

$$L_k = \begin{cases} 5.67z & z < 60 \text{ m} \\ 340.2 & z \geq 60 \text{ m} \end{cases} \tag{16}$$

where z is the height above sea level.

Figure 8 shows the time history of 100 s of fluctuating wind, where the turbulence intensity is 0.1, and the sampling frequency is 100 Hz. In the actual calculation, the wind speed is multiplied by a certain coefficient to obtain the required value of wind load.

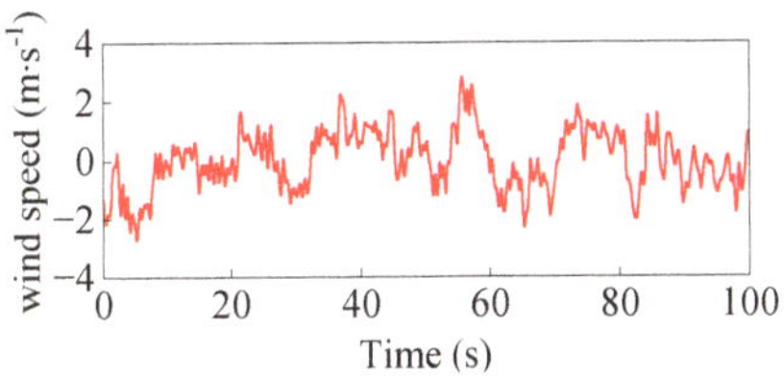

Figure 8. Time history of 100 s of fluctuating wind.

4.4. Verification of the Model

The rationality of the theoretical model was verified by comparing its dynamic response with the FEM model. To reduce the diversity of the two models, the mode shape of the theoretical model is an approximate fitting of the FEM model, considering the linear deformation properties of the transition part. In this article, only the first-order mode is considered, and the mode shape is formulated with Equation (17). It can be seen from Figure 9 that the two mode shapes are proximal, and the mode shape defied by Equation (17) is rational.

$$\begin{aligned}
&c = \{\ c_1\ \ c_2\ \ c_3\ \ c_4\ \ c_5\ \ c_6\ \}^{\mathrm{T}} = \{\ -0.3114\ \ -0.3755\ \ 0.1777\ \ 0.3902\ \ 0.0652\ \ 0.0004\ \}^{\mathrm{T}} \\
&\lambda_1 = 1.54922;\ \lambda_2 = 1.49869 \\
&\varphi_{1t}\left(\overline{h}\right) = \begin{cases} c_1 \sin\left(\lambda_1\overline{h}\right) + c_2\cos\left(\lambda_1\overline{h}\right) + c_3\sinh\left(\lambda_2\overline{h}\right) + c_4\cosh\left(\lambda_2\overline{h}\right) & \frac{H}{L} < \overline{h} \le 1 \\ c_5\overline{h} + c_6 & 0 \le \overline{h} \le \frac{H}{L} \end{cases}
\end{aligned} \tag{17}$$

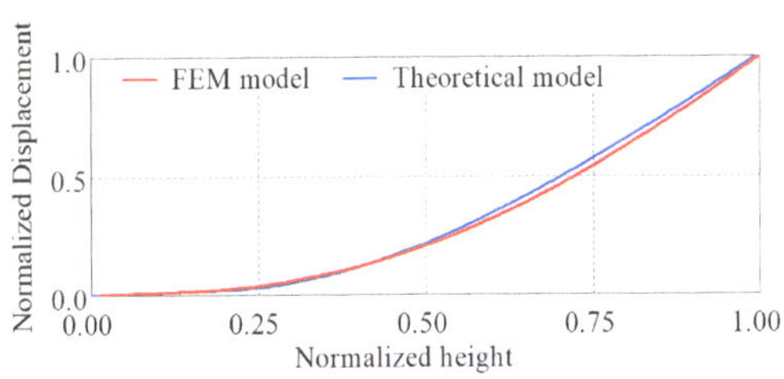

Figure 9. The mode shape of the tower.

Based on the fluctuating wind in Figure 9, the wind load of the root mean square (RMS) value 10 kN was generated and added to the tower top of the FEM model and the theoretical model, respectively. The absolute displacements of the head and foundation were extracted, as shown in Figure 10. As can be seen from the figure, the displacement time history of both models is basically the same; only the gap around 50 s is obvious. The RMS and maximum displacement values of the two models were calculated, as listed in Table 5. The head displacements of the theoretical model are 2.80 mm and 6.80 mm, and the foundation displacements of the theoretical model are 0.66 mm and 1.67 mm. The relative errors of the two models are from 3.78% to 5.03%, indicating that the theoretical model in this paper meets the accuracy requirement and is reasonable.

Table 5. The displacements and errors.

Displacement		FEM (mm)	Theoretical Model (mm)	Error (%)
Head	RMS	2.91	2.80	3.78
	maximum	7.13	6.80	4.63
Foundation	RMS	0.69	0.66	4.28
	maximum	1.76	1.67	5.03

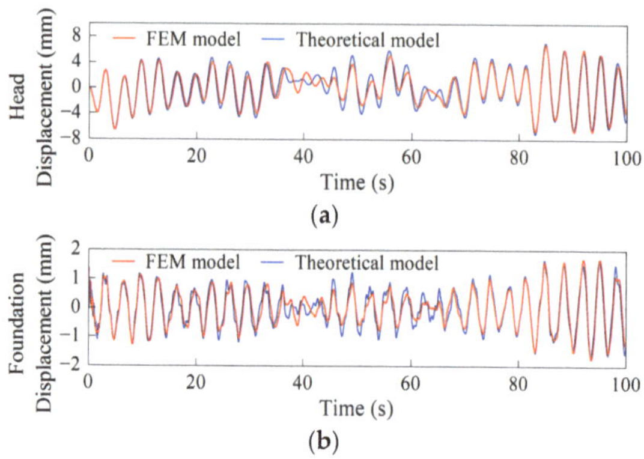

Figure 10. The displacement time history of the two models: (**a**) tower top; (**b**) foundation.

5. Analysis of the Influence of the Transition Part Parameters on the CBF Structure

First, a CBF structure model was built. The tower length was h = 75 m, the diameter was D = 4 m, the wall thickness was 30 mm and the section bending stiffness EI was 290 GPa. The head mass was set to 100 t. The height of the transition part H = 25 m, total mass $m_f = 7.34 \times 10^4$ kg, moment of initial is $I_f = 1.54 \times 10^7$ kg·m^2 and mass center height h_f = 12.5 m. The original natural frequency targets were f_{np} = [0.3 2.3 7.8] Hz, and the foundation stiffness obtained by GA was 1.92×10^7 N·m^{-1}, 3.62×10^{10} N·m·rad^{-1}. The influence of transition part parameters (m_f, I_f, h_f) was analyzed with the theoretical model based on the above values.

5.1. Influence on the Natural Frequencies of the Structure

Figure 11 shows the influence of transition part mass on the first three orders of natural frequencies of the CBF structure. The value of the right axis is the ratio of each point to the initial value. The mass ranges from 7.34×10^4 kg to 1.0×10^6 kg. It can be seen that the increase in the transition part mass causes the decrease in the natural frequencies, in which the first order shows a linear trend, and the second and third order show a gradually slowing trend. Comparing the ratios, the first and third order decrease by about 6%, but the second order decreases by 60%, indicating that the transition part mass mainly influences the second-order natural frequency.

Figure 12 shows the influence of the transition part moment of inertia on the first three orders of natural frequencies of the CBF structure. The moment of inertia ranges from 1.54×10^7 kg·m^2 to 3.0×10^7 kg·m^2. It can be seen that the increase in the moment of inertia causes the decrease in the third natural frequency by 23%, but the first- and second-order frequencies stay unchanged, indicating that the transition part moment of inertia has no effect on the first two natural frequencies; it only causes the third-order frequency reduction.

Figure 13 shows the influence of the transition part mass center height on the first three orders of natural frequencies of the CBF structure. The height ranges from 0 m (transition part bottom) to 25 m (transition part top). It can be seen that the increase in the mass center height causes the decrease in the first-order frequency by 0.7% and causes the increase in the second and third-order frequencies by about 8%, indicating that the transition part mass center height has a slight influence on the natural frequencies of the CBF structure.

5.2. Influence on the Foundation Stiffness

When the transition parameters change, the foundation stiffness provided below the CBF mudline should also be changed to meet the natural frequency design requirements of the wind turbine. In the following comparisons, the natural frequencies [0.3 2.3 7.8] Hz remain unchanged.

Figure 14 shows the influence of the transition part mass on the foundation stiffness of the CBF structure. As can be seen, the mass grows from 7.34×10^4 kg to 1.0×10^6 kg, the horizontal stiffness increases linearly by 6.97 times and the increase in rotational stiffness gradually slows down, with an increase rate of 11.14%, indicating that the transition part mass mainly affects the horizontal stiffness.

Figure 15 shows the influence of the transition part moment of inertia on the foundation stiffness of the CBF structure. As can be seen, the moment of inertia grows from 1.54×10^7 kg·m^2 to 3.0×10^7 kg·m^2, and the horizontal stiffness maximally increases by only 0.06%, while the rotational stiffness increases linearly by 96%. Therefore, it can be considered that the transition part moment of inertia mainly affects the rotational stiffness and has no effect on the horizontal stiffness.

Figure 16 shows the influence of the transition part mass center height on the foundation stiffness of the CBF structure. It can be seen that the two stiffnesses show a similar linear decrease as the mass center height rises from 0 m to 25 m. The change rate compared with the original height is about 17%, indicating that the height change in the mass center has the same influence on the two stiffnesses.

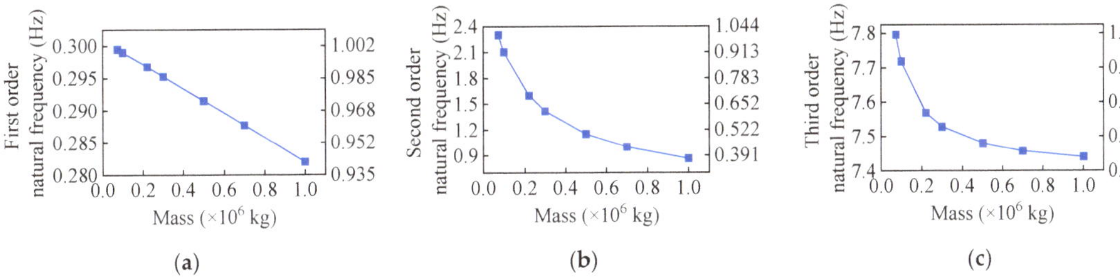

Figure 11. The influence of the transition part mass on the natural frequencies: (**a**) first order; (**b**) second order; (**c**) third order.

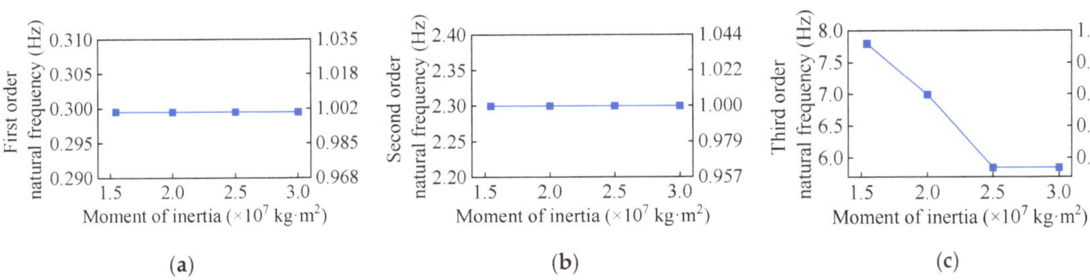

Figure 12. The influence of the transition part moment of inertia on the natural frequencies: (**a**) first order; (**b**) second order; (**c**) third order.

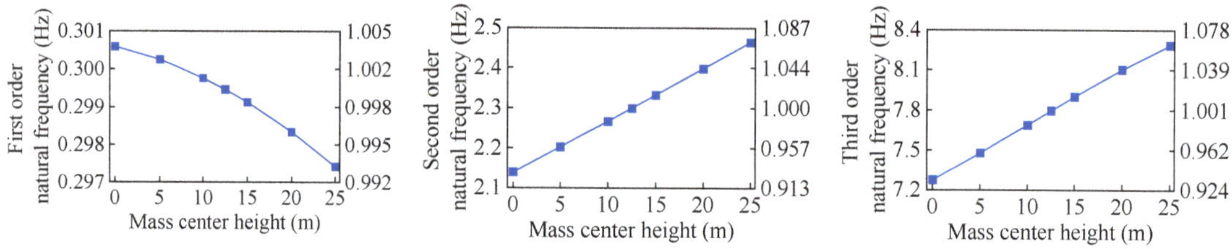

(**a**) (**b**) (**c**)

Figure 13. The influence of the transition part mass center height on the natural frequencies: (**a**) first order; (**b**) second order; (**c**) third order.

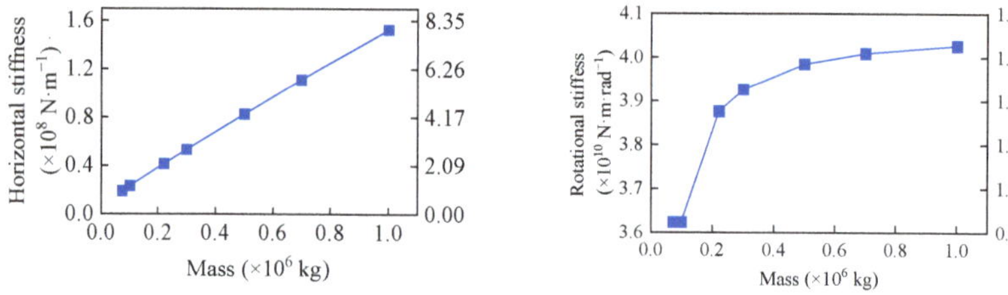

(**a**) (**b**)

Figure 14. The influence of the transition part mass on the foundation stiffness: (**a**) horizontal stiffness; (**b**) rotational stiffness.

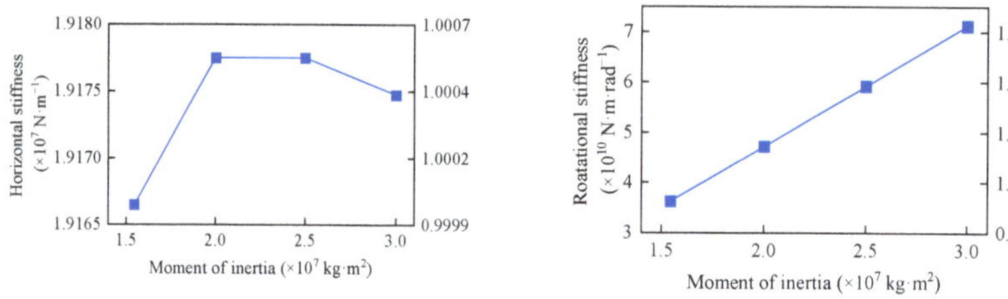

(**a**) (**b**)

Figure 15. The influence of the transition part moment of inertia on the foundation stiffness: (**a**) horizontal stiffness; (**b**) rotational stiffness.

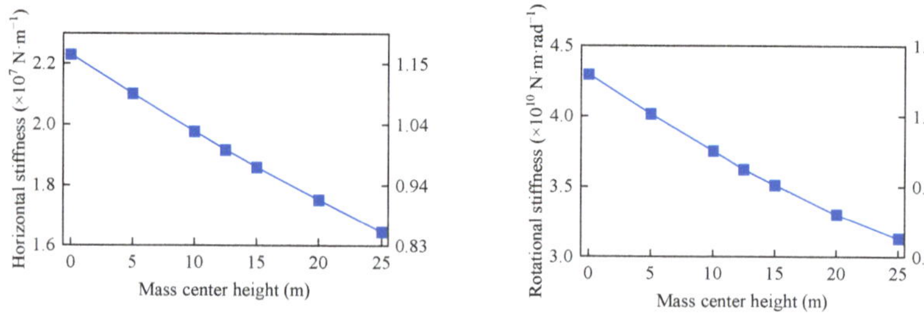

(**a**) (**b**)

Figure 16. The influence of the transition part mass center height on the foundation stiffness: (**a**) horizontal stiffness; (**b**) rotational stiffness.

5.3. Influence on the Vibration Properties

When analyzing the influence of transition part parameters on the vibration properties of the CBF structure, the input load is also the fluctuating wind load of the RMS, 10 kN, and the calculation time is 100 s. The head displacement RMS value of the wind turbine was taken to represent the vibration response. The results are shown in Figure 17.

Figure 17a is the influence of the transition part mass on the vibration properties. It can be seen that the displacement increases first and then decreases with the mass of the transition part, and the peak 3.67 mm appears at 2.20×10^5 kg. It was found that the displacement is largest when the transition part mass is equal to the tower mass; if the transition part mass is different from the tower mass, the displacement is rapidly reduced. Generally, the CBF has a huge transition part, whose mass is much greater than the tower mass. Accordingly, enlarging the transition part mass can reduce the vibration of the structure.

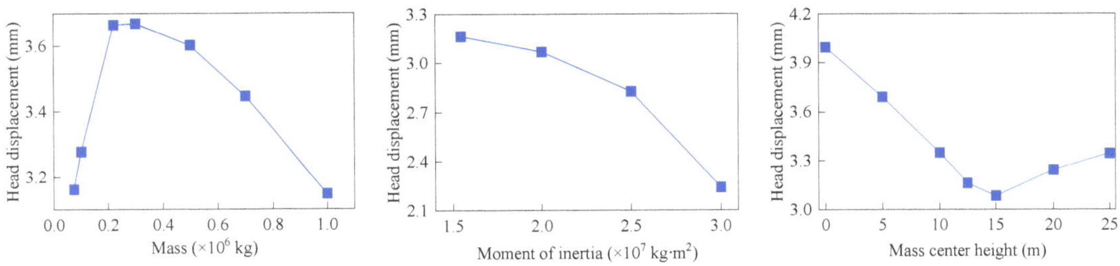

(a) (b) (c)

Figure 17. The influence of transition part parameters on the vibration properties: (**a**) mass; (**b**) moment of inertia; (**c**) height of mass center.

Figure 17b shows the influence of the transition part moment of inertia on the vibration properties. It can be seen that with the increase in the moment of inertia, the displacement gradually decreases from 3.16 mm to 2.24 mm, which may be caused by the increased utilization of wind energy for the rotation of the CBF. So, enlarging the transition part moment of inertia can also reduce the vibration of the structure.

Figure 17c shows the influence of the transition part mass center height on the vibration properties. It can be seen that the displacement presents the "V"-type with the elevation of the mass center height, and the minimum value 3.16 mm appears when the mass center height is 15 m. According to the analysis, the mass center height deviation from the mudline produces an additional moment of inertia $m_f h_f^2$; when it is equal to the original moment of inertia I_f, the displacement is the minimum, so the deviation is $\sqrt{I_f/m_f} = 14.48$ m, which is consistent with the figure. Therefore, it is better for the transition part mass center height to be designed near the deviation $\sqrt{I_f/m_f}$ to reduce the vibration.

6. Vibration Properties under Different Operation Loading Conditions

Also, the load RMS value was set to 10 kN. Figure 18 shows the tower-top displacements under the 1P/wind/3P load conditions of the theoretical model. It can be seen that the 1P/3P displacements are positively and negatively correlated with rotation frequency. When rotation frequency nears 0.1 Hz, the 3P frequency is close to 0.3 Hz (the natural frequency of the model), so 3P resonance [33] occurs. The 3P displacement is obviously higher than 1P and wind displacements. Therefore, close scrutiny is required when the 3P frequency is close to the natural frequency of the structure.

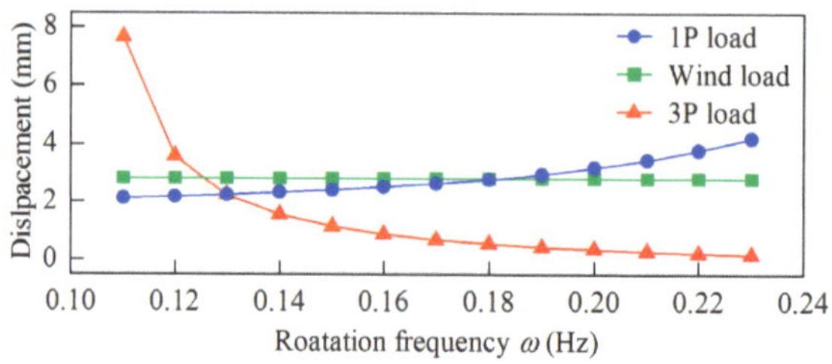

Figure 18. The displacements under different load conditions.

For the CBF structure mentioned in Section 2, the main frequencies of all 100 s data were extracted using the spectral analysis method and combined with the corresponding operation status, The distribution of the main frequency with rotation speed is presented in Figure 19.

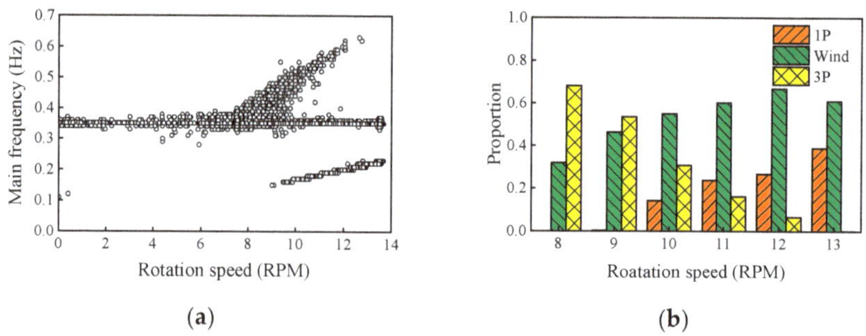

Figure 19. The distribution of the main frequency with rotation speed: (**a**) scatter; (**b**) histogram.

As shown in Figure 19a, the main frequency is concentrated in the first-order natural frequency of the wind turbine if rotation speed is less than 7.5 RPM. When rotation speed is higher than 7.5 RPM, the 1P/3P and natural frequencies are all reflected in the main frequency. The 3P frequency aggregates at the rotation interval near 7.5 RPM, which results from the 3P resonance caused by the 3P frequency 0.375 Hz approaching the natural frequency 0.35 Hz of the structure.

The occurrence frequencies of the main frequencies were recorded according to the speed range, and their proportion was calculated and plotted as shown in Figure 19b. It can be seen that the proportion of the 1P frequency increases, and the 3P proportion decreases gradually with respect to the increase in rotational speed. The natural frequency increases first and then decreases with the rotational speed. The 3P frequency is the most prominent main frequency at rotation intervals of less than 9 RPM, indicating that the 3P load has an important influence on this structure. Furthermore, the 1P load is less remarkable compared with 3P and wind loads. Dong XF et al. [35] tested the first full-scale CBF model and found that the 3P load has nearly no effect on the structure, which is quite different from the conclusion in this article. Therefore, the 3P load should not be ignored; it must be taken into account, lest excessive vibration or damage occurs.

7. Conclusions

A CBF prototype and monitoring system is introduced. By simplifying the CBF and whole wind turbine structure, a theoretical model was established and verified, the influence of the transition part parameters on the dynamic characteristics of the structure was studied and the vibration responses of operation loads were compared. The main conclusions are as follows:

(1) The transition part of the CBF can be regarded as a rigid body. Based on this simplification, the theoretical model of the CBF structure is accurate and reliable, and the errors of the displacement RMS and maximum values compared to the FEM results are 3.78% to 5.03%.

(2) The transition part mass mainly has a negative effect on the second-order natural frequency of the CBF structure, and the transition part moment of inertia only causes the third-order frequency reduction, while the mass center height has a slight influence on the natural frequencies.

(3) The transition part mass and moment of inertia mainly affect the horizontal and rotational stiffness of the CBF structure, respectively, with linear positive trends. In contrast, the influence of mass center height on the two stiffness of the foundation is consistent but limited.

(4) Enlarging both the transition part mass and moment of inertia can reduce the vibration of the CBF structure. The mass center shows a "V"-type trend for the vibration response, and the extreme value appears at the height of $\sqrt{I_f/m_f}$.

(5) From the theoretical analysis and the in situ measurement results, the 3P load has a great influence on the wind turbine if the 3P frequency is close to the natural frequency, and it should not be ignored.

With this work, the influence of transition part parameters on the CBF structure is clear and helpful for CBF design, and new knowledge on the vibration response difference between operation loads can guide the operation strategy optimization for wind turbines to reduce structural vibration.

Author Contributions: Conceptualization, J.L.; methodology, X.D. and H.Z.; validation, X.D. and J.L.; formal analysis, H.Z.; investigation, J.L., X.D. and H.Z.; resources, J.L.; data curation, X.D. and H.Z.; writing—original draft preparation, H.Z.; writing—review and editing, X.D.; visualization, H.Z.; supervision, X.D.; project administration, J.L.; funding acquisition, J.L. All authors have read and agreed to the published version of the manuscript.

Funding: This research was funded by the National Natural Science Foundation of China, grant number U21A20164.

Institutional Review Board Statement: Not applicable.

Informed Consent Statement: Not applicable.

Data Availability Statement: Data are contained within the article.

Conflicts of Interest: The authors declare no conflicts of interest.

References

1. Global Wind Report. 2023. Available online: https://gwec.net/globalwindreport2023/ (accessed on 27 March 2023).
2. Zhou, L.; Li, Y.; Liu, F.S.; Jiang, Z.Q.; Yu, Q.X.; Liu, L.J. Investigation of dynamic characteristics of a monopile wind turbine based on sea test. *Ocean Eng.* **2019**, *189*, 106308. [CrossRef]
3. Tu, Z.S.; Zhang, C.L.; Liu, H.Q.; Zhu, R.H. Hydrodynamic analysis of a multi-pile-supported offshore wind turbine integrated with an aquaculture cage. *J. Mar. Sci. Eng.* **2023**, *11*, 1830. [CrossRef]
4. Ku, C.Y.; Chien, L.K. Modeling of load bearing characteristics of jacket foundation piles for offshore wind turbines in Taiwan. *Energies* **2016**, *9*, 625. [CrossRef]
5. Oha, K.Y.; Namb, W.; Ryuc, M.S.; Kimc, J.Y.; Epureanu, B.I. A review of foundations of offshore wind energy convertors: Current status and future perspectives. *Renew. Sust. Energ. Rev.* **2018**, *88*, 16–36. [CrossRef]
6. Kapitanov, L.R.; Duroska, P.; Garcia, C.A.; Puigvert, M.M.; Ibsen, L.B.; Vandatirad, M.J. Literature review on cyclic lateral loading effects of mono-bucket foundations. In Proceedings of the 6th International Conference on Structural Engineering, Mechanics and Computation (SEMC), Cape Town, South Africa, 5–7 September 2016.
7. Penner, N.; Griessmann, T.; Rolfes, R. Monitoring of suction bucket jackets for offshore wind turbines: Dynamic load bearing behaviour and modelling. *Mar. Struct.* **2020**, *72*, 102745. [CrossRef]
8. Lian, J.J.; Sun, L.Q.; Zhang, J.F.; Wang, H.J. Bearing capacity and technical advantages of composite bucket foundation of offshore wind turbines. *Trans. Tianjin Univ.* **2011**, *17*, 132–137. [CrossRef]
9. Houlsby, G.T.; Kelly, R.B.; Huxtable, J. Field trials of suction caissons in clay for offshore wind turbine foundations. *Geotechnique* **2005**, *55*, 287–296. [CrossRef]

10. Houlsby, G.T.; Kelly, R.B.; Huxtable, J. Field trials of suction caissons in sand for offshore wind turbine foundations. *Geotechnique* **2006**, *56*, 3–10. [CrossRef]
11. Nielsen, S.D.; Ibsen, L.B.; Nielsen, S.A. Performance of a mono bucket foundation—A case study at Dogger Bank. *Int. J. Offshore Polar* **2017**, *27*, 326–332. [CrossRef]
12. Wang, B.; Qin, T.Q.; Yuan, C.F.; Li, L.; Yuan, M.H.; Li, Y. Analysis of bearing performance of monopile and single suction bucket foundation for offshore wind power under horizontal load. *Geofluids* **2022**, *2022*, 4163240. [CrossRef]
13. Zhang, P.Y.; Li, J.Y.; Le, C.H.; Ding, H.Y. Seismic responses of two bucket foundations for offshore wind turbines based on shaking table tests. *Renew. Energ.* **2022**, *187*, 1100–1117. [CrossRef]
14. Liu, X.X.; Chen, X.G.; Liu, J.Z.; Yu, T.S.; Sun, B.S. Scour effects on cyclic response of bucket foundations under waves and currents. *J. Coastal Res.* **2021**, *37*, 670–682. [CrossRef]
15. Song, H.Y.; Yu, T.S.; Yu, D.Y.; Zhao, Z.S. Hydrodynamic behavior of composite bucket foundation with random waves. *Ocean Eng.* **2023**, *281*, 114774. [CrossRef]
16. Liu, G.W.; Zhang, Q.H.; Zhang, J.F. Wave forces on the composite bucket foundation of offshore wind turbines. *Prog. Ind. Civil Eng. II* **2013**, *405–408*, 1420–1426. [CrossRef]
17. Ding, H.Y.; Pan, C.; Zhang, P.Y.; Wang, L.; Xu, Y.L. Shaking table tests and seismic response of three-bucket jacket foundations for offshore wind turbines. *J. Ocean Univ. China* **2022**, *21*, 719–736. [CrossRef]
18. Ding, H.Y.; Li, J.Y.; Le, C.H.; Pan, C.; Zhang, P.Y. Shaking table tests of four-bucket jacket foundation for offshore wind turbines. *China Ocean Eng.* **2022**, *36*, 849–858. [CrossRef]
19. Jalbi, S.; Nikitas, G.; Bhattacharya, S.; Alexander, N. Dynamic design considerations for offshore wind turbine jackets supported on multiple foundations. *Mar. Struc.* **2019**, *67*, 102631. [CrossRef]
20. Ding, H.Y.; Feng, Z.T.; Zhang, P.Y.; Le, C.H.; Guo, Y.H. Floating Performance of a Composite Bucket Foundation with an Offshore Wind Tower during Transportation. *Energies* **2020**, *13*, 882. [CrossRef]
21. Lian, J.J.; Zhou, H.; Dong, X.F. Study on structural vibration behavior and vibration reduction of bucket foundation wind turbines. *J. Hydroe. Eng.* **2022**, *41*, 1–9. (In Chinese)
22. Kraemer, P.; Friedmann, H. Vibration-based structural health monitoring for offshore wind turbines—Experimental validation of stochastic subspace algorithms. *Wind Struct.* **2015**, *21*, 693–707. [CrossRef]
23. Sun, G.D.; Liu, R.; Li, Q.X.; Huang, G.Y.; Zhou, Z.B.; Zhang, C. Research on the Critical Buckling Pressure of Bucket Foundations with Inner Compartments for Offshore Wind Turbines. *Ocean Univ. China* **2023**, *22*, 419–428. [CrossRef]
24. Sun, C.; Jahangiri, V. Bi-directional vibration control of offshore wind turbines using a 3D pendulum tuned mass damper. *Mech. Syst. Signal Pr.* **2018**, *105*, 338–360. [CrossRef]
25. Clough, R.W.; Penzien, J. *Dynamics of Structures*, 3rd ed.; Computers & Structures, Inc.: Walnut Creek, CA, USA, 2003.
26. Brasil, R.M.L.R.; Feitosa, L.C.S.; Balthazar, J.M. A nonlinear and non-ideal wind generator supporting structure. *Appl. Mech. Mater.* **2006**, *5–6*, 433–442. [CrossRef]
27. Bao, W.B.; Bai, Q.; Lu, H.Y. *Fundamentals and MATLAB Applications of Vibration Mechanics*; Tsinghua University Press: Beijing, China, 2015.
28. Carswell, W.; Johansson, J.; Lovholt, F.; Arwade, S.R.; Madshus, C.; DeGroot, D.J.; Myers, A.T. Foundation damping and the dynamics of offshore wind turbine monopiles. *Renew. Energ.* **2015**, *80*, 724–736. [CrossRef]
29. Lei, Y.J.; Zhang, S.W. *Toolbox and Application of Genetic Algorithm in MATLAB*; Xidian University Press: Xi'an, China, 2014.
30. Ogunjuyigbe, A.S.O.; Ayodele, T.R.; Bamgboje, O.D. Optimal placement of wind turbines within a wind farm considering multi-directional wind speed using two-stage genetic algorithm. *Front. Energy* **2021**, *15*, 240–255. [CrossRef]
31. Poudel, P.; Kumar, R.; Narain, V.; Jain, S.C. Power optimization of a wind turbine using genetic algorithm. In Proceedings of the 4th International and 19th National Biennial Conferences on Machines and Mechanisms (iNaCoMM), Suran, India, 5–7 December 2019.
32. Tian, J.M.; Tan, X. The Optimization Design of Wind Turbine Gearbox Based on Improved Genetic Algorithm and Feasibility Analysis. In Proceedings of the 4th International Conference on Advances in Materials and Manufacturing (ICAMMP 2013), Kunming, China, 18–19 December 2013.
33. Lian, J.J.; Zhou, H.; Dong, X.F. A theoretical approach for resonance analysis of wind turbines under 1P/3P loads. *Energies* **2022**, *15*, 5787. [CrossRef]
34. *DNV-OS-J101*; Design of Offshore Wind Turbine Structures. Det Norske Veritas: Oslo, Norway, 2011.
35. Dong, X.F.; Lian, J.J.; Wang, H.J. Vibration source features of offshore wind power structures under operational conditions. *J. Vib. Shock* **2017**, *36*, 21–28. [CrossRef]

Journal of
Marine Science
and Engineering

MDPI

Article

An Evaluation of Treatment Effectiveness for Reclaimed Coral Sand Foundation in the South China Sea

Ting Yao [1] and Wei Li [2,*]

[1] Sate Key Laboratory of Geomechanics and Geotechnical Engineering, Institute of Rock and Soil Mechanics, Chinese Academy of Sciences, Wuhan 430071, China; tyao@whrsm.ac.cn
[2] Faculty of Engineering, China University of Geosciences, Wuhan 430074, China
* Correspondence: liwei35@cug.edu.cn

Abstract: Mega land reclamation projects have been carried out on the coral reefs in the South China Sea. Coral sand was used as a backfill material through hydraulic filling, with fill heights ranging from 6 to 10 m. To enhance foundation stability, vibro-flotation and impact rolling have been employed. However, the uneven distribution of coral sand, irregular particle shape, lower single-particle strength, and paucity of engineering cases for reference have posed challenges in evaluating the effectiveness of these foundation treatments. In this study, the effectiveness of vibro-flotation and impact rolling on the densification and bearing capacity of coral sand foundations has been investigated. In situ tests, including the plate load test, California Bearing Ratio (CBR) test, density measurements, dynamic penetration test (DPT), and settlement monitoring, were conducted at four distinct zones: an untreated zone, a vibro-flotation zone at a 5 m depth, a vibro-flotation zone at a 10 m depth, and an impact rolling zone. The findings suggest that coral sand exhibits promising characteristics for foundation construction. Seepage and self-weight consolidation following land reclamation formation significantly enhance the compaction degree of the coral sand foundation, thereby meeting the requirements for areas with lower bearing capacity demands. Both vibro-flotation and impact rolling techniques could significantly enhance the foundation-bearing capacity, with marginal differences between them. Since the machinery is simple and construction speed is quick, the impact rolling method is considered to be the most efficient for the treatment of coral sand foundation. The DPT results suggest that the reinforcement effect of both vibro-flotation and impact rolling on the deep foundation is not as substantial as the surface layers. This study provides valuable insights into optimizing foundation treatments for land reclamation projects on the coral reefs.

Keywords: land reclamation; hydraulic filling; coral sand; foundation treatment; vibro-flotation; impact rolling

Citation: Yao, T.; Li, W. An Evaluation of Treatment Effectiveness for Reclaimed Coral Sand Foundation in the South China Sea. *J. Mar. Sci. Eng.* **2023**, *11*, 2288. https://doi.org/10.3390/jmse11122288

Academic Editor: Tom Spencer

Received: 24 October 2023
Revised: 26 November 2023
Accepted: 29 November 2023
Published: 1 December 2023

1. Introduction

Coral reefs are predominantly distributed in tropical oceans between N 30° and S 30° and are formed by *Scleractinia*, algae, and various marine organisms through biological processes [1]. In response to the growing demand for space and resources, extensive land reclamation projects have been undertaken on the coral reefs of the South China Sea. This involves excavating coral sand and debris from lagoons and reef flats using cutter suction dredgers, followed by the pumping of these materials through the pipeline onto the reef flat [2]. Through this method, the construction cost can be reduced and the construction period can be shortened significantly [3,4].

Coral sand is primarily composed of calcium carbonate, with distinct differences from terrigenous sediments formed through physical, biochemical, and chemical processes [5]. It has been observed that coral sands exhibit rich intra-particle pores, irregular particle shapes, and high susceptibility to breakage [6–8]. However, they possess higher peak and critical state friction angles, along with higher liquefaction resistance compared to quartz sand [9–13]. The authors of the *Hydraulic Fill Manual for Dredging and Reclamation Works* [14] designated coral sand as a specialized filling material, emphasizing its advantageous engineering performance and the potential for enhanced strength and stiffness over time due to particle bonding.

Traditionally, hydraulic-filled marine soils are characterized by their low-density and high-water content, as well as their void ratio, rendering them unsuitable as backfill material due to their low bearing capacity, high compressibility, and prolonged consolidation time [15,16]. Foundation treatment becomes imperative to enhance the bearing capacity of hydraulic-filled soils. However, there is currently no consensus on reinforcement methods for coral sand foundations. In the land reclamation areas within coral reefs, the filling layer is typically thicker than 5 m [4,17]. The vibro-flotation method has traditionally been employed to densify deeply buried hydraulic filling materials due to its commendable reinforcement effects; however, it is time-consuming and economically inefficient. Sand foundations are also commonly reinforced via impact rolling [18], which combines impact and rolling functions, providing a fast construction speed and cost-effectiveness [19]. Nevertheless, it falls short of meeting the requirements for deep subgrade reinforcement. Dynamic compaction, which utilizes instantaneous impact loads, has been utilized to compact soil, reduce soil compressibility, and enhance soil strength [20,21]. However, Ke et al. [22] declared that a coral sand foundation treated using dynamic compaction remains susceptible to liquefaction and that the reinforcement effect of dynamic compaction on the subgrade below the groundwater is minimal. Following on-site observation, it was found that the compaction degree of hydraulic-filled coral sand foundations nearly meets the criteria for use as the foundation for airports or other engineering structures with a straightforward foundation treatment [17].

In the case of quality control for a hydraulic-filled coral sand foundation, Van Impe et al. [23] highlighted that both the discharge methods of the dredgers, i.e., rainbowing or tilting, and ground improvement techniques, including dynamic compaction, rolling compaction, or vibro-compaction, can induce different initial stress an-isotropy and stress histories in coral sand. Therefore, in practical applications reinforcing hydraulic-filled coral sand foundation, relying solely on the tip resistance-relative density (q_c-D_r) relationship is insufficient and should be abandoned. Giretti et al. [24] also pointed out that empirical correlations developed with silica sands cannot be applied in the presence of crushable materials such as carbonate sands. Furthermore, Giretti et al. [25] performed a cone penetration test (CPT) calibration using a centrifuge model with supplementary large-scale calibration chamber tests, and a relationship between the q_c of CPT and state parameter was established, which serves as a pivotal benchmark for evaluating the reinforcement efficacy., Wang et al. [17] emphasized that the hydraulic-filled coral soils, in the land reclamation projects on the coral reefs in the South China Sea, exhibited uneven particle size distributions, ranging from silt-sized particles to blocks. Notably, the CPT sensors often faced damage when encountering these blocks, leading to test failures and inaccurate results. Consequently, the more resilient and cost-effective approach of utilizing the dynamic penetration test (DPT) was chosen to assess the density and bearing capacity of the coral soil foundation. Wang et al. [17] also found that there is an approximately linear relationship between the bearing capacity of the coral soil foundation and the blow count of DPT ($N_{63.5}$). Based on the above considerations, the DPT was also adopted for evaluating the performance of the current hydraulic-filled coral sand foundation.

In this study, the reclaimed foundation has a thick layer of hydraulic-filled coral sand, typically ranging from 6 to 10 m. It is crucial to carry out in situ experimental research on the treatment effectiveness of the coral sand foundation as it helps in exploring the optimal treatment scheme and technical construction parameters of coral sand foundations, allowing us to determine the quality control standard of foundation treatments and ensure the quality of the subsequent construction. The reclamation of reef islands in the South China Sea is mainly for the construction of airports. The requirement for an airport foundation bearing capacity is around 200 kPa [26], and a shallow foundation is normally enough to bear the overlying structures. Since previous experiences were limited and the construction time was short, two common methods of treating inland foundations, vibro-flotation and impact rolling, were employed to treat the hydraulic-filled coral sand foundation. After treatment, the plate load test, California Bearing Ratio (CBR) test, density measuring, and settlement monitoring were performed to evaluate the compaction degree, bearing capacity, and settlement of the surface layer, which is the main bearing layer; moreover, dynamic penetration test (DPT) was adopted for assessing the bearing capacity and compactness of deeper foundation soils.

2. Site Conditions

The study area is on a reef island in the South China Sea (Figure 1a), and the stratigraphy of the studied reef island, as depicted in Figure 1b, comprises six distinct layers. The uppermost layer represents the hydraulic filling layer, primarily consisting of uncemented coral soils with particle sizes ranging from silt to blocks. The second layer is the original reef flat, which is covered by a layer of beach rocks that extend to approximately 2 m in thickness, cemented by algae and laying about 1–2 m below sea level. Below the reef flat, the third to fifth layers consist of lightly cemented coral debris, predominantly composed of coral soil, shell remnants, and reef blocks. The sixth layer comprises the cemented reef limestone stratum. Comprehensive topographical and subsurface information about the reef island can be found in Sun et al. [27]. Figure 2 shows the hydraulic filling process for creating foundations for the construction of infrastructures, mainly including airport runways, drainage works, pipe gallery construction, oil tanks, and associated buildings. In this method, sedimentary coral sands near the reef platform are dredged by a large-scale cutter suction dredger and pumped through pipelines along with seawater onto the reef platform. Through the self-weight consolidation of the coral sand, excess water drains away, and a certain stable elevation is achieved.

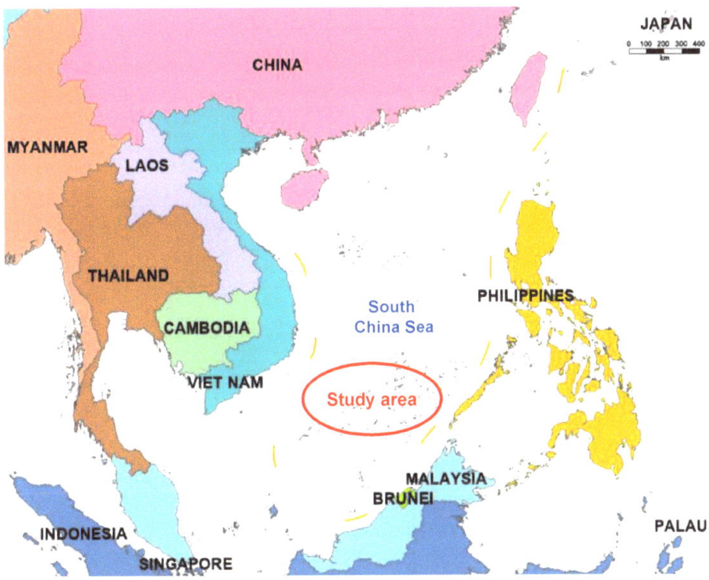

(a)

Figure 1. *Cont.*

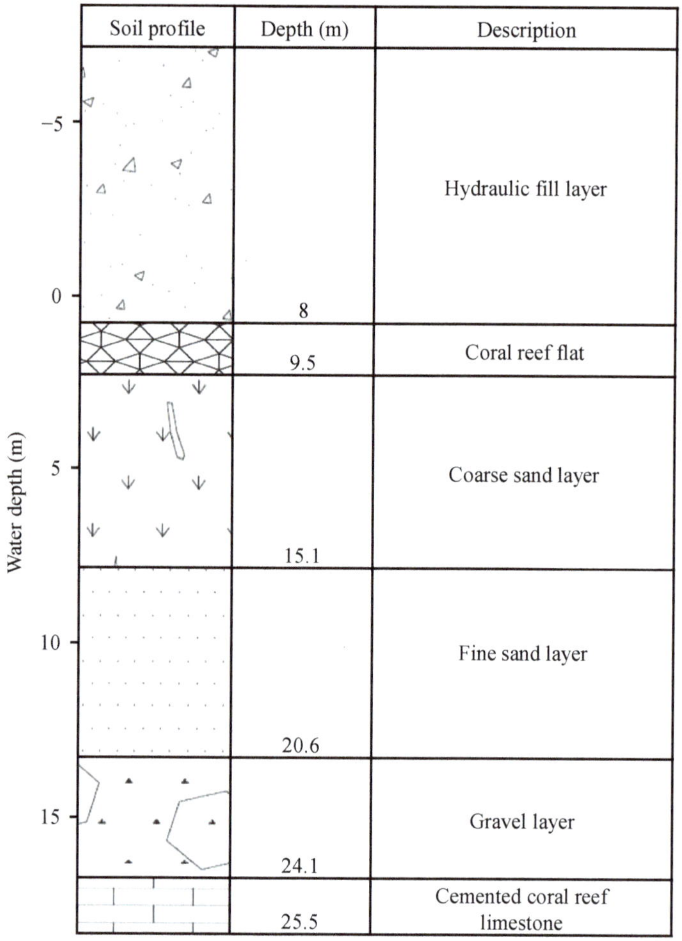

(b)

Figure 1. (a) Location of study area; (b) profile of the stratigraphy of the studied reef island.

Figure 2. Formation of coral sand foundation.

The hydraulic-filled coral sand is mainly formed through the accumulation of the skeletal remains of marine organisms like shells and corals. The soil particles often exhibit irregular branching, spindle-like, flaky, and blocky shapes, among others, with a higher content of branching and spindle-like particles. Coarser particles often show more irregular shapes with numerous intra-particle pores, while the finer particles have relatively regular shapes but still contain a significant number of intra and surface pores, as shown in Figure 3, which is also observed in Yao and Li [12]. Sieving tests were performed on five sets of coral sand samples collected from the surface of the hydraulic fill, where particles coarser than 4.75 mm (about 2–4%) were removed. The grading curves of the coral sand samples are presented in Figure 4. It can be seen that the particles with sizes ranging from 0.3 to 0.6 mm have the highest content, accounting for approximately 28.8% of the total mass. The median particle size (d_{50}) is around 0.44 to 0.68 mm, the coefficient of uniformity (C_u) falls within the range from 4.00 to 5.62, and the coefficient of curvature (C_c) is in the range from 0.85 to 1.05. According to the ASTM [28] D2487, among the five samples, 1, 4, and 5 are well graded, while 2 and 3 are poorly graded, indicating that there is a certain degree of variability in the distribution of soil materials at the site. The fundamental physical properties of the coral sands are listed in Table 1. The minimum dry density of the coral sand samples was determined according to ASTM [29] D4254. In the case of measuring the maximum dry density, the vibration hammering method was used to avoid significant particle damage [30]. Three parallel tests were performed for each measurement to obtain an average value.

(a)

Figure 3. *Cont.*

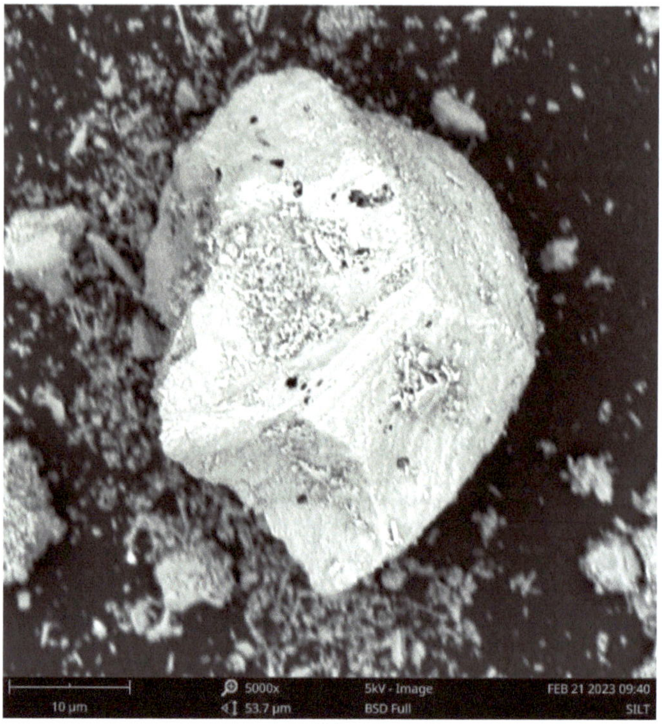

(b)

Figure 3. SEM images of coral sand: (a) coarse particle; (b) fine particle.

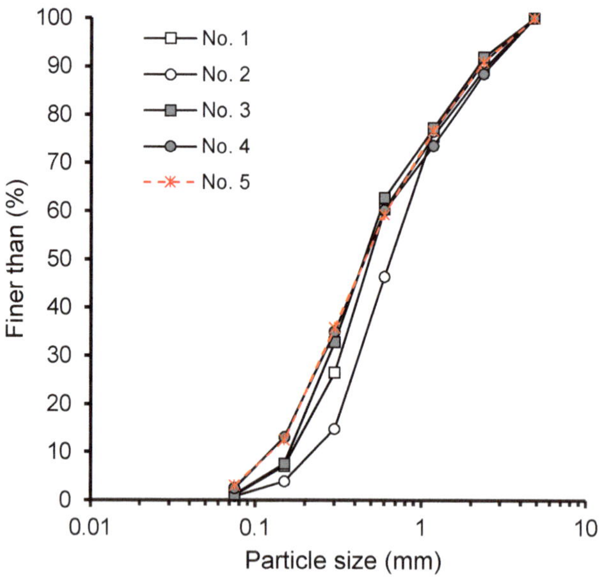

Figure 4. Particle size distribution curves of the hydraulic-filled coral sand.

Table 1. Fundamental physical properties of the calcareous soils.

G_s	e_{max}	e_{min}	w_i (%)	ρ_{di} (g/cm^3)	$\rho_{d,max}$ (g/cm^3)	$\rho_{d,min}$ (g/cm^3)
2.8	1.31	0.69	16.3–21.2	1.43–1.48	1.66	1.21

Note: G_s, specific gravity; e_{max}, maximum void ratio; e_{min}, minimum void ratio; w_i, water content after self-weight consolidation; ρ_{di}, dry density after self-weight consolidation; $\rho_{d,max}$, maximum dry density; $\rho_{d,min}$, minimum dry density.

3. Foundation Treatment Schemes

The study area was divided into four experimental zones, including one zone (A1) without post-filling treatment, two zones (A2 and A3) with vibro-flotation treatment, and one zone (A4) with impact rolling treatment. The purpose was to assess the treatment effects of coral sand foundations under different treatment techniques. The thickness of the hydraulic-filled coral sand layer in each experimental zone was approximately 7 to 8 m, and the depth of groundwater was around 2 m. The specific dimensions and layout of each zone are shown in Figure 5.

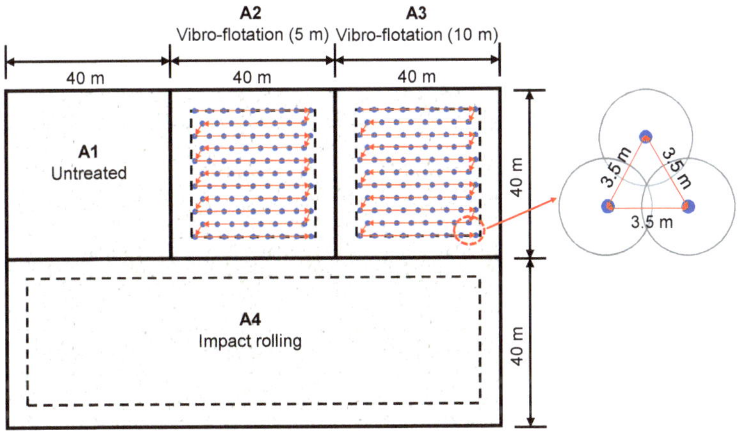

Figure 5. Dimensions and layout of each experimental zone.

During on-site implementation, hydraulic filling was first performed to create the initial onshore foundation. After the filling was completed, the site was left for one month to allow seepage and drainage. Once the seepage and drainage processes were stable and met the conditions for foundation treatment, the on-site foundation treatment was initiated. The treatment methods for each experimental zone were as follows:

(1) Untreated Zone A1: After the formation of the coral sand foundation through hydraulic filling, no further foundation treatment was conducted. The improvement of the foundation relied solely on self-weight consolidation.

(2) Vibro-flotation Zones A2 and A3: A 180 kW dynamic vibrator was used for the foundation treatment. The zones were divided into A2 and A3 based on different treatment depths, with compaction depths of 5 m and 10 m, respectively. The vibro-flotation points were arranged in an equilateral triangle and the distance between two treatment points was 3.5 m. According to in situ observation, it was found that the effective vibration zone of the 180 kW dynamic vibrator has a diameter of about 4 m, beyond which the soil density was almost unaffected. The layouts of the treatment points as well as the construction sequence can be observed in Figure 5.

(3) Impact Rolling Zone A4: A 25 kJ triangular roller was used to impact and roll the surface layer of the hydraulic-filled coral sand. In total, 20 rolling passes were performed for the treatment.

During the treatment, to prevent the mutual interference of seepage effects in different experimental zones after foundation treatment and to expedite the drainage and self-weight consolidation process, a 2 m deep drainage trench was excavated around each experimental zone after the completion of filling. Additionally, a 5 m wide untreated platform was left around each experimental zone during the treatment process.

4. On-Site Evaluation Tests

After the foundation treatment was completed, a series of evaluation tests were conducted on site, including a plate load test, California Bearing Ratio (CBR) test, soil density measurement, and dynamic penetration test (DPT), to identify the extent of the increase in the foundation density and bearing capacity, assess the deformation characteristics, and analyze the reinforcement depth. Finally, the monitoring data of the on-site settlement of the hydraulic-filled coral sand foundation were discussed.

4.1. Normal Shallow Plate Load Test

The shallow plate load test is a direct means of measuring the bearing capacity and deformation modulus of the foundation and typically involves recording the foundation settlement under incremental loading, which could reflect the foundation's resistance to deformation [31]. The maximum load applied is generally twice the design load of the foundation. In this study, the maximum design load is 300 kPa; therefore, a maximum load of 600 kPa was employed. A square steel plate with sides of 1000 mm and a thickness of 50 mm was used as the loading plate. Counterweights in conjunction with a high-pressure hydraulic pump were employed to apply the load incrementally, and the displacement of the load plate at various time points and under different pressures was recorded using a dial gauge. A curve illustrating displacement as a function of pressure was plotted, i.e., the p–s curve. According to GB 50007-2011 [31], when there is a proportional limit on the p–s curve, the load value corresponding to the proportional limit is taken as the characteristic value of the bearing capacity; however, when the ultimate load is less than the load value of the corresponding proportional limit, half of the ultimate load should be taken as the characteristic value of the bearing capacity. The deformation modulus is calculated according to the following equation [32]:

$$E = I_0\left(1 - \mu^2\right)\frac{pb}{s} \tag{1}$$

where E represents the deformation modulus of the foundation; p denotes the vertical stress in the linear segment of the p–s curve; s corresponds to the settlement associated with p; b is the side length or diameter of the bearing plate; I_0 stands for the shape factor of the bearing plate (0.886 for a square plate); μ is the Poisson's ratio, and 0.3 is adopted for coral sands according to He et al. [33]. The shallow plate load tests were performed immediately after completing the foundation treatment and leveling the surface. For each experimental zone, three points near the center were randomly tested.

4.2. Plate Load Test for Determining the Modulus of the Subgrade Reaction

As a design parameter for pavement structures, the modulus of the subgrade reaction is typically used to determine the strength of the airport pavement foundation, reflecting the foundation's resistance to deformation. The modulus of the subgrade reaction can be determined by a specialized plate load test. In this test, an increasing load was applied to a bearing plate with a diameter of 760 mm. Deformation values under varying loads were measured to obtain a p–s curve. Based on Winkler's model, the modulus of the subgrade reaction can be calculated using the following equation [34]:

$$K_u = \frac{p_B}{0.00127} \tag{2}$$

where K_u represents the modulus of the subgrade reaction and p_B denotes the vertical stress corresponding to a settlement of 1.27 mm. After the treatment, plate load tests were conducted at three random points on the leveled surface in each experimental zone.

4.3. CBR Test

The CBR test is mainly for determining the stiffness of road subgrades. A penetration rod with a 50 mm metal cylinder was pressed into the subgrade at a rate of 1 mm/min. The pressure at a specific depth (usually 2.5 mm) was compared to the standard pressure for that penetration depth, which gave the measured *CBR* value for the subgrade [35]. The formula for calculating the *CBR* value is as shown below:

$$CBR = \frac{P_1}{P_0} \times 100\% \tag{3}$$

where P_1 is the measured vertical stress at the corresponding penetration depth and P_0 is the standard pressure for a 2.5 mm penetration depth (7 MPa). Similar to the plate load tests, in situ CBR tests were carried out at three randomly chosen locations within each experimental zone.

4.4. Soil Density Measurement

The compaction degree is one of the most important indicators for the quality control of the soil foundation, which refers to the ratio of the dry density of the compacted material to its maximum dry density. Only through adequate compaction of the foundation can the safety and stability of the upper structure construction be ensured. The density of coral sand was determined using the sand-filling method [35]. In each experimental zone, the density of the surface layer of the coral sand foundation (0–1.0 m) was assessed at 8 randomly chosen positions.

4.5. Dynamic Penetration Test

The dynamic penetration (DPT) test is a quick and cost-effective method for indirectly assessing the bearing capacity and compactness of deeper foundation soils. In this study, a heavy dynamic penetration test was employed using a 63.5 kg heavy hammer dropped freely from a fixed height of 76 cm to drive the probe rod into the soil. The number of blows required for the probe rod to penetrate 10 cm was recorded as $N_{63 \cdot 5}$, thus obtaining DPT values at different depths [36]. The DPT tests were performed at nine randomly selected positions within each experimental zone after the foundation treatment. Considering the hydraulic filling depth on the reef island ranging from 5 to 10 m, the DPT examination depth was set as 6 m, with $N_{63 \cdot 5}$ values recorded every 10 cm of penetration, commencing at a depth of 0.5 m.

4.6. Settlement Monitoring

The elevation of each experimental zone was measured before and after the treatment to calculate the construction settlement. After the treatment, three surface settlement markers were buried in each zone to observe the post-construction settlement. For the untreated zone, settlement markers were set just after the reclaimed land formation.

5. Results and Discussion

5.1. Bearing Capacity and Deformation Modulus

The representative p–s curves of the four experimental zones are shown in Figure 6a. It can be observed that, except for the A1 zone, there are no obvious inflection points on the p–s curves for the other zones. While on the site, it was found that, even when the tests reached their maximum loading of 600 kPa, there was no significant lateral extrusion of coral sand or a sudden increase in the settlement around the bearing plate for the A2, A3, and A4 zones. This suggests that the characteristic value of the bearing capacity must be at least 300 kPa for the three zones. Analyzing the p–s curve of the A1 zone, an inflection point appears at a load of 375 kPa, and a significant lateral extrusion of coral sand occurred when the load reached 450 kPa, indicating the onset of soil failure. This might be related to the particle breakage of the coral sand. Therefore, the load just before failure, 375 kPa, is taken as the ultimate load, and half of the ultimate load, 187.5 kPa, is considered the characteristic value of the bearing capacity of the A1 zone.

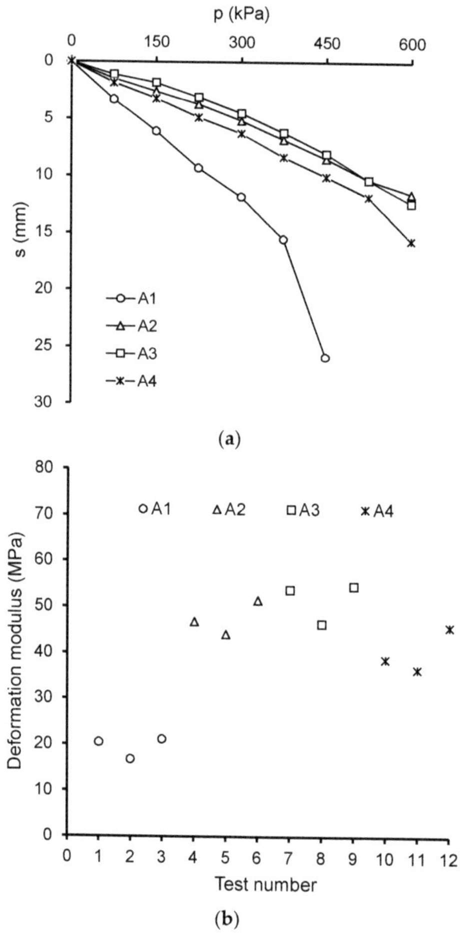

Figure 6. Shallow plate load testing results: (**a**) p–s curve; (**b**) deformation modulus.

It can be seen that the untreated coral sand foundation already has a relatively high bearing capacity that can meet the requirements of some light buildings and structures. At the same time, according to the p–s curves of the A2 and A3 zones, their differences are minimal, with the foundation bearing capacity and deformation modulus varying within the same range, indicating that there is no need to further extend the treatment depth to 10 m in vibro-flotation. The p–s curve of the A4 zone lies only slightly below those of A2 and A3, which suggests that surface impact rolling could effectively improve the bearing capacity of the coral sand foundation.

The deformation moduli of the four experimental zones, with three testing data for each zone, are shown in Figure 6b. It can be seen that the deformation modulus of the A1 zone is around 20 MPa, which is higher than common untreated terrestrial sands [37,38]. For the A4 zone, after the impact rolling treatment, the deformation modulus increases to about 40 MPa, essentially double that of the untreated zone. Firstly, this is because the soil particles rearrange and the void ratio reduces under the impact. Secondly, the impact force causes significant particle breakage, especially in the surface layer, further filling voids and increasing the degree of compaction. In contrast, although the deformation modulus of the A2 and A3 zones increases further under the vibro-flotation compaction treatment, the increase is marginal in comparison to the A2 zone.

5.2. Modulus of Subgrade Reaction

The representative p–s curves of the four experimental zones for measuring the modulus of the subgrade reaction are shown in Figure 7a. It can be observed that, for the A1 zone without treatment, the vertical stress when the bearing plate settlement reaches 1.27 mm is 56 kPa. Therefore, the modulus of the subgrade reaction for the A1 zone is determined to be 44.0 MN/m^3. The value is relatively high and can essentially meet the design requirements for the clay and silt subgrades of an airport runway [34]. This indicates that seepage consolidation has a certain compaction effect on the coral sand foundation. Similarly, the modulus of the subgrade reaction for the A2, A3, and A4 zones after vibro-flotation and impact rolling treatments are found to be 91.8 MN/m^3, 97.2 MN/m^3, and 78.5 MN/m^3, respectively. The modulus of subgrade reaction after impact rolling increases by nearly a factor of two compared to the untreated state. The foundation capacity essentially meets the design requirements for the coarse-grained soil subgrades of an airport runway [34].

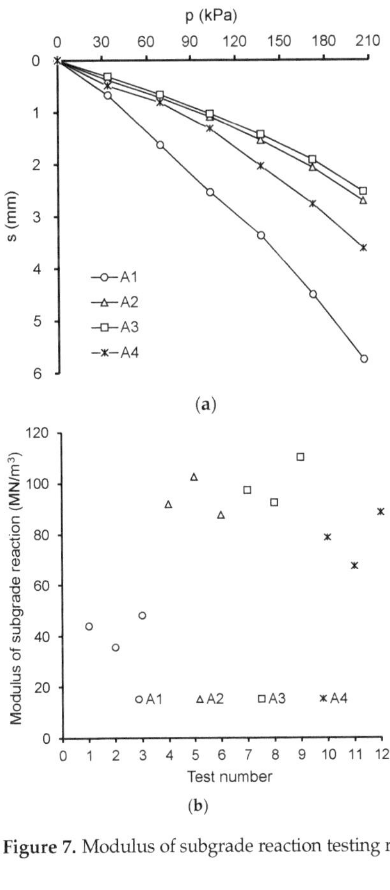

Figure 7. Modulus of subgrade reaction testing results: (**a**) p–s curve; (**b**) modulus of subgrade reaction.

The moduli of the subgrade reaction of the four experimental zones, with three testing data for each zone, are shown in Figure 7b. The average modulus of subgrade reaction for the A1, A2, A3, and A4 zones was found to be 42.6 MN/m^3, 94.0 MN/m^3, 99.8 MN/m^3, and 78 MN/m^3, respectively, indicating that the overall trend remains consistent with those described above.

5.3. CBR Value

The p–s curves from the representative location of each experimental zone are presented in Figure 8a. For the untreated A1 zone, the vertical stress at a penetration depth of 2.5 mm was 1092 kPa, and the corresponding CBR value was 15.6%. In accordance with AASHTO [39] T 193, a CBR value of 16% suggests that the soil may have a moderate-to-fair bearing capacity. The CBR values for the three treated zones are 66.5%, 66.0%, and 53.8%, respectively, indicating a significant improvement in the bearing capacity after vibro-flotation or impact rolling. Although the CBRs of the vibro-flotation-treated zones surpass those of impact rolling, considering the soil variability, both treatment methods are considered to be efficient.

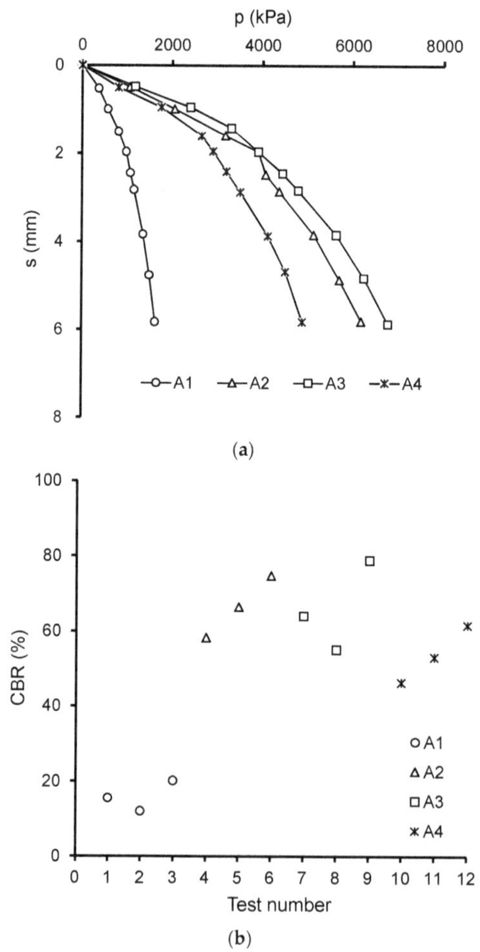

Figure 8. CBR testing results: (**a**) p–s curve; (**b**) CBR value.

A more detailed analysis of the CBR values for different experimental zones is shown in Figure 8b, where slightly scattered results could be observed within each experimental zone. Notably, the CBRs for the three tested positions within the untreated zone A1 consistently exceed 15%, with one reaching as high as 20%. This signifies that the stiffness of the hydraulic-filled coral sand foundation is already high enough for airport runway construction. Much higher CBR values are observed in the treated areas, which can be attributed to the substantial impact of both the vibro-flotation and impact rolling treatments on particle rearrangement and breakage, resulting in a much denser structure [4].

5.4. Soil Compactness

In each area, eight positions were randomly selected for the density measurement. The compaction degree of the hydraulic filling coral soils before and after treatment was determined by comparing the field-measured dry density, obtained through the sand replacement method, with the maximum dry unit weight of the coral sand achieved using the vibration hammering method (1.66 g/cm^3). The corresponding compaction degree for each point is summarized in Figure 9. The results show that the compaction degree within the A1 zone, which did not undergo any treatment, ranged from 86% to 90%, with a mean value of approximately 88% after hydraulic filling and self-weight consolidation. This suggests that the compaction degree of the hydraulic-filled coral sand foundation complies with the design requirements for areas with lower bearing capacity demands [34]. Following vibro-flotation and impact rolling treatment, due to the breakage of coarser coral sand particles, the voids are further filled. As a result, the compaction degree in the treated areas increases from 93% to 96%, essentially meeting the design requirements for airport construction [34]. The difference in enhancement between vibro-flotation and impact rolling, as seen in the comparison of the compaction degree, is negligible, with the difference in average values being less than 1%.

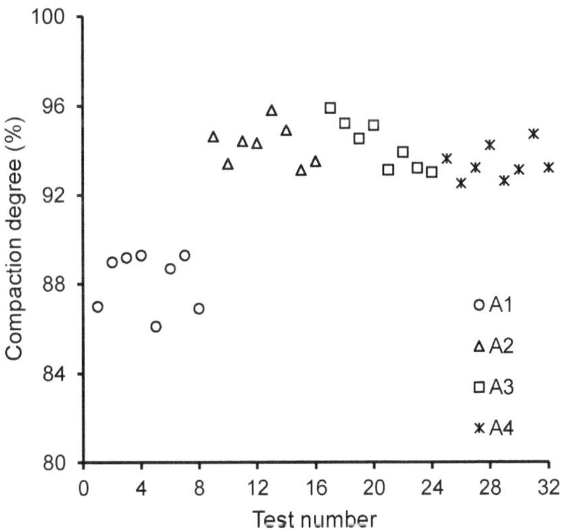

Figure 9. Compaction degree of the surface coral sand in each experimental zone.

Additionally, a comprehensive examination of the compaction degree was conducted at different depths and time intervals in a specific location within the A1 zone. This investigation aimed to understand the effects of seepage and self-weight consolidation on the densification of the coral sand. The evaluation spanned time periods of 1 week, 2 weeks, and 2 months following the reclaimed land formation. Figure 10 illustrates the compaction degree at this specific site. The data reveal a notable enhancement in the density of the coral sand ground after increased time following hydraulic filling, particularly during the first week. Subsequently, it stabilizes at around 14 days. Notably, one month later, the compaction degree within the top 1.0 m of the soil reaches 88.6%, aligning with the data shown in Figure 9. A comparison of the soil compaction degree at different depths indicates that the compactness improves with increasing burial depth, reaching as high as 90% within the 2.0–3.0 m range, and, in some instances, increasing up to 94%. This meets the essential requirement for airport subgrade quality. Self-weight consolidation plays a significant role in the densification of hydraulic-filled materials. Additionally, the relatively faster seepage velocity and the facilitated movement of finer particles within inter-particle voids are a consequence of the irregular particle shape of coral sand [40,41]. Both factors contribute to the densification of the coral sand.

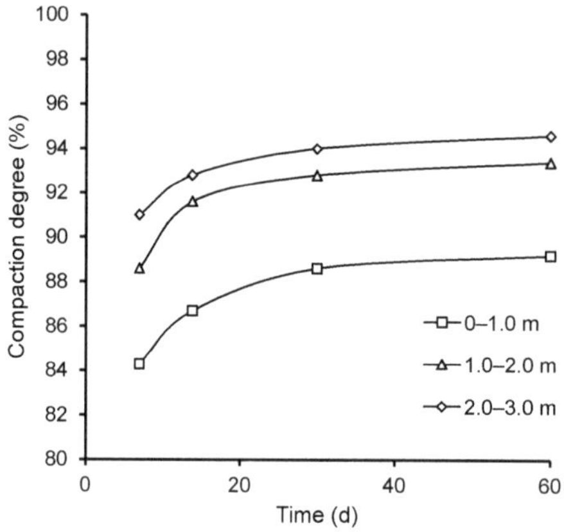

Figure 10. Compaction degree of the coral sand in the untreated zone at different depths and time intervals.

5.5. DPT Result

The test data for the four representative positions within each experimental zone in terms of the $N_{63.5}$ against depth are displayed in Figure 11. In the untreated A1 zone, the $N_{63.5}$ ranges from 5 to 12 within a 6 m depth. The scatteredness is mainly due to the uneven structure of the coral sand caused by the hydraulic filling process [42], with the maximum value having been found at a depth of 1.5–2.0 m. Beyond a depth of 3 m, the $N_{63.5}$ decreases due to the influence of groundwater. After foundation treatment, $N_{63.5}$ is generally greater than 10, with a maximum value of 23, indicating a significant improvement in foundation compactness. The most pronounced increase in $N_{63.5}$ occurs after vibro-flotation at a depth of 1–3 m. However, for foundation depths exceeding 3 m, the increase in $N_{63.5}$ is considerably less, suggesting that the reinforcement effect of both vibro-flotation and impact rolling on the deep foundation is not as substantial as the surface layers.

In each area, nine positions were randomly selected for the DPT evaluation. Figure 12 presents the average $N_{63.5}$ along the foundation depth for each test position. In the untreated area, the mean values of the $N_{63.5}$ range from approximately 7 to 11. After vibro-flotation treatments at depths of 5 m and 10 m, the $N_{63.5}$ values increase significantly, falling within the range from 14 to 24. However, the difference between the two tested zones within the depth of 6 m is not significant. In the impact rolling zone, the mean $N_{63.5}$ values increase to a range from 12 to 20. According to Wang et al. [17], when the $N_{63.5}$ of the coral sand foundation exceeds 10, it indicates a compaction degree higher than 85% and a bearing capacity greater than 500 kPa. In summary, the effectiveness of impact rolling meets the requirement for infrastructure construction.

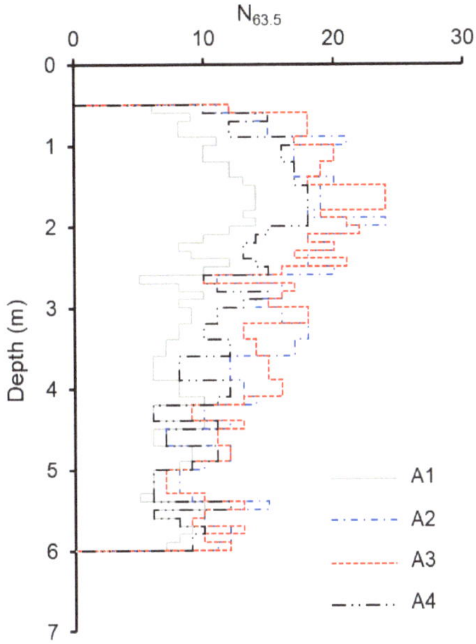

Figure 11. $N_{63.5}$ against depths of representative positions within each experimental zone.

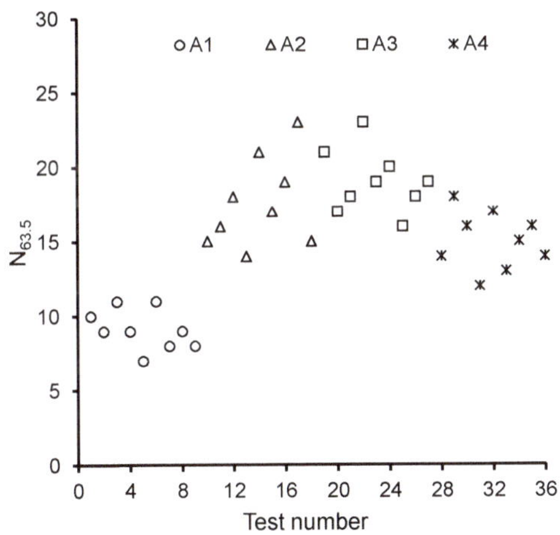

Figure 12. Average $N_{63.5}$ along the foundation depth for each test position.

5.6. Settlement Analysis

The settlement of the untreated A1 zone over time is shown in Figure 13. It is evident that the coral sand foundation experiences significant settlement over time, and the settlement rate initially increases rapidly. As the drainage process continues, the settlement gradually stabilizes, reaching a relatively stable state after one month, with a maximum settlement of approximately 30 cm. As pointed out by Wang et al. [41], the coral sand particles are porous and have an angular shape, resulting in higher permeability than common terrigenous sand and rapid self-weight consolidation characteristics.

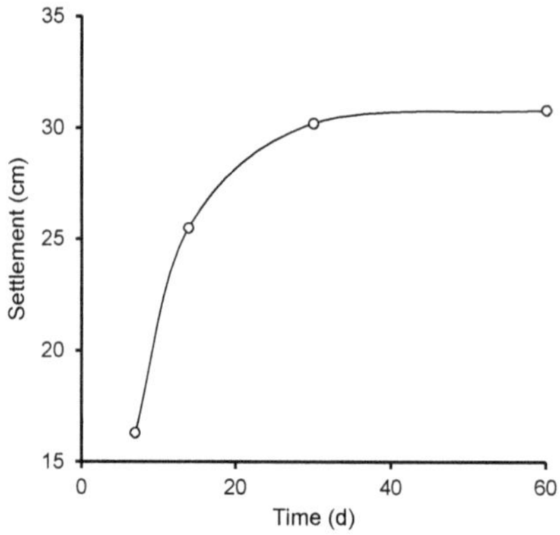

Figure 13. Settlement of the untreated A1 zone over time.

The additional construction and post-construction settlements of the A2, A3, and A4 zones are shown in Table 2. It can be observed that the surface coral sand undergoes further compression and consolidation because of particle breakage and rearrangement after the vibro-flotation and impact rolling treatments. Specifically, the settlement ranges from 21.0 to 24.1 cm for the vibro-flotation treatment and from 19.5 to 21.6 cm for the impact rolling treatment. In conjunction with the post-treatment settlement analysis, it is found that the post-vibro-flotation treatment settlement after 30 days ranges from 3.5 to 5.1 mm, and the post-impact rolling treatment settlement ranges from 5.2 to 6.8 mm. Both treatments result in a very low post-treatment settlement, indicating a more stable state; moreover, those values are much lower than the allowable post-construction settlement for an airport runway, which is 0.2–0.3 m [34]. The results also coincide with the findings by Wang et al. [4] that the ground settlement of the hydraulic filling coral sand treated using vibroflotation tends to be stable after 21 days, and the average settlement is 4.3 mm after 61 days. This might be due to the relatively low compressibility of coral sand compared to quartz sand at low stress levels [12].

Table 2. Settlement observation results of each test point.

Experimental Zone	Construction Settlement (cm)			Post-Construction Settlement after 30 Days (mm)		
	Point 1	Point 2	Point 3	Point 1	Point 2	Point 3
A2	23.7	21.2	21.0	5.1	4.7	4.2
A3	23.6	22.9	24.1	3.5	4.3	3.6
A4	19.5	20.3	21.6	5.2	6.8	5.5

6. Conclusions

Through a series of on-site evaluation tests, this study examined the reinforcement effectiveness of various foundation treatment processes on a reclaimed coral sand foundation, considering aspects including the bearing capacity, deformation modulus, modulus of the subgrade reaction, CBR, DPT value, and settlement. The main findings are as follows:

(1) The untreated hydraulic-filled coral sand foundation exhibits a bearing capacity exceeding 150 kPa and a deformation modulus of 20 MPa. This high-bearing capacity and deformation modulus could satisfy the requirements of some light buildings and structures. After impact rolling treatment, the bearing capacity increases to larger than 300 kPa, and the deformation modulus reaches 40 MPa. Vibro-flotation could result in a slightly higher bearing capacity and deformation modulus than when using the impact rolling treatment. Since both treatment methods could make the bearing capacity of the foundation meet the maximum requirement of design load (300 kPa),

impact rolling is considered to be the most efficient in this study because of the simple machinery and quick construction speed;

(2) Comparing the modulus of the subgrade reaction and CBR values of the coral sand foundation before and after the impact rolling treatment, the untreated foundation exhibits a modulus reaction of 40 MN/m^3 and CBR values exceeding 15%. These values are relatively higher compared to common terrestrial fill soils, indicating the beneficial effect of seepage densification in increasing the foundation's strength. After the impact rolling treatment, the modulus of the reaction and CBR values further increased. Impact rolling caused particle breakage and rearrangement in the surface layer of the coral sand, enhancing compaction and deformation resistance. The finer particles filling the voids result in a greater interlocking effect, improving the strength and stability of the foundation to meet airport engineering design standards. Vibro-flotation could result in a slightly higher modulus of reaction and CBR values than the impact rolling treatment;

(3) Due to the high permeability resulting in rapid self-weight consolidation, the untreated coral sand foundation exhibits a relatively high degree of compaction. After impact rolling or vibro-flotation treatment, coral sand particles undergo breakage and rearrangement, leading to denser structures. The compaction degrees after treatment essentially meet the technical requirements for airport runways;

(4) The $N_{63.5}$ obtained from the DPT of the untreated coral sand foundation ranges from 5 to 12 within 6 m of depth, with the maximum value found at a depth of 1.5–2.0 m. After impact rolling or vibro-flotation treatment, the $N_{63.5}$ increases markedly. The most pronounced increase in $N_{63.5}$ occurs after vibro-flotation at a depth of 1–3 m. However, for foundation depths exceeding 3 m, the increase in $N_{63.5}$ is considerably less, suggesting that the reinforcement effect of both vibro-flotation and impact rolling on the deep foundation is not as substantial as in the surface layers;

(5) The untreated coral sand foundation, due to the effects of seepage densification and self-weight consolidation, experiences settlement during the drainage process. The consolidation occurs rapidly, with the foundation essentially stabilizing after one month, resulting in total settlements ranging from 27.5 to 32.8 cm. After impact rolling or vibro-flotation treatment, the foundation is further densified, with post-construction settlements ranging from 3.5 to 6.8 mm, indicating a more stable state.

Author Contributions: Conceptualization, T.Y. and W.L.; Methodology, W.L.; Formal analysis, T.Y. and W.L.; Investigation, T.Y.; Writing—original draft, T.Y. and W.L. All authors have read and agreed to the published version of the manuscript.

Funding: This research was funded by the National Key Technologies R&D Program of China [2022YFC3102101], the National Natural Science Foundation of China [42177154, 42102327, 42377176] and the Open Fund of the Key Laboratory of Special Environment Road Engineering of Hunan Province, Changsha University of Science & Technology, China [kfj220503]. The APC was funded by [42177154].

Institutional Review Board Statement: Not applicable.

Informed Consent Statement: Not applicable.

Data Availability Statement: The data presented in this study are available on request from the corresponding author.

Conflicts of Interest: The authors declare no conflict of interest.

References

1. Darwin, C. *The Structure and Distribution of Coral Reefs*; Smith, Elder and Co.: London, UK, 1842.
2. Wang, X.; Weng, Y.; Wei, H.; Meng, Q.; Hu, M. Particle obstruction and crushing of dredged calcareous soil in the Nansha Islands, South China Sea. *Eng. Geol.* **2019**, *261*, 105274. [CrossRef]
3. Shen, J.H.; Wang, X.; Cui, J.; Wang, X.Z.; Zhu, C.Q. Shear characteristics of calcareous gravelly sand considering particle breakage. *Bull. Eng. Geol. Environ.* **2022**, *81*, 130. [CrossRef]

4. Wang, X.; Ding, H.; Wen, D.; Wang, X. Vibroflotation method to improve silt interlayers of dredged coral sand ground–a case study. *Bull. Eng. Geol. Environ.* **2022**, *81*, 472. [CrossRef]
5. Coop, C.A. Origin of carbonate sand. *Sedimentology* **1990**, *37*, 503–515.
6. Kong, D.; Fonseca, J. Quantification of the morphology of shelly carbonate sands using 3D images. *Géotechnique* **2018**, *68*, 249–261. [CrossRef]
7. Zhou, B.; Ku, Q.; Wang, H.; Wang, J. Particle classification and intra-particle pore structure of carbonate sands. *Eng. Geol.* **2020**, *279*, 105889. [CrossRef]
8. Wu, Y.; Li, N.; Wang, X.; Cui, J.; Chen, Y.; Wu, Y.; Yamamoto, H. Experimental investigation on mechanical behavior and particle crushing of calcareous sand retrieved from South China Sea. *Eng. Geol.* **2021**, *280*, 105932. [CrossRef]
9. Coop, M.R. The mechanics of uncemented carbonate sands. *Géotechnique* **1990**, *40*, 607–626. [CrossRef]
10. Wang, X.; Cui, J.; Wu, Y.; Zhu, C.; Wang, X. Mechanical properties of calcareous silts in a hydraulic fill island-reef. *Mar. Georesources Geotechnol.* **2021**, *39*, 1–14. [CrossRef]
11. Gao, R.; Ye, J. Mechanical behaviors of coral sand and relationship between particle breakage and plastic work. *Eng. Geo.* **2023**, *316*, 107063. [CrossRef]
12. Yao, T.; Li, W. Effect of initial fabric from sample preparation on the mechanical behaviour of a carbonate sand from the South China Sea. *Eng. Geol.* **2023**, *326*, 107311. [CrossRef]
13. Yao, T.; Li, W. Quantifying the particle shape and surface roughness of sands. *Bull. Eng. Geol. Environ.* **2023**, *82*, 135. [CrossRef]
14. Van't Hoff, J.; van der Kolff, A.N. *Hydraulic Fill Manual for Dredging and Reclamation Works*; CRC Press, Taylor and Francis Group: Boca Raton, FL, USA, 2012.
15. Dong, Y.; Wang, D.; Randolph, M.F. Quantification of impact forces on fixed mudmats from submarine landslides using the material point method. *Appl. Ocean Res.* **2020**, *102*, 102227. [CrossRef]
16. Fan, N.; Jiang, J.; Nian, T.; Dong, Y.; Guo, L.; Fu, C.; Tian, Z.; Guo, X. Impact action of submarine slides on pipelines: A review of the state-of-the-art since 2008. *Ocean Eng.* **2023**, *286*, 115532. [CrossRef]
17. Wang, X.; Ding, H.; Meng, Q.; Wei, H.; Wu, Y.; Zhang, Y. Engineering characteristics of coral reef and site assessment of hydraulic reclamation in the South China Sea. *Constr. Build. Mater.* **2021**, *300*, 124263. [CrossRef]
18. Chen, Y.; Jaksa, M.B.; Kuo, Y.L.; Airey, D.W. Experimental analysis of rolling dynamic compaction using transparent soils and particle image velocimetry. *Can. Geotech. J.* **2022**, *59*, 254–271. [CrossRef]
19. Ranasinghe, R.A.T.M.; Jaksa, M.B.; Pooya Nejad, F.; Kuo, Y.L. Predicting the effectiveness of rolling dynamic compaction using genetic programming. *Proc. Inst. Civ. Eng.-Ground Improv.* **2017**, *170*, 193–207. [CrossRef]
20. Xiao, Y.; Liu, H.; Chen, Q.; Long, L.; Xiang, J. Evolution of particle breakage and volumetric deformation of binary granular soils under impact load. *Granul. Matter* **2017**, *19*, 71. [CrossRef]
21. Xiao, Y.; Yuan, Z.; Chu, J.; Liu, H.; Huang, J.; Luo, S.N.; Wang, S.; Lin, J. Particle breakage and energy dissipation of carbonate sands under quasi-static and dynamic compression. *Acta Geotech.* **2019**, *14*, 1741–1755. [CrossRef]
22. Ke, J.; Ma, L.; Luo, Z.; Shi, X.; Song, Y.; Tan, Y. Compressibility and crushing of coral sands under one-dimensional dynamic compaction. *Granul. Matter* **2021**, *23*, 95. [CrossRef]
23. Van Impe, P.O.; Van Impe, W.F.; Manzotti, A.; Mengé, P.; Van den Broeck, M.; Vinck, K. Compaction control and related stress-strain behaviour of off-shore land reclamations with calcareous sands. *Soils Found.* **2015**, *55*, 1474–1486. [CrossRef]
24. Giretti, D.; Fioravante, V.; Been, K.; Dickenson, S. Mechanical properties of a carbonate sand from a dredged hydraulic fill. *Géotechnique* **2018**, *68*, 410–420. [CrossRef]
25. Giretti, D.; Been, K.; Fioravante, V.; Dickenson, S. CPT calibration and analysis for a carbonate sand. *Géotechnique* **2018**, *68*, 345–357. [CrossRef]
26. Wang, X.Z. Study on Engineering Geological Properties of Coral Reefs and Feasibility of Large Project Construction on Nansha Islands. Ph.D. Thesis, Institute of Rock and Soil Mechanics, The Chinese Academy of Sciences, Wuhan, China, 2008.
27. Sun, Z.; Zhan, W.; Zhu, J. Rockmass structure and engineering geological zones of coral reefs, Nansha Islands of South China Sea. *J. Trop. Oceanogr.* **2004**, *23*, 11–20.
28. *ASTM. D2487*; Standard Practice for Classification of Soils for Engineering Purposes (Unified Soil Classification System). American Society for Testing and Materials: West Conshohocken, PA, USA, 2017.
29. *ASTM. D4254-16*; Standard Test Methods for Minimum Index Density and Unit Weight of Soils and Calculation of Relative Density. American Society for Testing and Materials: West Conshohocken, PA, USA, 2016.
30. Lunne, T.; Knudsen, S.; Blaker, Ø.; Vestgården, T.; Powell, J.J.M.; Wallace, C.F.; Krogh, L.; Thomsen, N.V.; Yetginer, G.; Ghanekar, R.K. Methods used to determine maximum and minimum dry unit weights of sand: Is there a need for a new standard? *Can. Geotech. J.* **2019**, *56*, 536–553. [CrossRef]
31. *GB 50007-2011*; Code for Design of Building Foundation. National Standard of the People's Republic of China: Beijing, China, 2011.
32. Reznik, Y.M. Comparison of results of oedometer and plate load tests performed on collapsible soils. *Eng. Geol.* **1995**, *39*, 17–30. [CrossRef]
33. He, H.; Li, W.; Senetakis, K. Small strain dynamic behavior of two types of carbonate sands. *Soils Found.* **2019**, *59*, 571–585. [CrossRef]
34. *MH/T 5027-2013*; Code for Geotechnical Engineering Design of Airport. China Civil Aviation Press: Beijing, China, 2013.

35. *JTG 3450-2019*; Field Test Methods of Highway Subgrade and Pavement. Standard of the Ministry of Transport of the People's Republic of China: Beijing, China, 2019.
36. *GB 50021-2001*; Code for Investigation of Geotechnical Engineering. National Standard of the People's Republic of China: Beijing, China, 2009.
37. Wang, X.Z.; Wang, R.; Meng, Q.S.; Liu, X.P. Study of plate load test of calcareous sand. *Rock. Soil. Mech.* **2009**, *30*, 147–152.
38. Wang, X.Z.; Wang, X.; Liu, H.F.; Meng, Q.S.; Zhu, C.Q. Field test study of engineering behaviors of coral reef foundation. *Rock. Soil. Mech.* **2017**, *38*, 2065–2070.
39. *AASHTO T 193*; Standard Method of Test for the California Bearing Ratio. American Association of State Highway and Transportation Officials: Washington, DC, USA, 2022.
40. Hu, M.J.; Jiang, H.H.; Zhu, C.Q.; Weng, Y.L.; Ruan, Y.; Chen, W.J. Discussion on permeability of calcareous sand and its influencing factors. *Rock. Soil. Mech.* **2017**, *38*, 2895–2900.
41. Wang, X.Z.; Wang, X.; Chen, J.W.; Wang, R.; Hu, M.J.; Meng, Q.S. Experimental study on permeability characteristics of calcareous soil. *Bull. Eng. Geol. Environ.* **2018**, *77*, 1753–1762. [CrossRef]
42. Dong, Y.; Liu, C.; Wei, H.; Meng, Q.; Zhou, H. Size segregation mechanism of reclamation fill sand due to rainbowing operations in hydraulic dredging activities. *Ocean Eng.* **2021**, *242*, 109957. [CrossRef]

Journal of
Marine Science and Engineering

Article

Experimental Study on Mechanical Properties of Marine Mud Slurry Treated by Flocculation-Solidification-High Pressure Filtration Combined Method

Chao Han [1], Hongping Xie [1], Bin Bai [1], Rongjun Zhang [2,*], Yingchao Gao [2,*] and Zhekun Zhao [2]

[1] Construction Branch of State Grid Jiangsu Electric Power Co., Ltd., Nanjing 210000, China
[2] School of Civil Engineering, Wuhan University, Wuhan 430072, China
* Correspondence: ce_zhangrj@whu.edu.cn (R.Z.); ycg1995@whu.edu.cn (Y.G.)

Abstract: For the massive quantities and negative impacts of dredged mud slurry, its disposal and utilization have become one of the most noticeable issues in the world. In this paper, the flocculation-solidification-high pressure filtration combined method is proposed to effectively dispose of marine mud slurries. The advantages of this method are demonstrated herein in the following three aspects: dewatering performance, material savings, and the shear strength of the treated marine mud slurry. Then, the effects of the anionic polyacrylamide (APAM) dose, composite solidification agent dose, initial water content of marine mud slurries, and initial thickness of geo-bags on the mechanical properties of the marine mud slurry treated by the flocculation-solidification-high pressure filtration combined method are studied. Experimental results show that with increasing doses of APAM, the structures of mud slurries become more stable, and the optimal dose of APAM is determined as 0.16%. Moreover, the increase in the composite solidification agent dose and initial water content of the marine mud slurry, and the decrease in the initial thickness of geo-bags both contribute to the increase in the shear strength of the marine mud slurry treated by the flocculation-solidification-high pressure filtration combined method.

Keywords: marine mud slurry; flocculation; composite solidification agent; high-pressure filtration; shear strength

Citation: Han, C.; Xie, H.; Bai, B.; Zhang, R.; Gao, Y.; Zhao, Z. Experimental Study on Mechanical Properties of Marine Mud Slurry Treated by Flocculation-Solidification-High Pressure Filtration Combined Method. *J. Mar. Sci. Eng.* **2023**, *11*, 2270. https://doi.org/10.3390/jmse11122270

Academic Editors: José António Correia and José-Santos López-Gutiérrez

Received: 19 October 2023
Revised: 15 November 2023
Accepted: 28 November 2023
Published: 29 November 2023

1. Introduction

In the projects of navigation construction, enormous amounts of marine mud slurry (MS) are hydraulically dredged from ports, harbors, and channels around the world [1]. Globally, several hundred million cubic meters of marine mud slurries with poor engineering characteristics are inevitably produced. For the massive quantities and negative impacts of dredged mud slurry, its disposal and utilization have become one of the most noticeable issues in the world [2–7]. Traditional solutions, such as inland deposits and ocean dumping, are increasingly unpopular for their deleterious influence on surroundings [8]. Considering the shortage of construction materials in civil engineering, a practically feasible solution is to convert large volumes of marine mud slurry into construction materials for reuse, such as embankment [9], subbase materials [10,11] and brick production [12]. Therefore, it is of great significance to investigate the efficient treatment and utilization of dredged marine mud slurry.

Scholars have demonstrated that the vacuum preloading method can significantly improve the mechanical properties of slurries [13–17]. The vacuum preloading method is used in the treatment of soft clays for the advantages of simple construction technology and low cost. However, the slurry has a high fine particle content and low permeability, and can easily block the drainage board. The vacuum preloading method has low drainage efficiency and long drainage times for slurry with high water content [18]. Furthermore, the limited vacuum load inevitably results in the treated soil still belonging to soft soil, and can not directly convert marine mud slurry into filling material.

It has been widely accepted that the stabilization/solidification (S/S) techniques play a significant role in dredging sediments [19–22]. By adding a solidification agent, such as ordinary Portland cement (OPC), ground granulated blast furnace slag (GGBS) or metakaolin, the strength of slurries can be significantly increased [23]. Previous studies have gained a general understanding on the mechanism of sediment solidification [24]. It should be noted that the treatment efficiency of solidification agents on slurries is related to the water content. With the increase in water content, the curing efficiency of the solidification agent significantly decreases [25–27]. For marine mud slurry with high water content, the solidification treatment is either ineffective or extremely costly. The technique of flocculation and pressure of flocculated sediments is used in dewatering treatment of slurries. Flocculants have the function of promoting the dewatering rate of treated slurries [28,29]. Pressure filtration is a process of solid–liquid separation [30]. The mechanical properties of slurries treated by flocculation and pressure filtration will deteriorate after encountering water, making it difficult to reuse them.

In recent years, the combined physical and chemical method that integrates physical dewatering and chemical solidification has been proposed. This method can transform marine mud slurry into construction materials and apply them in reclamation, coastal embankment filling, and coastal slope protection [31]. However, as a new method, its fundamental principles and influencing factors are not well-understood. During the treatment process, the effects of the process parameters on the mechanical properties of the marine mud slurry are still unclear. This paper focuses on the flocculation-solidification-high pressure filtration combined method (FSHCM) to effectively process marine mud slurry. The effects of flocculant dose, composite solidification agent dose, initial water content and initial thickness of geo-bags on the mechanical properties of treated marine mud slurry are studied through laboratory experiments.

2. Laboratory Experiments

2.1. Materials

The materials employed in the experiments consisted of marine mud, solidification agent, and flocculants. The mud sample belonged to a marine deposit collected from an actual construction site located in Wenzhou, China. The basic physical and engineering properties of the mud sample are listed in Table 1.

Table 1. Basic properties of the marine mud slurries used in laboratory experiments.

Specific Density Gs	Liquid Limit w_L	Plastic Limit w_P	Plasticity Index	Organic Content (Ignition Loss) mo	Sand Fraction (0.075–2 mm)	Silt Fraction (0.002–075 mm)	Clay and Colloid Fraction (<0.002 mm)
2.69	56.1%	26.7%	29.4	4.41%	14.9%	79.5%	5.6%

The particle size of the mud was determined using a laser particle size analyzer, and the particle size distribution curve is shown in Figure 1. Based on the Unified Soil Classification System [32], the mud could be categorized as fat clay (CH). Additionally, the mud composition was analyzed using the 09 Empyrean X-ray diffractometer. This mud was composed of quartz, kaolinite, illite, smectite and mica. The composite solidification agents chosen for this experiment were the 425# OPC and GGBS. The flocculants used in the experiments were AN926SH anionic polyacrylamide (APAM) solution and $Ca(OH)_2$. APAM is a widely employed substance in various industrial fields like water treatment, sludge dewatering, and the paper industry [33–35]. In this experiment, the APAM solution was prepared at a concentration of 1:500 (mass ratio of dry APAM powder to water). The APAM dose was defined as the dry weight ratio of APAM powder to the dry soil particles. $Ca(OH)_2$ can be consumed by GGBS; the hydration products improve the solidification effect [36].

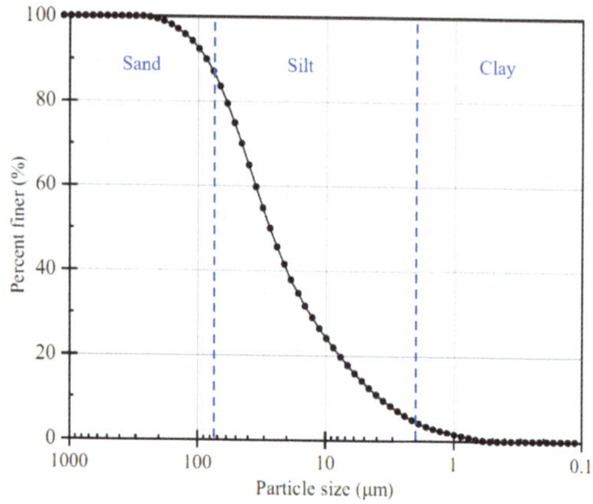

Figure 1. Particle size distribution curve for the mud sample.

2.2. Testing Procedures

The procedures of two types of tests, i.e., sedimentation tests and pressure filtration tests, are illustrated.

(1) Sedimentation tests

Many scholars use settlement tests to investigate the sedimentation behavior of mud slurries [37–39], which provided a reference for our experiments. The detailed procedures for the sedimentation tests are listed below.

(a) Set the initial dry weight of the mud sample (300 g) and the initial water content of the marine mud slurry (300%), and calculate the amount of experimental materials required.
(b) Sequentially add water and $Ca(OH)_2$ to the marine mud slurry, and mix them at a constant speed using a multi-function mixer for 5 min. Then, manually stir the slurry for half a minute (to avoid insufficient mechanical mixing) to obtain a homogeneous slurry mixture.
(c) Add the APAM solution to the slurry mixture and stir thoroughly to obtain a composite flocculant mixture.
(d) Pour the composite flocculant mixture evenly into a 1.5 L capacity graduated cylinder, ensuring consistent initial liquid level height. In the initial stage of the experiment, record the sludge–water interface separation value every hour, and later, record it every 2 h until the value remains constant.

(2) Pressure filtration tests

(a) Use an electric agitator to uniformly mix the original marine mud slurry and take a sample to obtain its natural water content. Based on the natural water content of the mud slurry and the predetermined mix ratio, calculate the required masses of water, $Ca(OH)_2$, OPC, and GGBS for the experiment.
(b) Based on the calculation results from step (a), add the required amounts of water, $Ca(OH)_2$, OPC, and GGBS sequentially to the slurry. Use a multifunctional mixer to thoroughly stir the slurry to obtain the mixture. The mixing process includes 5 min machine mixing in the beginning, 1 min hand mixing in the middle, and 3 min machine mixing in the end.
(c) Add APAM solution to the slurry and use a stirring machine to mix it evenly. Slow down the stirring speed when agglomerates begin to form. Stop stirring when the supernatant liquid separates from the mixed slurry.
(d) A specially developed device is used, as shown in Figure 2. This device is similar to a piston, through which the mud slurry can be dewatered, as demonstrated in the relevant standard [40]. Pour the mixed slurry into a geo-bag using a funnel and close

the zipper. Then, place the geo-bag in trough I and secure the cover plate on top of the pressure chamber using high-strength bolts. The material chosen for the geo-bags in this test was polypropylene fiber, which effectively prevents the seepage of mud particles and exhibits good permeability. The geo-bags had a planar dimension of 30 cm × 30 cm, and the maximum thickness was 20 cm. The aperture diameter of the geo-bags was 48 um.

(e) A hydraulic jack is used to apply pressure to the geo-bag and maintain the hydraulic pressure gauge reading at 15 MPa for 12 min. Observe the pressure gauge readings and control the pressure level accordingly.

(f) Mud cake is obtained after the pressurization is completed. To ensure uniform initial water content for each sample, when taking samples with a ring cutter, avoid the edge of the mud cake. The samples are cylindrical, 61.8 mm in diameter and 20 mm in height. Place the sample into sealed bags and conduct the curing process. The curing temperature of the water bath incubator is 20 ± 3 °C, and the curing ages are 7 days, 14 days, 21 days, and 28 days, respectively. After the samples reach the curing age, conduct water content tests and direct shear tests [41,42].

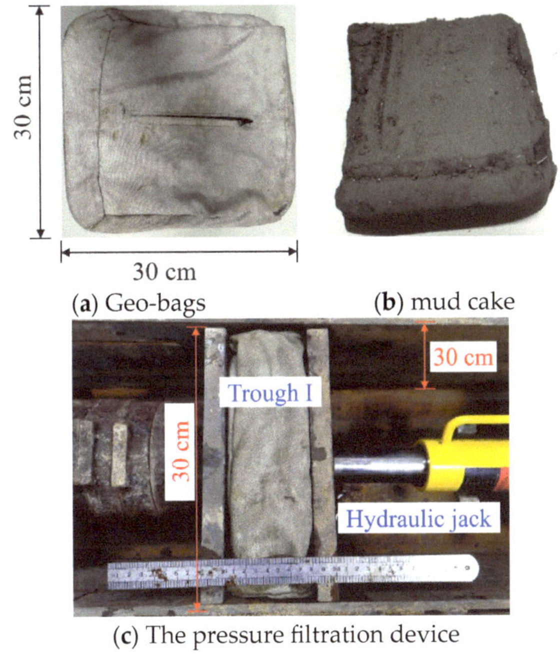

Figure 2. Pressure filtration tests.

2.3. Testing Program

The laboratory experiments included a total of 5 groups and 32 tests. Table 2 provides an overview of the mix proportions for all of these groups and tests. The curing conditions of samples for each group were the same. The curing temperature of the water bath incubator was 20 ± 3 °C, and the curing ages were 7 days, 14 days, 21 days, and 28 days. The definitions of the symbols given in the table are as follows: w_{ei} represents the equivalent initial water content, defined as the ratio of the sum of the mass of water contained in the slurry itself and in the APAM solution to the mass of solids; *M* represents the dry weight of the slurry; *C* represents the dose of composite solidification agent (mass fraction).

Table 2. Program for the laboratory experiments.

Group	Test	w_{ei} (%)	*M* (kg)	*C* (%)	APAM (%)	$Ca(OH)_2$ (%)
A	A1	200	2.5	6	0.16	1.5
	A2	200	0.3	0	0.16	1.5
	A3	200	2.5	20	0.16	1.5
B	B1	300	0.3	-	0	1.5
	B2	300	0.3	-	0.04	1.5
	B3	300	0.3	-	0.08	1.5
	B4	300	0.3	-	0.12	1.5
	B5	300	0.3	-	0.16	1.5
	B6	300	0.3	-	0.20	1.5
	B7	300	0.3	-	0.24	1.5
	B8	300	0.3	-	0.28	1.5
	B9	200	2.5	-	0	1.5
	B10	200	2.5	-	0.16	1.5
	B11	200	2.5	-	0.20	1.5
	B12	200	2.5	-	0.24	1.5
	B13	200	2.5	-	0.28	1.5
C	C1	200	2.5	3	0.16	1.5
	C2	200	2.5	5	0.16	1.5
	C3	200	2.5	7	0.16	1.5
	C4	200	2.5	9	0.16	1.5
D	D1	100	2.5	6	0.12	1.5
	D2	200	2.5	6	0.12	1.5
	D3	300	2.5	6	0.12	1.5
	D4	400	2.5	6	0.12	1.5
	D5	500	2.5	6	0.12	1.5
	D6	600	2.5	6	0.12	1.5
E	E1	200	2.5	6	0.16	1.5
	E2	200	3.0	6	0.16	1.5
	E3	200	3.5	6	0.16	1.5
	E4	200	4.0	6	0.16	1.5
	E5	200	4.5	6	0.16	1.5
	E6	200	5.0	6	0.16	1.5

Group A consisted of a feasibility exploration test for the FSHCM. This group included three tests (A1, A2 and A3) that adopted the FSHCM, flocculation-high pressure filtration combined method (FHCM), and flocculation-solidification combined method (FSCM), respectively. The feasibility of the FSHCM was verified by laboratory experiments. The strength of samples was obtained through direct shear tests. For each test, four samples were sheared under normal consolidation stresses of 50 kPa, 100 kPa, 150 kPa, and 200 kPa, respectively. The samples were cylindrical, 61.8 mm in diameter and 20 mm in height. The horizontal shear rate was set at 0.8 mm/min. In the three tests, the values of C were 0%, 6%, and 20%, respectively. A pressure filtration device was used in tests A1 and A2, with a pressure of 0.35 MPa and an action time of 12 min.

Group B included 13 tests with fixed w_{ei} (B1~B8: 300%; B9~B13: 200%), *M* (0.3 kg), and $Ca(OH)_2$ dose (1.5%) but different APAM doses, varying from 0 to 0.28% (mass fraction). Sedimentation tests (B1~B8) and pressure filtration tests (B9~B13) were used to explore the effects of the APAM dose on the dewatering efficiency of FSHCM. The pressure and action time of the pressure filtration tests were the same as those of Group A.

The Group C tests explored the effects of the composite solidification agent dose on the shear characteristics of mud cake. The initial water content of the marine mud slurry was 200%, and the mass of dry soil was 2.5 kg. The doses (mass fraction) of $Ca(OH)_2$ and APAM were 1.5% and 0.16%, respectively. In the four tests in this group, the composite solidification agent doses were s 3%, 5%, 7%, and 9% of the dry soil mass, respectively.

Group D included six tests, i.e., D1~D6, aiming to investigate the effects of initial water content on the shear characteristics of mud cake. The initial water contents of samples were 100%, 200%, 300%, 400%, 500%, and 600%, respectively.

Group E tests were conducted to study the effects of the initial thickness of geo-bags on the shear characteristics of mud cake. The initial thickness of geo-bags refers to the thickness of geo-bags when they were filled with mud slurries and placed in the pressure filtration device. The initial thickness of geo-bags included the thickness of the mud inserted in a geo-bag and the thickness of the geo-bag itself. This group involved five tests with different initial thicknesses of geo-bags (66 mm, 79 mm, 92 mm, 106 mm, 119 mm, and 132 mm).

3. Results and Discussion

3.1. Feasibility Exploration Tests for the FSHCM

The experimental results of Group A are illustrated in this section to demonstrate the feasibility of FSHCM. Figure 3 depicts the relationship between water content and curing age for the mud samples treated using FSHCM, FHCM, and FSCM, respectively. When the curing age was less than 1 day, FSHCM had the best effect of dewatering, with the lowest sample water content. FHCM yielded a slightly higher water content compared to the former. FSCM, on the other hand, had the poorest effect of dewatering, with water contents significantly higher than those of the first two methods. FSHCM-MS, FHCM-MS, and FSCM-MS represent mud slurries treated by flocculation-solidification-high pressure filtration combined method, flocculation-high pressure filtration combined method, and flocculation-solidification combined method, respectively. When the curing age was longer than 1 day, the water content of FSHCM-MS decreased slowly. FHCM-MS showed almost no change in water content, while FSCM-MS exhibited a faster decrease in water content. The higher the dose of composite solidification agent, the more pore water was consumed by chemical reactions, resulting in a more significant decrease in the water content of the samples. FSHCM can reduce the water content of the marine mud slurry to a relatively low level through flocculation and high-pressure filtration. During the curing process, the water content of marine mud slurry is further decreased by the consumption of pore water. FSHCM has advantages both in terms of dewatering and material savings.

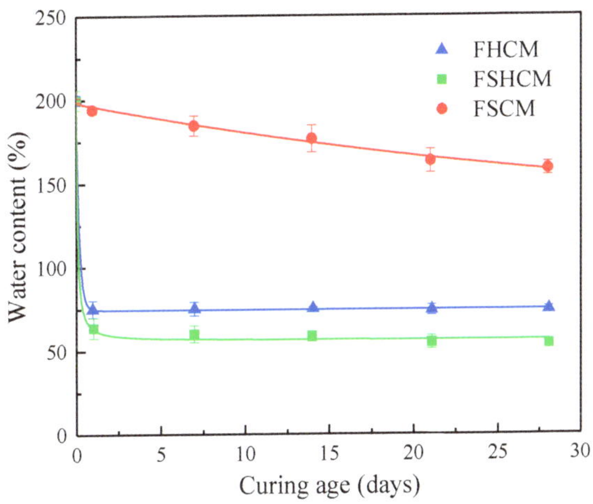

Figure 3. Relationship between water contents of mud samples and curing age.

Figure 4 illustrates the representative strength envelopes of samples treated by the three methods. It should be noted that the solid lines represent FSCM and the dashed lines represent FHCM in Figure 4b. With the increasing curing age, the shear strength envelopes of FSHCM-MS and FSCM-MS moved upward, indicating an increase in shear strength. Additionally, the shear strength envelope of FHCM-MS remained unchanged with curing age. The shear strength envelope of FSHCM-MS at 7 days was located above the shear strength envelope of FSCM-MS at 28 days, indicating that FSHCM-MS achieved greater shear strength in a shorter time. Compared to FSCM, FSHCM needed less composite solidification agent, and FSHCM-MS achieved higher shear strength, demonstrating the efficiency of FSHCM.

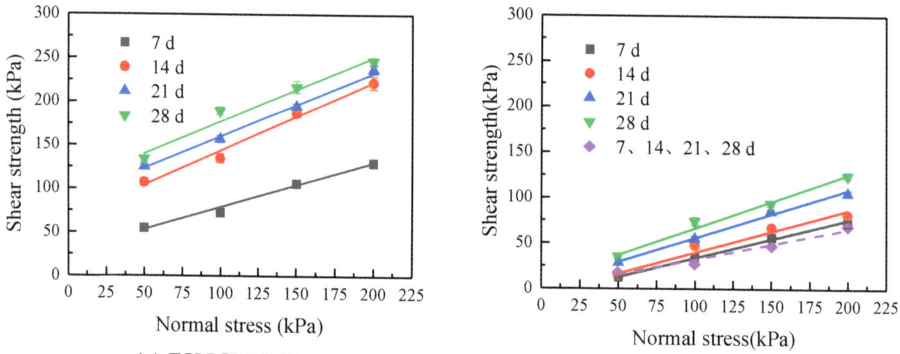

(**a**) FSHCM-MS (**b**) FSCM-MS and FHCM-MS

Figure 4. Strength envelopes of samples.

Figure 5 illustrates the effects of curing age on the cohesion and friction angle of samples treated by the three methods. It can be observed from Figure 5a that FHCM-MS exhibited no cohesion at 7 days, 14 days, 21 days, and 28 days of curing. This was attributed to the higher water content in FHCM-MS, which did not allow effective bonding between soil particles. FSCM-MS exhibited no cohesion at 7 days and 14 days of curing age, and it showed minimal cohesion at 21 days, which gradually increases with the curing period. After flocculation and dewatering treatment, there were still numerous pores between soil particles in the slurries. Under the influence of the composite solidification agent, it was challenging to establish effective bonds between soil particles in a short period. The cohesion of FSHCM-MS increased with curing age. Compared to FHCM-MS, the cohesion of FSHCM-MS increased by 104.89 kPa after 28 days of curing. Compared to FSCM-MS, FSHCM-MS saved 14% of the composite solidification agent dose and increased the cohesion by 94.59 kPa after 28 days of curing. From Figure 5b, it can be observed that the friction angle values for FSHCM-MS ranged from 20° to 38.5°, presenting significant advantages.

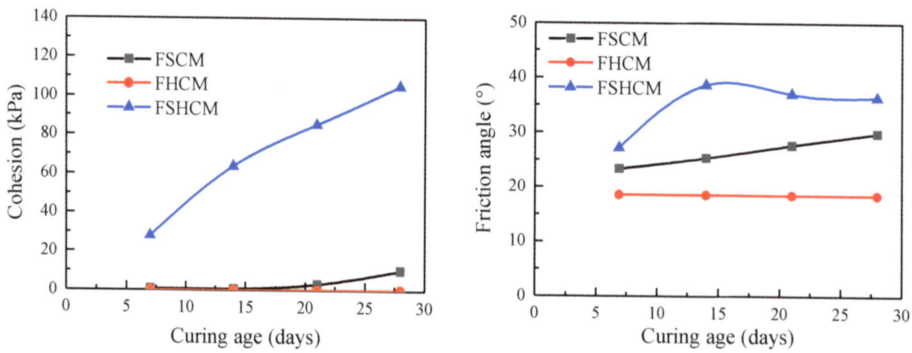

(**a**) Cohesion (**b**) Friction angle

Figure 5. Effects of curing age on the cohesion and friction angle of samples.

3.2. Effects of APAM Dose on the Dewatering Performance of FSHCM-MS

In sedimentation tests, the mixed slurry was poured into a 1.5 L capacity graduated cylinder, and the supernatant liquid height, H_S, was recorded every 1 or 2 h. Figure 6 shows variation curves of H_S against time at various APAM doses. The same dose of $Ca(OH)_2$, 1.5%, was adopted for all tests. As can be seen in Figure 6a, when the APAM content was 0%, 0.04%, 0.08% and 0.12%, the value of *Hs* increased significantly with time within 2000 min. The reason for this phenomenon is that the APAM dose was too small to form an effective flocculation structure between the clay particles. The mixed mud slurry was still a suspension.

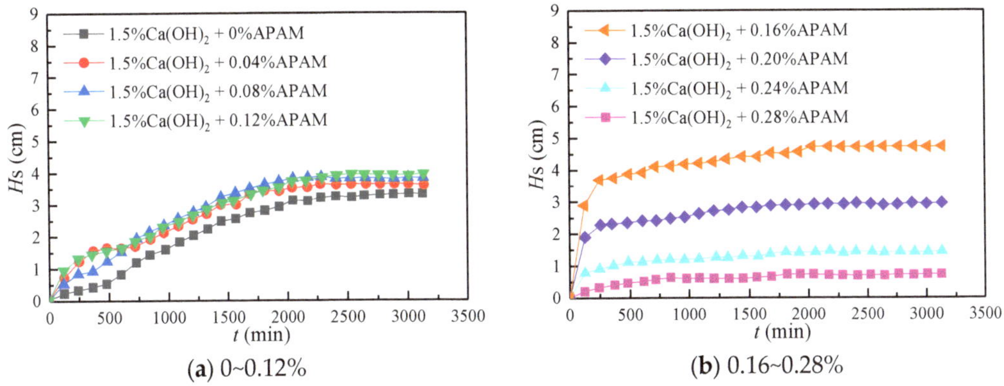

Figure 6. Variation curves of H_S against time at various APAM doses.

When the APAM dose was 0.16%, the value of *Hs* increased evidently, and the slope of the curve was larger within 200 min. Then, the curve entered a flatter state, and H_S approached a constant value after 2000 min, as shown in Figure 6b. As the APAM dose increases (0.20%, 0.24% and 0.28%), the H_S value gradually decreased, and the less time it took for the curve to reach a relatively stable state. The reason for this phenomenon is that with increasing APAM doses, the flocculation structure became larger, and the mixed slurry quickly approached a stable state. Therefore, a stable flocculation structure could be formed in the mud slurry when the APAM dose was 0.16%. As the APAM dose continued to increase, the flocculation structure became larger, the mixed slurry reached a stable state quickly, and the value of *Hs* changed less with time.

To investigate the effects of the APAM dose on the dewatering performance of FSHCM-MS, representative APAM doses (0%, 0.16%, 0.20%, 0.24%, and 0.28%) were selected for the analysis of the dewatering performance of the marine mud slurry. As shown in Figures 7 and 8, the mud cake without the addition of APAM was fragmented, and the filtrate was turbid. When APAM was added, the mud formed into a cake shape, and the filtrate was clear.

Figure 9 shows the effects of the APAM dose on the water content of the mud cake. When the APAM dose was 0.16%, the water content of the mud cake was only 65.56%. When APAM was not added, the water content of the mud cake was 150.6%, which was 2.3 times that of the mud cake without APAM, and the dewatering efficiency was significantly improved. Moreover, as the APAM dose increased, the water content of the mud cake decreased, but the rate became smaller.

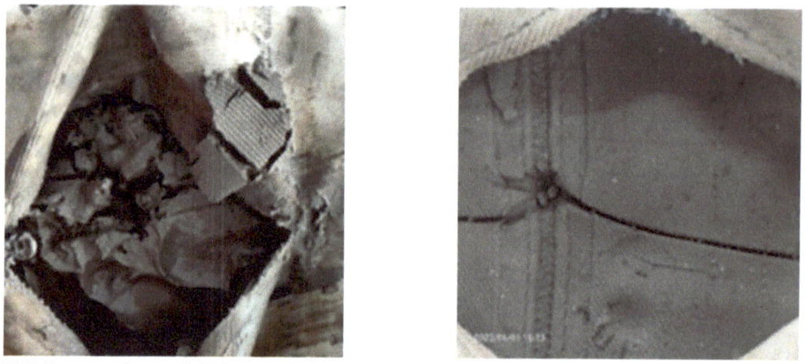

(**a**) Without the addition of APAM (**b**) With the addition of APAM

Figure 7. Mud cake.

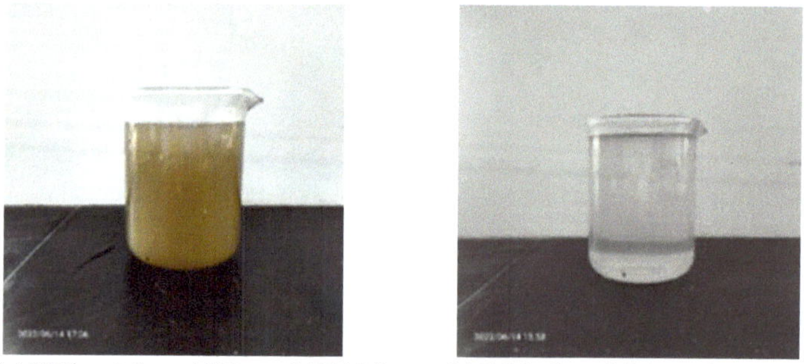

(**a**) Without the addition of APAM (**b**) With the addition of APAM

Figure 8. Filtrate.

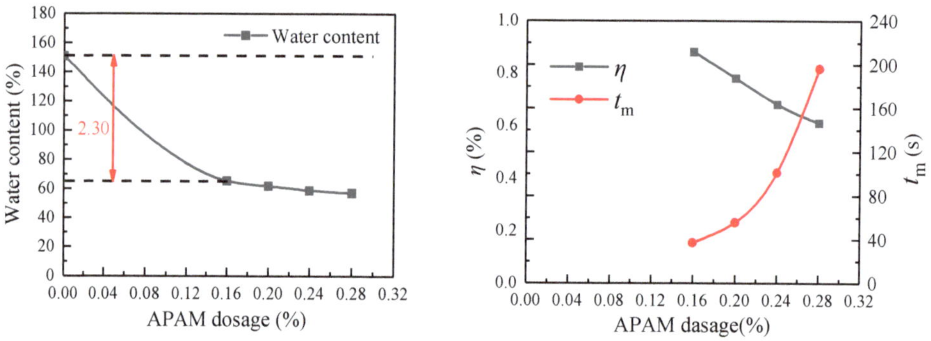

(**a**) Water content (**b**) Evaluation indicator

Figure 9. Effects of APAM dose on the water content and evaluation indicators.

The beneficial value of water consumption, η, was defined as shown in Equation (1). A smaller η value indicates a more significant wastage of water resources:

$$\eta = \frac{\omega_1 - \omega_2}{\omega_2} \times 100\% \quad (1)$$

where w_1 represents the mass of drainage during the filtration process, and w_2 represents the mass of water required to prepare the corresponding flocculant solution.

Additionally, mechanical stirring time was considered as another evaluation indicator. The mechanical stirring time was the mixing time of the agitator after adding the APAM solution to the mud slurries. The shorter the mechanical stirring time, the greater the mud volume that could be processed per unit time. The effects of APAM dose on evaluation indicators are shown in Figure 9b. It can be observed that with an increase in APAM dose, the value of η exhibited a generally linear decrease, while the mechanical stirring time (t_m), showed an exponential increase. The results indicate that as the APAM dose increased, the investment cost of FSHCM increased. Therefore, this study recommends an optimal APAM dose of 0.16%.

3.3. Effects of Composite Solidification Agent on the Shear Characteristics of FSHCM-MS

Figure 10 illustrates the effects of composite solidification agent dose on the water content of the mud cake. As seen in Figure 10a, the initial water content of the mud cake showed relatively small variations with changes in the dose of the composite solidification agent, and the dose of the composite solidification agent had little effect on the dewatering efficiency of the FSHCM. Figure 10b shows that with an increasing curing age, a higher solidification agent dose resulted in a faster decrease in the water content of the mud cake. For example, when the solidification agent dose was 3%, 5%, 7%, and 9%, the 28-day water content of the mud cake was 63.15%, 59.47%, 56.28%, and 52.84%, respectively.

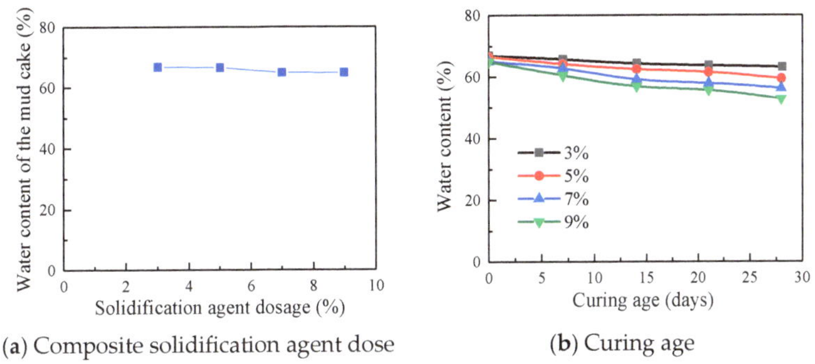

Figure 10. Effects of composite solidification agent dose on the water content of the mud cake.

Figure 11 illustrates the effects of the dose of the composite solidification agent on the shear strength of mud cakes. It can be observed from Figure 11a that samples with a higher dose of composite solidification agent had shear strength envelopes positioned above those with a lower dose. Under the same normal stress conditions, they exhibited greater shear strength.

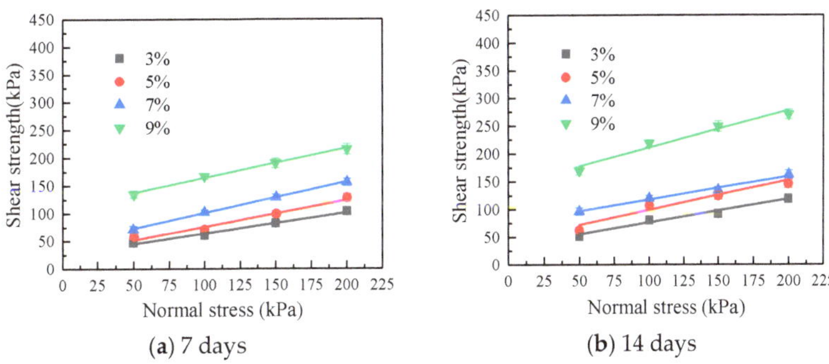

Figure 11. *Cont.*

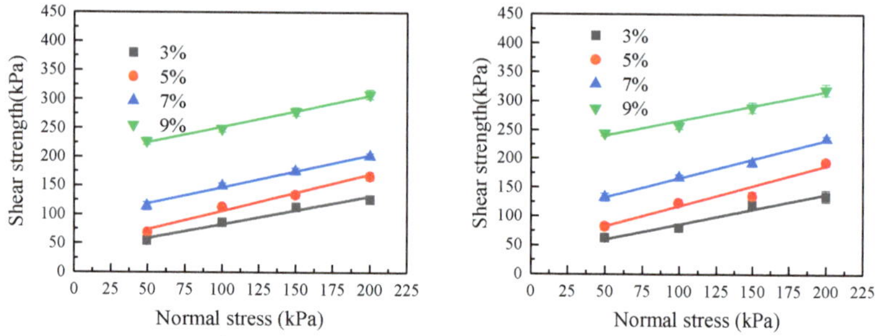

(**c**) 21 days (**d**) 28 days

Figure 11. Shear strength envelopes.

Additionally, it is evident that with an increase in the dose of composite solidification agent, the spacing between the shear strength envelopes gradually increased. The spacing between the envelopes was notably larger for samples with a dose between 7% and 9%, suggesting that a higher dose of composite solidification agent leads to a greater increase in shear strength of the samples.

It can be observed from Figure 11 that the shear strength of the samples increased with the curing age. Using the experimental results with a normal force of 200 kPa as an example, when the samples were cured from 7 days to 28 days, the shear strength of the samples increased by 30.2 kPa, 64.12 kPa, 76.13 kPa, and 104.45 kPa, respectively, for composite solidification agent doses of 3%, 5%, 7%, and 9%. The higher the dose of the composite solidification agent, the faster the increase in the shear strength of the samples with curing age, and the greater the ultimate shear strength.

The Mohr–Coulomb criterion can be used in direct shear tests. Figure 12 illustrates the effects of the dose of the composite solidification agent on the cohesion and friction angle of the samples. As shown in Figure 12a, the cohesion of samples increased with curing age, and a higher dose of composite solidification agent resulted in a more pronounced increase in cohesion. For the same curing age, samples exhibited greater cohesion with a higher dose of composite solidification agent. As can be seen from Figure 12b, the values of the friction angle ranged between 20° and 35°.

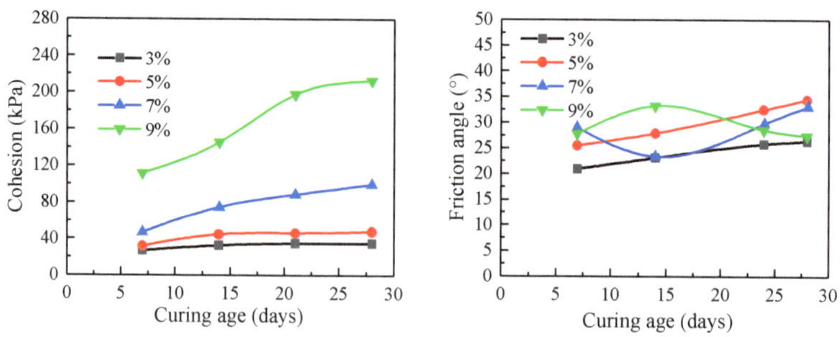

(**a**) Cohesion (**b**) Friction angle

Figure 12. Effects of the dose of composite solidification agent on the cohesion and friction angle of samples.

3.4. Effects of Initial Water Content on the Shear Characteristics of FSHCM-MS

Figure 13 illustrates the effects of the initial water content of the marine mud slurry on the water content of the mud cake. As can be seen in Figure 13a, changing the initial water content of the marine mud slurry had a significant effect on the water content of the mud cake. As the initial water content of the marine mud slurry increased, the water content of the mud cake decreased, and the rate of decrease became progressively smaller. When the initial water content of the marine mud slurry increased from 100% to 400%, the water content of the mud cake decreased by 19.6%. However, when the initial water content increased from 400% to 600%, the water content of the mud cake only decreased by 2.39%. During the experiment, it was observed that the higher the initial water content of the marine mud slurry, the larger the flocs formed by the flocculant, resulting in a faster sedimentation rate.

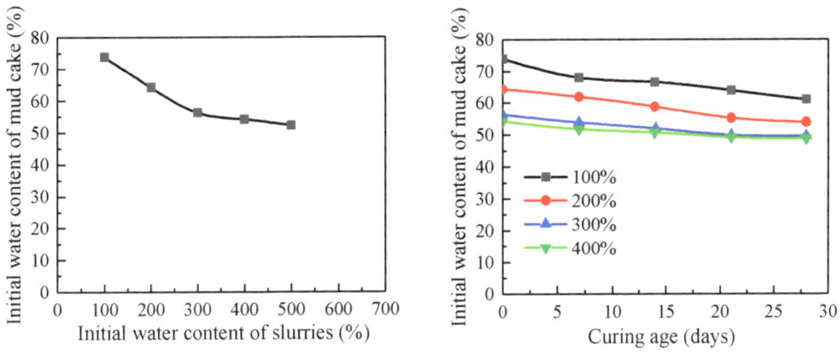

(**a**) Initial water content of slurries (**b**) Curing age

Figure 13. Effects of the initial water content of marine mud slurries on the initial water content of the mud cake.

It can be observed from Figure 13b that for a given dose of composite solidification agent, the higher the initial water content of the marine mud slurry, the lower the 28-day water content of the mud cake. Specifically, when the initial water content of the marine mud slurry was 100%, 200%, 300%, and 400%, the 28-day water content of the mud cake was 61.0%, 53.8%, 49.5%, and 48.7%, respectively. The reasons for this phenomenon include the following two points. The main reason is that as the initial water content of the marine mud slurry increased, there were fewer solid particles per unit volume, resulting in a higher initial porosity. This led to shorter drainage paths and higher drainage rates. Another possible reason is that the chosen APAM dose for this experiment (0.12%) was optimal for mud with high water content. At this dose, the dewatering performance was better in high-water-content samples compared to low-water-content samples.

Figure 14 illustrates the effects of the initial water content of the marine mud slurry on the shear strength of the mud cake. As can be seen from Figure 14a, it is evident that the shear strength envelope of the samples with a higher initial water content was positioned above that of the samples with a lower initial water content. This indicates that a higher initial water content of the marine mud slurry leads to greater shear strength of the filter-pressed mud cake under the same normal stress conditions.

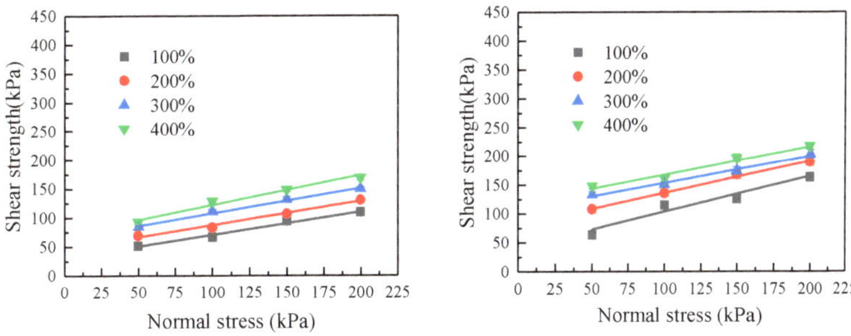

(**a**) 7 days (**b**) 14 days

Figure 14. *Cont.*

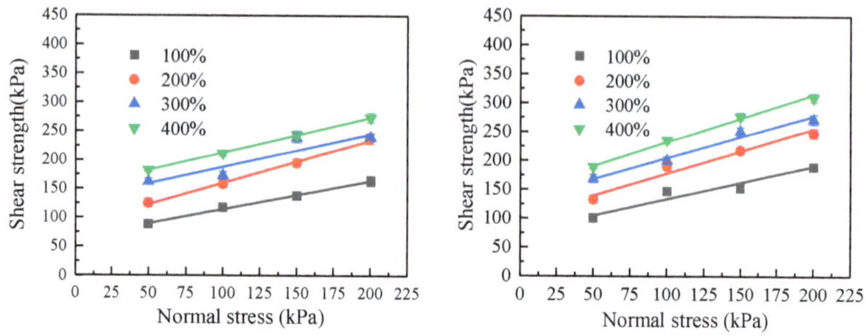

(**c**) 21 days (**d**) 28 days

Figure 14. Shear strength envelopes.

Furthermore, the shear strength of the samples increased with curing age. Taking the normal stress of 200 kPa as an example, for samples cured from 7 days to 28 days, the shear strength increased as follows for the marine mud slurry with the initial water content of 100%, 200%, 300%, and 400%: 79.67 kPa, 117.77 kPa, 123.2 kPa, and 130.75 kPa, respectively. The higher the initial water content of the marine mud slurry, the faster the rate of increase in shear strength of the samples with curing age, resulting in a higher final shear strength value.

Figure 15 shows the effects of the initial water content of marine mud slurry on the cohesion and friction angle of mud cake. For slurries with the same initial water content, their cohesion increased with curing age. For mud cake cured from 7 days to 28 days, the cohesion of the samples increased as follows for the marine mud slurries with initial water contents of 100%, 200%, 300%, and 400%: 49.76 kPa, 60.58 kPa, 67.69 kPa, and 86.90 kPa, respectively. Moreover, a higher initial water content in the marine mud slurry led to greater cohesion of the samples. Figure 15b illustrates the curve of the friction angle of the samples with curing age. The values of the friction angle ranged between 20° and 35°.

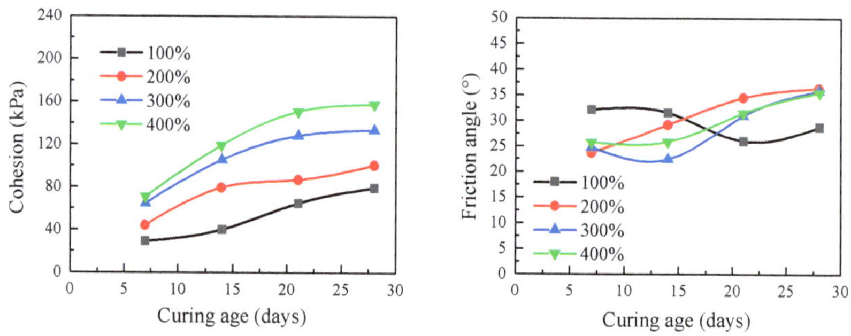

(**a**) Cohesion (**b**) Friction angle

Figure 15. Effects of initial water content of marine mud slurries on the cohesion and friction angle of mud cake.

3.5. Effects of the Initial Thickness of Geo-Bags on the Shear Characteristics of FSHCM-MS

Figure 16 illustrates the effects of the initial thickness of the geo-bags on the initial water content of the mud cake. It can be observed from Figure 16a that as the initial thickness of the geo-bags increased, the initial water content of the mud cake remained relatively constant at first and then gradually increased. Specifically, the water content of the mud cake rose by 14.8% due to the increase in the initial thickness of the geo-bags from 92 mm to 132 mm, while it remained almost unchanged from 66 mm to 92 mm. Figure 16b illustrates the effects of the curing age on the water content of mud cake. When the same dose of composite solidification agent was used, a smaller initial thickness of geo-bags resulted in a lower 28-day water content of the mud cake, mainly because the former led to shorter pore channels and better dewatering efficiency in the FSHCM.

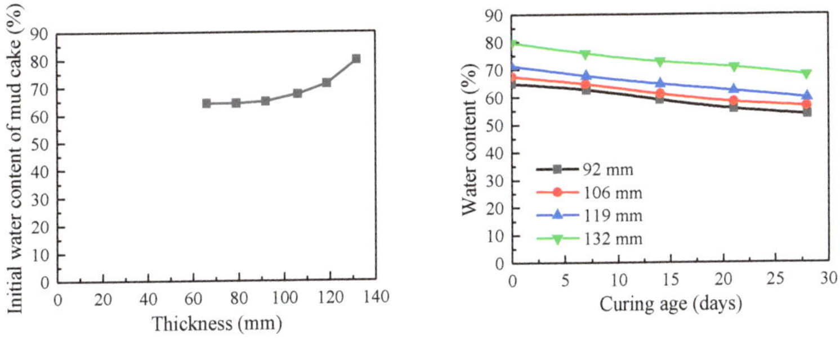

(**a**) Initial water content of marine mud slurry (**b**) Curing age

Figure 16. Effects of the initial thickness of geo-bags on the water content of the mud cake.

Figure 17 illustrates the effects of the initial thickness of geo-bags on the shear strength of the mud cake. It is obvious that the shear stress of the mud cake increases with increasing normal stress. Additionally, as shown in Figure 17a, the envelope of shear strength for mud cake with lower initial thickness of geo-bags was situated above that for mud cake with higher geo-bag thickness. This indicates that under the same normal stress conditions and curing age, a smaller initial thickness of geo-bags results in a larger shear strength of the mud cake. Taking a normal stress of 200 kPa as an example, the 7-day shear strength of the mud cake was 123.97 kPa, 118.59 kPa, 115.59 kPa, and 100.14 kPa for the thickness of 92 mm, 106 mm, 119 mm, and 132 mm, respectively.

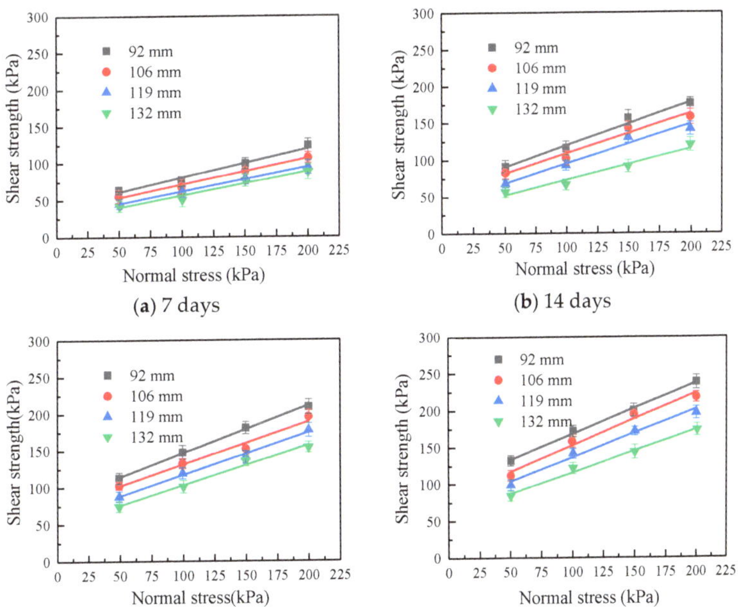

(**a**) 7 days (**b**) 14 days (**c**) 21 days (**d**) 28 days

Figure 17. Shear strength envelopes.

Furthermore, with the increase in the curing age, the shear strength envelope gradually shifted upward, indicating that the shear strength of the mud cake increased with the curing age. Taking the normal stress of 200 kPa as an example, as the curing period increased from 7 days to 28 days, the shear strength of the samples increased by 114.48 kPa, 109.35 kPa, 92.2 kPa, and 74.86 kPa under the geo-bag thickness of 92 mm, 106 mm, 119 mm, and 132 mm, respectively. The results show that under the same normal force, the greater the initial thickness of the geo-bags, the smaller the increase in shear strength of mud cake with the change in curing period.

Figure 18 illustrates the effects of the initial thickness of the geo-bags on the cohesion and friction angle of the samples. As seen in Figure 18a, under the same curing age, a smaller thickness resulted in greater cohesion of the mud cake. Under the same geo-bag thickness, the cohesion of the mud cake increased with the increase in curing time. When the curing period increased from 7 days to 28 days, the cohesion of samples increased by 62.05 kPa, 51.79 kPa, 46.91 kPa, and 36.38 kPa under the geo-bag thickness of 92 mm, 106 mm, 119 mm, and 132 mm, respectively. Figure 18b depicts the variation in the friction angle. Under the same geo-bag thickness, the friction angle increased with the increase in curing age. The maximum value of the friction angle was 35°, and the minimum value was 20°.

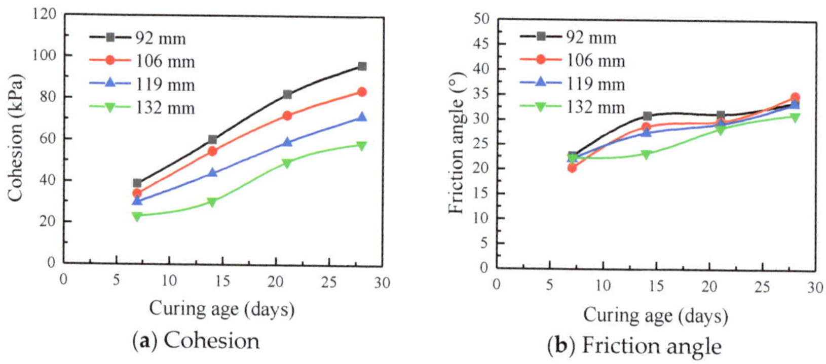

Figure 18. Effects of the initial thickness of geo-bags on the cohesion and friction angle of the samples.

4. Conclusions

In this study, a series of laboratory experiments were conducted to investigate the feasibility of FSHCM in marine mud slurries and the effects of APAM dose, composite solidification agent dose, the initial water content of marine mud slurries and the initial thickness of geo-bags on the mechanical properties of FSHCM-MS. The main conclusions are as follows:

1. Compared to the FHCM, the cohesion of the samples treated by the FSHCM decreased by 104.89 kPa at 28 days. Compared with FSCM, the FSHCM saved 14% of the composite solidification agent and increased cohesion by 94.59 kPa at 28 days. By comparing the water content and cohesion of mud cake treated by three methods, it was found that the FSHCM is more efficient.
2. When the APAM dose was 0.16%, FSHCM-MS exhibited a noticeable flocculation effect. As the APAM dose increased, the dewatering efficiency of FSHCM-MS improved. However, the water use efficiency value η decreased, and mechanical mixing time increased. Considering all factors, the optimal dose of APAM was determined as 0.16%.
3. The composite solidification agent dose had little effect on the dewatering efficiency of the FSHCM. The initial water content of the marine mud slurry had a significant effect on the dewatering efficiency of the combined method when it varied between 100% and 400%. Higher initial water content in the mud led to better dewatering

results. As the initial thickness of geo-bags increased, the initial water content of the mud cake remained constant at first, and then gradually increased.

4. An increase in composite solidification agent dose led to a higher rate of increase in the shear strength and cohesion of mud cake. As the initial water content of the marine mud slurry rose, the shear strength and cohesion of the mud cake increased. Moreover, an increase in the initial thickness of the geo-bags led to a decrease in the shear strength and cohesion of the mud cake.

Author Contributions: Conceptualization, C.H., H.X., B.B., R.Z. and Y.G.; methodology, C.H., H.X., R.Z. and Y.G.; validation, C.H., H.X. and B.B.; investigation, C.H., B.B., R.Z. and Y.G.; resources, C.H. and B.B.; data curation, H.X., B.B. and Y.G.; writing—original draft preparation, Z.Z., H.X. and C.H.; writing—review and editing, R.Z. and Y.G.; supervision, R.Z.; funding acquisition, C.H., H.X. and B.B. All authors have read and agreed to the published version of the manuscript.

Funding: This research was funded by the Science and Technology Project of State Grid Jiangsu Electric Power Co., Ltd. (Grant No. J2021047).

Institutional Review Board Statement: Not applicable.

Informed Consent Statement: Not applicable.

Data Availability Statement: All data have been provided in the paper.

Conflicts of Interest: Authors Chao Han, Hongping Xie, and Bin Bai were employed by the Construction Branch of State Grid Jiangsu Electric Power Co., Ltd. The remaining authors declare that the research was conducted in the absence of any commercial or financial relationships that could be construed as a potential conflict of interest.

References

1. Chrysochoou, M.; Grubb, D.G.; Drengler, K.L.; Malasavage, N.E. Stabilized Dredged Material. III: Mineralogical Perspective. *J. Geotech. Geoenviron. Eng.* **2010**, *136*, 1037–1050. [CrossRef]
2. Xu, G.; Gao, Y.; Hong, Z.; Ding, J. Sedimentation behavior of four dredged slurries in China. *Mar. Georesour. Geotechnol.* **2012**, *30*, 143–156. [CrossRef]
3. Zeng, L.; Hong, Z.; Tian, W.; Shi, J. Settling behavior of clay suspensions produced by dredging activities in China. *Mar. Georesour. Geotechnol.* **2018**, *36*, 30–37. [CrossRef]
4. Wang, S.Y.; Fu, J.Y.; Zhang, C.; Yang, J.S. *Shield Tunnel Engineering, Chapter 10 Slurry Treatment for Shield Tunnelling and Waste Slurry Recycling*; Elsevier: Amsterdam, The Netherlands, 2021. [CrossRef]
5. Foged, S.; Duerinckx, L.; Vandekeybus, J. An innovative and sustainable solution for sediment disposal problems. In Proceedings of the Eighteenth World Dredging Congress, Lake Buena Vista, FL, USA, 27 May–1 June 2007; pp. 1513–1528.
6. Spadaro, P.; Rosenthal, L. River and harbor remediation:"polluter pays," alternative finance, and the promise of a "circular economy". *J. Soils Sediments* **2020**, *20*, 4238–4247. [CrossRef]
7. Hussan, A.; Levacher, D.; Mezazigh, S. Investigation of natural dewatering of dredged sediments incorporating evaporation and drainage. *Dry. Technol.* **2023**, *41*, 2227–2240. [CrossRef]
8. Sheehan, C.; Harrington, J. Management of dredge material in the Republic of Ireland–A review. *Waste Manag.* **2012**, *32*, 1031–1044. [CrossRef] [PubMed]
9. Levacher, D.; Sanchez, M. Characterization of marine sediments for a reuse in land disposal and embankment. *Eur. J. Environ. Civ. Eng.* **2011**, *15*, 167–178. [CrossRef]
10. Zentar, R.; Dubois, V.; Abriak, N.E. Mechanical behaviour and environmental impacts of a test road built with marine dredged sediments. *Resour. Conserv. Recycl.* **2008**, *52*, 947–954. [CrossRef]
11. Achour, R.; Abriak, N.-E.; Zentar, R.; Rivard, P.; Gregoire, P. Valorization of unauthorized sea disposal dredged sediments as a road foundation material. *Environ. Technol.* **2014**, *35*, 1997–2007. [CrossRef]
12. Samara, M.; Lafhaj, Z.; Chapiseau, C. Valorization of stabilized river sediments in fired clay bricks: Factory scale experiment. *J. Hazard. Mater.* **2009**, *163*, 701–710. [CrossRef]
13. Sun, L.; Gao, X.; Zhuang, D.; Guo, W.; Hou, J.; Liu, X. Pilot tests on vacuum preloading method combined with short and long PVDs. *Geotext. Geomembr.* **2018**, *46*, 243–250. [CrossRef]
14. Sun, H.; Weng, Z.; Liu, S.; Geng, X.; Pan, X.; Cai, Y.; Shi, L. Compression and consolidation behaviors of lime-treated dredging slurry under vacuum pressure. *Eng. Geol.* **2020**, *270*, 105573. [CrossRef]
15. Wang, J.; Ni, J.; Cai, Y.; Fu, H.; Wang, P. Combination of vacuum preloading and lime treatment for improvement of dredged fill. *Eng. Geol.* **2017**, *227*, 149–158. [CrossRef]
16. Wang, J.; Cai, Y.; Ma, J.; Chu, J.; Fu, H.; Wang, P.; Jin, Y. Improved vacuum preloading method for consolidation of dredged clay-slurry fill. *J. Geotech. Geoenviron. Eng.* **2016**, *142*, 06016012. [CrossRef]

17. Zhou, Y.; Cai, Y.; Yuan, G.; Wang, J.; Fu, H.; Hu, X.; Geng, X.; Li, M.; Liu, J.; Jin, H. Effect of tamping interval on consolidation of dredged slurry using vacuum preloading combined with dynamic consolidation. *Acta Geotech.* **2021**, *16*, 859–871. [CrossRef]
18. Cui, Y.; Pan, F.; Zhang, B.; Wang, X.; Diao, H. Laboratory test of waste mud treated by the flocculation-vacuum-curing integrated method. *Constr. Build. Mater.* **2022**, *328*, 127086. [CrossRef]
19. Boutouil, M. Traitement des Vases de Dragage par Stabilisation/Solidification à Base de Ciment et Additifs (Treatment of Dredging Muds by Stabilization/Solidification with Cement and Additives). Ph.D. Thesis, University of Le Havre, Le Havre, France, 1998.
20. Lahtinen, P.; Forsman, J.; Kiukkonen, P.; Kreft-Burman, K.; Niutanen, V. Mass stabilisation as a method of treatment of contaminated sediments. In Proceedings of the South Baltic Conference on Dredged Materials in Dike Construction, Rostock, Germany, 10–12 April 2014; pp. 145–155.
21. Lirer, S.; Liguori, B.; Capasso, I.; Flora, A.; Caputo, D. Mechanical and chemical properties of composite materials made of dredged sediments in a fly-ash based geopolymer. *J. Environ. Manag.* **2017**, *191*, 1–7. [CrossRef]
22. Yoobanpot, N.; Jamsawang, P.; Poorahong, H.; Jongpradist, P.; Likitlersuang, S. Multiscale laboratory investigation of the mechanical and microstructural properties of dredged sediments stabilized with cement and fly ash. *Eng. Geol.* **2020**, *267*, 105491. [CrossRef]
23. Horpibulsuk, S.; Rachan, R.; Suddeepong, A. Assessment of strength development in blended cement admixed Bangkok clay. *Constr. Build. Mater.* **2011**, *25*, 1521–1531. [CrossRef]
24. Sollars, C.; Perry, R. Cement-based stabilization of wastes: Practical and theoretical considerations. *Water Environ. J.* **1989**, *3*, 125–134. [CrossRef]
25. Yamashita, E.; Cikmit, A.A.; Tsuchida, T.; Hashimoto, R. Strength estimation of cement-treated marine clay with wide ranges of sand and initial water contents. *Soils Found.* **2020**, *60*, 1065–1083. [CrossRef]
26. Vu, M.C.; Satomi, T.; Takahashi, H. Influence of initial water, moisture, and geopolymer content on geopolymer modified sludge. *Constr. Build. Mater.* **2020**, *235*, 117420. [CrossRef]
27. Su, Y.; Luo, B.; Luo, Z.; Xu, F.; Huang, H.; Long, Z.; Shen, C. Mechanical characteristics and solidification mechanism of slag/fly ash-based geopolymer and cement solidified organic clay: A comparative study. *J. Build. Eng.* **2023**, *71*, 106459. [CrossRef]
28. Arjmand, R.; Massinaei, M.; Behnamfard, A. Improving flocculation and dewatering performance of iron tailings thickeners. *J. Water Process Eng.* **2019**, *31*, 100873. [CrossRef]
29. Maraschin, M.; Ferrari, K.F.H.; Carissimi, E. Acidification and flocculation of sludge from a water treatment plant: New action mechanisms. *Sep. Purif. Technol.* **2020**, *252*, 117417. [CrossRef]
30. Wakeman, R.; Tarleton, S. *Solid/Liquid Separation: Principles of Industrial Filtration*; Elsevier: Amsterdam, The Netherlands, 2005.
31. Zhang, R.; Dong, C.; Lu, Z.; Pu, H. Strength characteristics of hydraulically dredged mud slurry treated by flocculation-solidification combined method. *Constr. Build. Mater.* **2019**, *228*, 116742. [CrossRef]
32. *ASTM D2487-11*; Standard Practice for Classification of Soils for Engineering Purposes (Unified Soil Classification System). ASTM International: West Conshohocken, PA, USA, 2017.
33. Amuda, O.; Amoo, I. Coagulation/flocculation process and sludge conditioning in beverage industrial wastewater treatment. *J. Hazard. Mater.* **2007**, *141*, 778–783. [CrossRef] [PubMed]
34. Jin, L.; Zhang, P.; Zhang, G.; Li, J. Study of sludge moisture distribution and dewatering characteristic after cationic polyacrylamide (C-PAM) conditioning. *Desalin Water Treat.* **2016**, *57*, 29377–29383. [CrossRef]
35. Kang, J.; McLaughlin, R.A. Simple systems for treating pumped, turbid water with flocculants and a geotextile dewatering bag. *J. Environ. Manag.* **2016**, *182*, 208–213. [CrossRef]
36. Preetham, H.; Nayak, S. Geotechnical investigations on marine clay stabilized using granulated blast furnace slag and cement. *Int. J. Geosynth. Groun* **2019**, *5*, 28. [CrossRef]
37. Bertrand, G.; Roy, P.; Filiatre, C.; Coddet, C. Spray-dried ceramic powders: A quantitative correlation between slurry characteristics and shapes of the granules. *Chem. Eng. Sci.* **2005**, *60*, 95–102. [CrossRef]
38. Du, Y.; Yang, Y.; Fan, R.; Wang, F. Effects of phosphate dispersants on the liquid limit, sediment volume and apparent viscosity of clayey soil/calcium-bentonite slurry wall backfills. *KSCE J. Civ. Eng.* **2016**, *20*, 670–678. [CrossRef]
39. He, J.; Chu, J.; Tan, S.K.; Vu, T.T.; Lam, K.P. Sedimentation behavior of flocculant-treated soil slurry. *Mar. Georesour. Geotechnol.* **2017**, *35*, 593–602. [CrossRef]
40. *EN 14701-2:2013 (E)*; Characterisation of Sludges-Filtration Properties-Part 2: Determination of the Specific Resistance to Filtration. CEN—UNI EN: Brussels, Belgium, 2013.
41. *ASTM D2216-19*; Standard Test Methods for Laboratory Determination of Water (Moisture) Content of Soil and Rock by Mass. ASTM International: West Conshohocken, PA, USA, 2019.
42. *ASTM D7702/D7702M-14*; Standard Guide for Considerations When Evaluating Direct Shear Results Involving Geosynthetics. ASTM International: West Conshohocken, PA, USA, 2021.

Journal of
Marine Science
and Engineering

MDPI

Article

Vertical Monotonic and Cyclic Responses of a Bucket in Over-Consolidated Clay

Jun Jiang [1], Dong Wang [1,2,*] and Dengfeng Fu [1]

1 MOE Key Laboratory of Marine Environment and Ecology, Ocean University of China, 238 Songling Road, Qingdao 266100, China; junjiang@stu.ouc.edu.cn (J.J.); dengfeng.fu@ouc.edu.cn (D.F.)
2 Laboratory for Marine Geology, National Laboratory for Marine Science and Technology, Qingdao 266061, China
* Correspondence: dongwang@ouc.edu.cn; Tel.: +86-13165325292

Abstract: Bucket foundations, especially multi-bucket foundations, have become an alternative for large offshore wind turbines. Vertical responses of a single bucket are critical for the serviceability design of tripod or tetrapod bucket foundations. Centrifuge tests are conducted to investigate the responses of a single bucket under monotonic and symmetric cyclic loading in over-consolidated clay. The strength of clay is obtained by cone penetration tests. The monotonic vertical capacity measured in the centrifuge tests are compared with the finite element results, with errors less than 6%. The effects of the ratio of cyclic loading amplitude to vertical capacity (ranging between 0.37 and 0.64) and the number of cycles on the accumulation of vertical displacement and evolution of stiffness are explored. Simplified functions are proposed to predict the evolutions of dimensional and dimensionless stiffness.

Keywords: vertical capacity; cyclic loading; bucket foundations; clay; offshore wind turbine; centrifuge testing

check for updates

Citation: Jiang, J.; Wang, D.; Fu, D. Vertical Monotonic and Cyclic Responses of a Bucket in Over-Consolidated Clay. *J. Mar. Sci. Eng.* **2023**, *11*, 2044. https://doi.org/10.3390/jmse11112044

Academic Editor: M. Dolores Esteban

Received: 26 September 2023
Revised: 22 October 2023
Accepted: 23 October 2023
Published: 25 October 2023

1. Introduction

With the expansion of offshore wind energy, bucket foundations have been developed and considered as an economic option to support wind turbines [1,2]. Compared to conventional piles or large diameter mono-piles, the bucket foundations are with smaller aspect ratios and can be installed conveniently into the seabed by jacking and assisted suction [3,4]. The bucket foundations used in the practical applications can be categorized as the mono-bucket and multi-bucket. The latter is usually composed of three or four buckets, designed to resist large overturning moments caused by the horizontal loading applied on the wind turbine and supporting structure [5,6]. The interactions between single buckets may be neglected given that the bucket spaces are sufficiently large [7,8], then the overturning moment is resisted by the compression of buckets downwind and tension of buckets upwind, as Figure 1 shows [9,10]. The responses of a single bucket under monotonic and cyclic vertical forces are critical for the design of multi-bucket foundations [11,12]. The geotechnical conditions at several sites of offshore wind farms (e.g., Houhu in China and Dogger Bank in England) are mainly clay over sand layers, while the clay layers are dominant over the depth between the mudline and the bucket tip [13,14]. The concern of this paper is the single bucket of the multi-bucket foundation in clay.

For the bucket foundations in clay, the monotonic compressive and tensile capacities have been studied through finite element (FE) simulations [15,16] and model tests [11,17,18]. The monotonic vertical capacity factors depended on the aspect ratio of the bucket, the soil non-homogeneity, and the bucket–soil adhesion factor [16,19]. Unlike the end-bearing mechanism for the compressive capacity as Figure 2a shows, the soil failure mechanism caused by the tension load became complicated due to suction developed at the base of the bucket foundation. As Figure 2b–d show, the mechanisms observed in the tensile loading tests included the reverse end-bearing, pull-out of bucket, and pull-out of bucket and soil plug inside [3]. The tensile capacities were usually assumed to be lower than the compressive ones, for example, by 20% for sealed buckets and by even up to 30% for the unsealed [18].

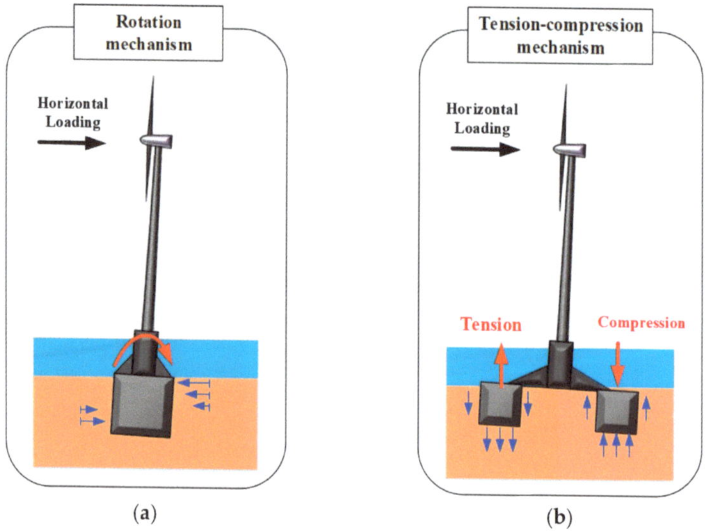

Figure 1. Response of bucket foundation subjected to horizontal loading: (**a**) Mono-bucket foundation; (**b**) Multi-bucket foundation.

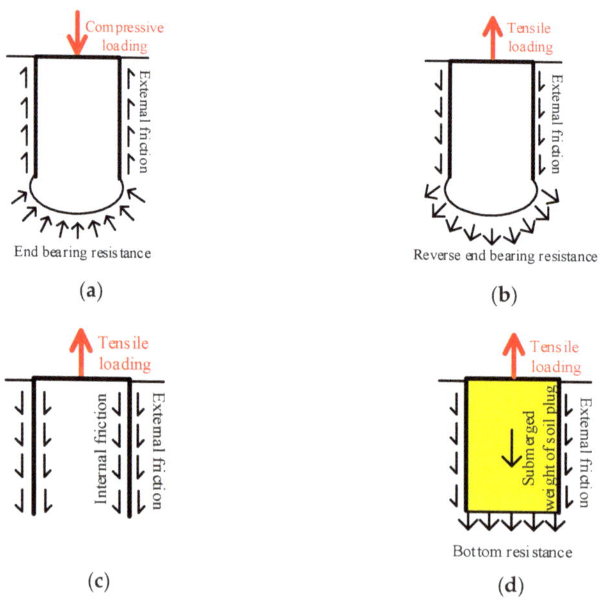

Figure 2. Resistances under compressive and tensile loadings: (**a**) End-bearing mechanism; (**b**) Reverse end-bearing mechanism; (**c**) Siding mechanism; (**d**) Pulling-out mechanism.

The magnitude and direction of vertical cyclic loading affected the displacement and stiffness of the bucket, and were quantified by ζ_b, the ratio of the maximum loading

V_{max} to the monotonic vertical capacity V_0, and ζ_c, and the ratio of the minimum loading V_{min} to V_{max}, as Figure 3 shows [20]. Figure 3 also demonstrates the definitions of the loading amplitude V_c and average loading V_a. The displacements caused were defined, respectively, as the maximum w_{max}, the minimum w_{min}, and the average w_a, and the displacement amplitude was w_c. The studies on the vertical cyclic response of the bucket in clay were relatively limited, compared to those in sand and sand over clay [6,21–25]. For the bucket in sand, purely compressive loading ($\zeta_c \geq 0$) and cycles with sufficiently large V_a on the compressive side leaded to downwards residual displacement which was approximately equal to w_a [24,26,27]. The load–displacement response became stiffer with N due to the soil hardening for one-way compressive loading, and the enhancement of stiffness is more obvious at higher V_a [6]. In contrast, upwards w_a occurred under zero and tensile average loading, independent of the V_c value [24]. The decrease in stiffness with N was more severe under zero average loading ($\zeta_c = -1$) than that under one-way tensile loading [6,28], indicating loading with $\zeta_c = -1$ may be more dangerous. For the bucket in sand over clay, the direction of w_a was also governed by V_a [25]. As for the bucket in clay, a few conventional [11,29] and centrifuge [30,31] model tests have been conducted. Given that the bucket was displaced under undrained conditions, the direction of w_a and the failure mechanism were a function of the direction of V_a: zero and tensile V_a leaded to upwards w_a, while compressive V_a larger than a small level, for example, 9% of the monotonic vertical capacity, may always generate downward w_a [31]. These phenomena of displacements were related to the positive excess pore pressure under compressive average loading and negative excess pore pressure under zero average loading [11]. At higher loading magnitudes, the accumulation of displacement was larger due to the more severe disturbance on soil. However, for loading magnitudes below a certain threshold, the bucket may be at a stable state without obvious residual displacement [31,32]. The displacement accumulated was observed under force-controlled loading, while the vertical resistance was degraded by about 35% in the first ~10 cycles and became stable at ~20 cycles if one-way cyclic loading with a small amplitude of ~0.009D was applied [33]. The stiffness was lower at higher V_c under purely compressive loading with $\zeta_c = 0$ and a logarithmic function was used to describe the relationship between stiffness and the number of cycles [29]. For the bucket in clay subjected to cyclic loading, most existing studies were focused on the responses under horizontal cyclic loading [34–37]. The bucket under cyclic vertical loading was concerned in a limited number of recent studies only, such as by [29,33], however, their model tests were conducted at 1 g. Since the stress levels of soil are much lower than those in practical applications, the centrifuge test is preferred to provide more reliable data, especially for the loading with larger V_c.

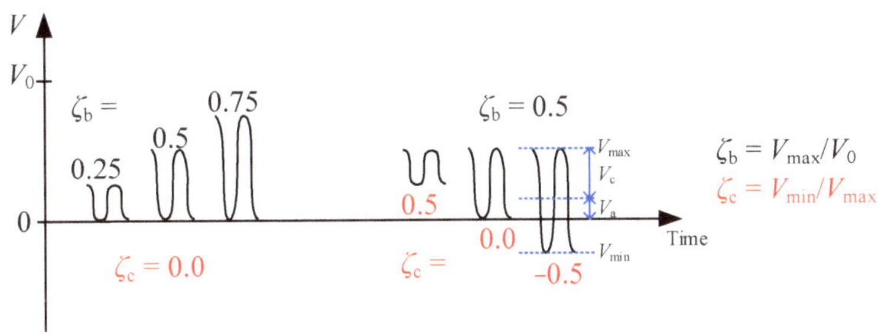

Figure 3. Definitions of V_{max}, V_{min}, ζ_b, ζ_c, V_c, and V_a.

In this paper, the load–displacement responses of the single bucket in clay under monotonic and symmetric cyclic vertical loading ($\zeta_c = -1$) are investigated by centrifuge tests. The vertical capacities under monotonic loading are measured, validated by comparison with the finite element analyses. Then the accumulation of vertical displacement and the evolution of stiffness of the bucket are explored against various loading amplitudes. As a result, simplified functions are proposed to predict the stiffness of the bucket under different loading amplitudes and number of cycles. Figure 4 shows the process of the methodology.

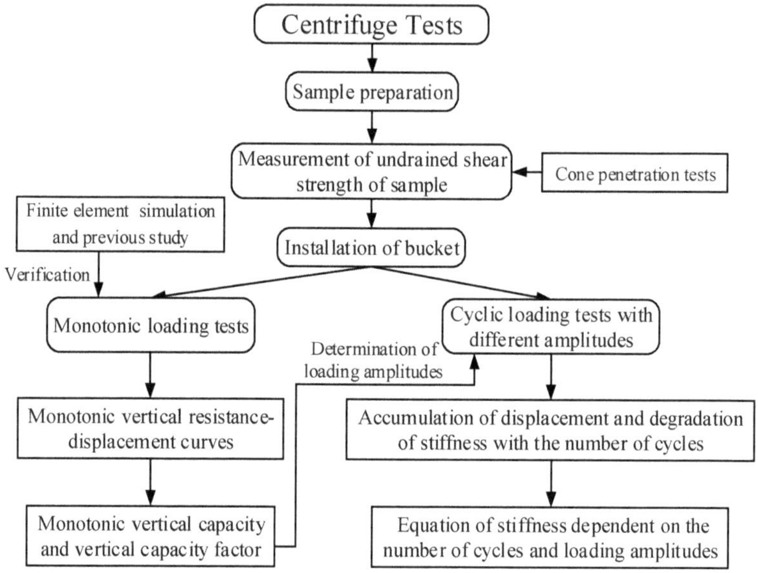

Figure 4. Process of the methodology.

2. Experimental Equipment and Soil Preparation

2.1. Experimental Equipment and Model Bucket

The tests were performed in a drum centrifuge with 1.4 m diameter at the Dalian University of Technology, China. Monotonic or cyclic vertical loads were applied on a bucket in over-consolidated kaolin clay. As Figure 5a,b show, the model bucket was made of aluminum alloy, with a diameter D of 40 mm, skirt length L of 40 mm, and skirt thickness t_s of 1 mm. Then the sizes in prototype were $D = 4$ m, $L = 4$ m, and $t_s = 0.1$ m at an acceleration of 100 g. The skirt thickness ratio t_s/D of the model bucket was 0.025 which was larger than 0.005–0.008 in practice [38] to avoid buckling during installation in centrifuge. A vent was set on the bucket cap with a thickness of 3 mm. A load cell with a measurement range of 300 N and a laser displacement sensor with a precision of 0.01 mm, were used to measure the load and the vertical displacement of the bucket, as Figure 5c shows.

2.2. Sample Preparation and Strength Profile

To prepare the soil sample, dry clay powder was mixed with water in a vacuum tank for at least 4 h to form a slurry at a moisture content of twice the liquid limit. Table 1 demonstrates the properties of the clay. The slurry was then poured into a strongbox of 310 mm × 290 mm × 230 mm, followed by one-dimensional compression under 1 g conditions. The final pressure σ_v' was 90 kPa for Samples 1–3 and 60 kPa for Sample 4, with the compressions lasting about 38 and 36 d, respectively. The sensitivities, S_t, of clay samples in each strongbox were measured by vane shear tests with a range of 2.0–2.3 and an average value of 2.1. After consolidation, the samples were maintained wet throughout the tests by spraying water on the sample surfaces.

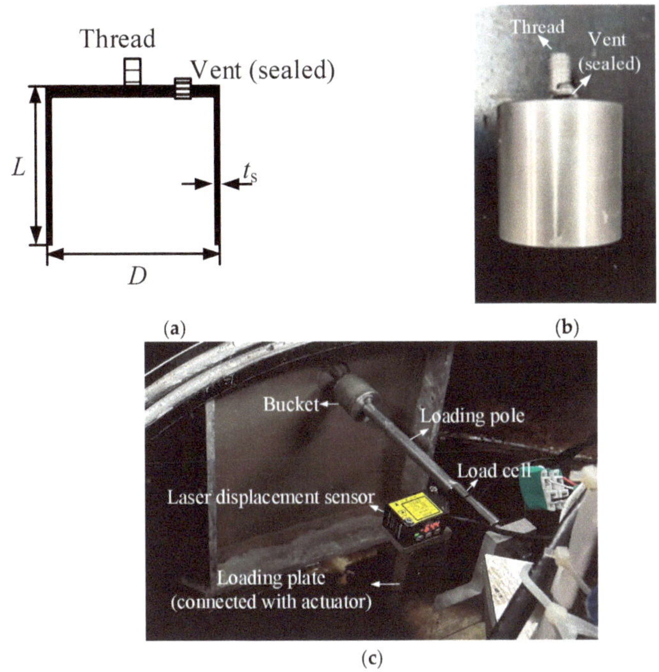

Figure 5. Model bucket and experimental apparatus: (**a**) Schematic model bucket; (**b**) Image of model bucket; (**c**) Experimental apparatus.

Table 1. Properties of kaolin clay.

Parameters	Values
Specific gravity, G_s	2.70
Effective unit weight, γ' (kN/m^3)	6.97
Liquid limit (%)	42.8
Plastic limit (%)	20.8
Sensitivity, S_t	2.1
Vertical coefficient of consolidation, c_v (mm^2/s) ($\sigma_v{}'$ = 60 kPa)	0.11
Vertical coefficient of consolidation, c_v (mm^2/s) ($\sigma_v{}'$ = 90 kPa)	0.14

To characterize the strength profile of clay, cone penetrometer tests (CPTs) were conducted prior to installation of the strongbox into the centrifuge, using a probe with a diameter d of 10 mm. The distances between the cone penetrometer and the strongbox boundaries or the circumferences of buckets were at least 6d, to avoid the boundary effect. The penetration velocity v of cone was 4 mm/s, leading to dimensionless velocity $V = vd/c_v$ = 363.6 or 285.7 for $\sigma_v{}'$ = 60 or 90 kPa, respectively, where c_v is the vertical coefficient of consolidation. Refs. [39–41] suggested that undrained responses occurred at $V > 30$. With the cone resistance profiles measured during the CPTs, the undrained shear strength s_u was calculated as:

$$s_u = (q_t - \sigma_{v0})/N_{kt}, \tag{1}$$

where q_t is the cone resistance; σ_{v0} the total overburden pressure; and N_{kt} the cone factor, which ranged between 9 and 18 for typical clays [42–44]. N_{kt} was taken as 15 in this study. Figure 6 demonstrates the strength profiles recorded during the CPTs at 1 g condition (solid lines) and the corresponding fitting curves (dash lines). The strength of the soil sample was typically increased with soil depth [45,46] and can be described as $s_u = s_{um} + kz$, where s_{um}

is the undrained shear strength at clay surface and k is the gradient of strength with soil depth z. The soil depth z at 1 g condition was multiplied by 100 (as centrifuge tests at an acceleration of 100 g) to obtain strength profiles in the prototype scale. As a result, s_u (in the unit of kPa) in centrifuge was determined as 6.5 + 0.55z, 11.6, 9.0 + 0.4z, and 6.0 + 0.18z for Samples 1–4, where z was in the unit of m.

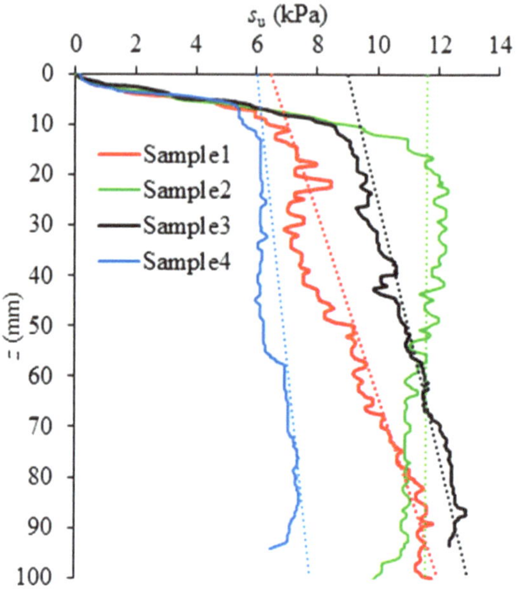

Figure 6. Strength profiles of the clay samples (the fitting curves of the strength profiles are shown as the dotted lines, N_{kt} = 15, model scale).

3. Experimental Arrangements

The model bucket was subjected to monotonic or cyclic loading after installation. For each strongbox, three locations were designed with one monotonic test and two cyclic tests. The distances between the bucket skirts and the strongbox sides were at least 1.75D, while the distance from one skirt of a bucket to another was at least 1.95D.

After the centrifuge was spun up to an acceleration level of 100 g, the bucket was jacked into soil at a velocity v of 1 mm/s. As the soil sample was consolidated at 60 or 90 kPa, the corresponding normalized velocity was $V = vD/c_v$ = 363.6 or 285.7. The undrained response during installation was guaranteed since V was larger than 30. The vent on the cap of the bucket was open during installation, allowing the air/water inside the bucket to be expelled. When the bucket cap reached the soil surface, the penetration resistance increased quickly, indicating that the installation was completed. Then the centrifuge was stopped and the vent was sealed manually. This was to maintain the potential suction developed inside as the bucket was subjected to cyclic loading, which was similar to the operation in most practical applications. The centrifuge was spun up again to 100 g for the subsequent loading process.

In the monotonic loading tests, the bucket was penetrated at v = 1 mm/s until the displacement reached a relatively large value of 10 mm (0.25L), where the corresponding penetration resistance was defined as the monotonic capacity V_0.

In the cyclic loading tests, the loading amplitude V_c of the sinusoidal cyclic load was selected as a specified percentage of V_0, as Table 2 shows. Tests 2-3 and 4-3 were not presented here due to the unqualified precision of loading control. Only 6 cyclic loading tests were reported, with the ratios of loading amplitude to monotonic capacity, V_c/V_0, ranging between 0.37 and 0.64. It was expected to apply symmetric two-way cyclic loading since this was the most dangerous type of cyclic loading [3,47]. However, the average loading ratio, V_a/V_0, obtained were in the range of 0.01–0.05 due to limitations of the controlling system. The frequencies were controlled in the range of 0.5–0.84 Hz. When the frequency was 0.5 Hz, the corresponding dimensionless time factor $T = c_v t/D^2$ over one cycle was 1.4×10^{-4} and 1.8×10^{-4} at $c_v = 0.11$ and $0.14\ \text{mm}^2/\text{s}$, respectively. T became reduced with the frequency larger than 0.5 Hz. The response of clay over one cycle might be undrained given that $T < 1.3 \times 10^{-3}$, suggested by [48]. However, partial drainage may occur over dozens or hundreds of cycles, since $T = 1.0$ was suggested to be associated with at least 90% consolidation [49,50]. Partial drainage was accompanied by an increase in the shear strength of soil. The cyclic tests were shut down after at least 650 cycles, or, when the maximum vertical displacement reached 0.25L. Note that the compressive vertical loads and downward vertical displacements were taken positive in the following discussions.

Table 2. Experimental arrangements involving monotonic and cyclic tests.

Test No.	Load Type	V_0 (N)	V_a/V_0	V_c/V_0
1-1	Monotonic	129.3		
1-2	Cyclic		0.02	0.42
1-3	Cyclic		0.03	0.53
2-1	Monotonic	168.0		
2-2	Cyclic		0.01	0.58
3-1	Monotonic	160.6		
3-2	Cyclic		0.01	0.37
3-3	Cyclic		0.01	0.51
4-1	Monotonic	107.8		
4-2	Cyclic		0.05	0.64

4. Installation

During installation, the penetration resistance is increased slowly prior to displacement of ~35 mm and then is enhanced rapidly, as Figure 7 shows. The abrupt change in the penetration resistance indicates the touchdown of the bucket cap. At the touchdown moments (solid points in Figure 7), the penetration resistance is changed from the sum of the friction resistances along both sides of the bucket skirt and the tip resistance to the sum of the friction resistance along the outside of the bucket and the end resistance. If the penetration resistance Q_{tot} at the touchdown moment is calculated as the sum of the friction resistance along both sides of the bucket skirt and the tip-bearing resistance, then $Q_{tot} = z\alpha s_{ua}\pi(D + D_i) + (\gamma' z + s_{utip}N_c)A_{tip}$ [51], where α is the adhesion factor; D_i is the internal diameter of bucket; and s_{ua} and s_{utip} are the average undrained shear strength along the bucket skirt and the undrained shear strength at the bucket tip, respectively; N_c is the bearing factor, usually taken as 7.5; A_{tip} is the cross-sectional area of the bucket tip. Figure 7a–d show Q_{tot} as 39.0, 58.1, 47.3, and 33.0 N, respectively, agreeing well with the results by [51] of 32.0, 50.4, 41.6, and 27.3 N with errors less than 18%. The depths of the touchdown points are slightly less than 37 mm (the bucket length minus the thickness of bucket cap), which is due to the soil plug formed during installation. The penetration resistance curves measured in each strongbox are close to each other, indicating that the prepared soil samples were uniform.

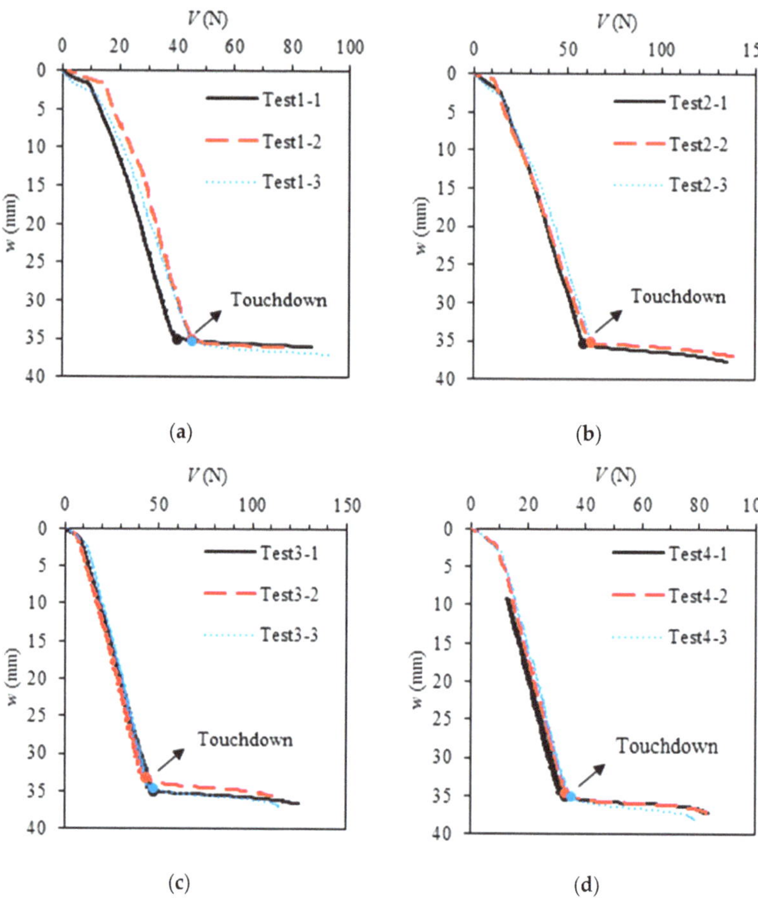

Figure 7. Penetration resistance–depth curves at different point locations (model scale): (**a**) Strongbox1; (**b**) Strongbox2; (**c**) Strongbox3; (**d**) Strongbox4.

5. Monotonic Loading

Under monotonic vertical loading, the bucket is penetrated to around 10 mm (0.25L) deeper than the touchdown point to obtain the vertical capacities V_0; Figure 8 shows them as hollowed points. V_0 measured at four strongboxes are 129.3, 168.0, 160.6, and 107.8 N, respectively.

To testify the reliability of the centrifuge tests, the vertical capacity is determined using the commercial FE package Plaxis 3D CE V20 [52], in which the bucket is assumed to be wished-in-place, i.e., the bucket is at the touchdown position (Figure 8). The effect caused by installation on the following monotonic loading is considered by reducing the shear stress along the skirt–soil interface. Only half of the bucket and corresponding soil are simulated due to the symmetry of the foundation. To avoid a boundary effect, the soil bottom is 4L away from the bucket tip and the soil sides are 3.4D away from the bucket skirt, as Figure 9 shows. The bucket and soil are discretized with ten-node wedge elements with full integration and the bucket–soil interfaces are composed of twelve-node interface elements. To satisfy the numerical convergence and accuracy, the coarseness factors of mesh are chosen as 1 for the far-field soil, 0.3 for the soil near the bucket (0.5D horizontally and L vertically away from the bucket), and 0.1 for the bucket and soil inside the bucket. Clay is regarded as a Tresca material under undrained conditions. The parameters of undrained shear strength s_u in each of the monotonic loading tests, including s_{um} and k, are deduced from CPTs, as Section 2.2 and Figure 6 describe. Specifically, $s_u = 6.5 + 0.55z$, 11.6, $9.0 + 0.4z$, and $6.0 + 0.18z$ for Test 1-1 to 4-1, where z is in the unit of m. The installation effect is considered by reducing the shear stress along the skirt–soil interfaces to αs_u, where the adhesion factor α is taken as $1/S_t$ [8,19]. The value of the adhesion factor is taken as $\alpha = 0.5$ since the values of S_t in each strongbox are averaged as 2.1. The typical Young's modulus of clay is ranged between (200 and 800)s_u, with 400s_u adopted. The effective unit weight of clay is 6.97 kN/m^3 (see Table 1) and Poisson's ratio is 0.495 to the approximate constant volume under undrained conditions. The bucket is simplified as a rigid body since the stiffness of the bucket is much higher than that of soil. The center of the bucket top at the mudline level is taken as a reference point, while the load/displacement at the reference point represents that of the whole bucket. The vertical displacement is applied at the reference point, such that the corresponding vertical reaction force of the bucket can be obtained. Similar to that in the tests, V_0 is the reaction force when the vertical displacement reaches 0.25L.

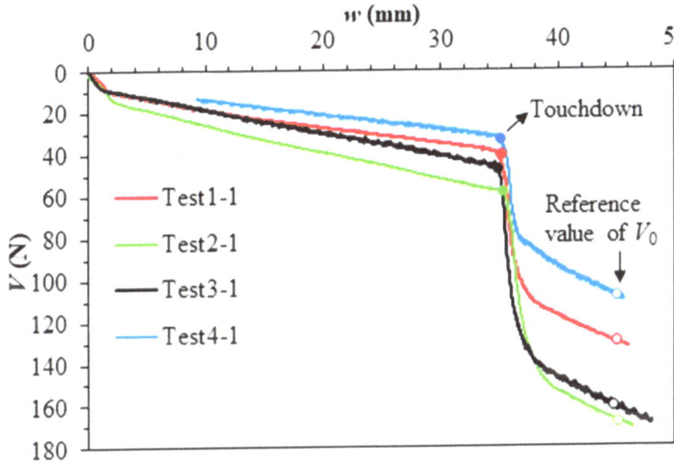

Figure 8. Load–displacement curves under monotonic vertical loading.

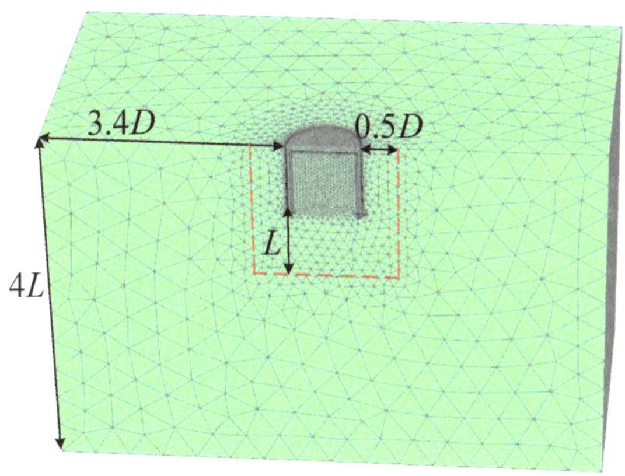

Figure 9. Mesh strategy for the bucket and soil.

Figure 10 presents the monotonic vertical reaction force–displacement curves ($V - w/L$ curves) by FE. Since the bucket is wished in place in the FE simulation, w in Figure 10 represents the displacement from the touchdown point. Figure 10 also plots the experimental curves recorded in the tests. The FE values of V_0 at four strongboxes are 132.1, 170.7, 159.8, and 100.8 N, respectively, agreeing well with centrifuge tests with errors less than 6%. The vertical capacity factor of bucket N_{cv} can be calculated as $(V_0 - \alpha \pi L D s_{uav})/A s_{utip}$, where A is the cross-sectional area of the bucket. The corresponding N_{cv} in four tests are 10.1, 9.5, 10.2, and 10.9, respectively. The errors between N_{cv} in this study and N_{cv} = 10.3 calculated by the equation proposed by [53] were less than 7%, indicating the reliability of monotonic loading results by centrifuge.

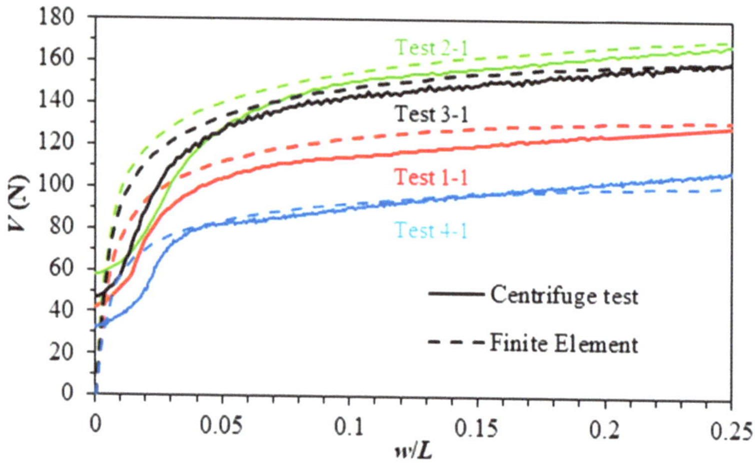

Figure 10. Vertical reaction force–displacement curves ($V - w/L$ curves) under monotonic loading by FE and centrifuge test.

6. Cyclic Loading

6.1. Evolution of Vertical Displacement

Figure 11 shows the relationships between the normalized vertical displacement w/L and the number of cycles N. In Test 3-2 (V_c/V_0 = 0.37, V_a/V_0 = 0.01), the peak data points near N = 300, 650, and 900 are not recorded due to signal loss, causing missing parts in the load curves, as Figure 11a shows. The minimum displacements w_{min} are accumulated upwards under V_a/V_0 = 0.01 in Figure 11a but are accumulated downwards under V_a/V_0 = 0.02–0.05 in Figure 11b. A possible reason is that the evolution of displacements depends on V_a/V_0, although V_a/V_0 is limited to less than 0.05. The small increase in V_a/V_0 may cause a relatively larger average displacement towards the compression side at the beginning of the tests, affecting the accumulations of average displacements and displacement amplitudes. At the end of the tests, normalized average displacements w_a/L are between −0.024 and −0.109. The negative w_a represents the uplift of the bucket. Similar phenomena were reported by [11].

The negative w_a under symmetric loading represents that the residual tensile displacements are larger than the compressive ones. The reason is that the tensile capacity of the bucket is lower than the compressive. When cyclic loading into compression is applied, the resistance is composed of the friction along the outside skirt and the end bearing, as Figure 2a shows. The friction along the inside skirt is not mobilized since the soil plug is moved along with the bucket. However, when the cyclic loading into tension is applied, there are three potential failure mechanisms as Figure 2b–d show. In centrifuge tests, the vent of the bucket is sealed after installation, so the negative excess pore pressure is generated inside the bucket, which may prevent the relative movements between the soil plug and bucket skirt. As a result, the reverse end-bearing mechanism which Figure 2b shows, composed of external friction and reverse end-bearing resistance, occurs.

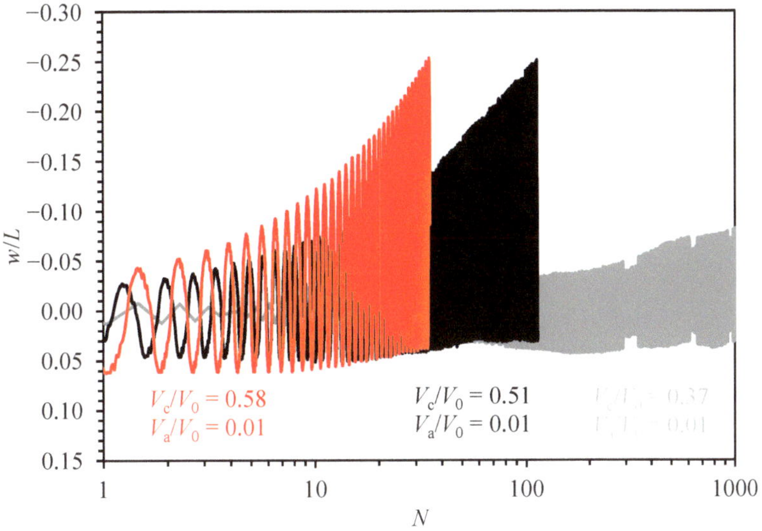

(**a**)

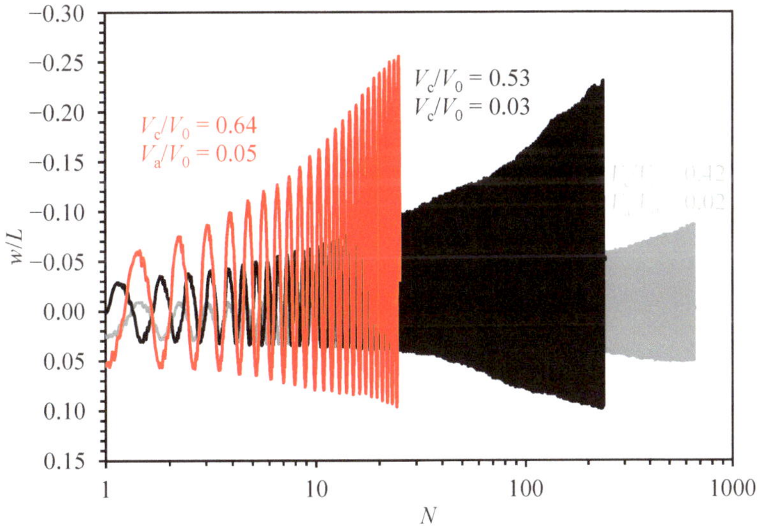

(**b**)

Figure 11. Normalized vertical displacement and cyclic number relationship: (**a**) V_c/V_0 = 0.37, 0.51, and 0.58; (**b**) V_c/V_0 = 0.42, 0.53, and 0.64.

The external friction of the bucket under compressive and tensile loading are similar to each other, since the soil along the outside of the skirt is roughly under a direct simple shear and the shear stress mobilized is not affected by the loading direction. The contributions of the end-bearing resistance and reverse end-bearing resistance are corresponding to the states of triaxial compression and triaxial tension, respectively. The former is usually larger due to higher soil strength at the triaxial compression state, i.e., the compressive capacity of the bucket is higher than the tensile. The residual upward displacement of the bucket is thus accumulated gradually even as the symmetrical vertical cyclic loading is applied.

The reverse end-bearing mechanism as Figure 2b shows is proofed further by the soil plug inside the bucket and the shallow pit left on the soil surface after the bucket is pulled out (Figure 12). Under cyclic loading into tension, the soil plug and the soil beneath the bucket tip are mobilized and move upwards along with the bucket due to the negative pressure. Ref. [11] proved the existence of suction by measuring negative excess pore pressure under the bucket lid.

(a)

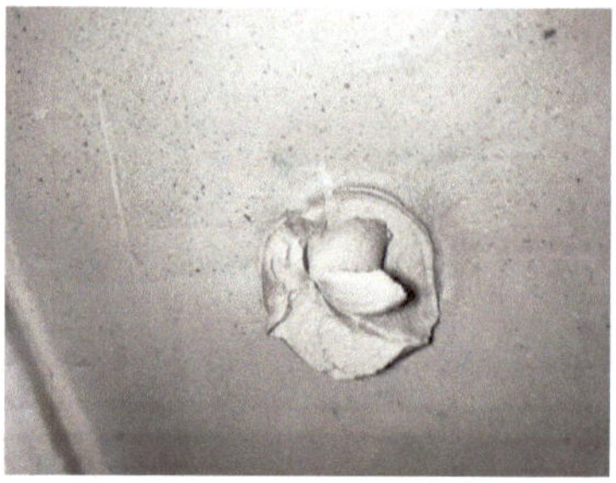

(b)

Figure 12. Soil plug inside bucket and soil surface after pull-out of bucket: (**a**) Soil plug inside bucket; (**b**) Soil surface after pull-out of bucket.

The accumulation rates of w_a and w_c are larger under higher V_c/V_0, as Figure 11 shows. For example, normalized average displacement w_a/L reaches -0.11 and normalized displacement amplitude w_c/L reaches 0.148 over 35 cycles under $V_c/V_0 = 0.58$. As the opposite, w_a/L is only -0.024 and w_c/L is 0.056 with 996 cycles of $V_c/V_0 = 0.37$. The reason may be that the pore pressure is accumulated more under higher V_c/V_0, causing lower effective stress. So the soil strength is lower under higher V_c/V_0, leading to larger displacements for the same number of cycles.

Figure 13 shows the hysteresis loop, the relationship between the normalized cyclic loading, and the normalized vertical displacement. To demonstrate the changes of hysteresis loops clearly, Figure 13 demonstrates only the hysteresis loops at typical loading stages, and hysteresis loops without peak data in Test 3-2 ($V_c/V_0 = 0.37$, $V_a/V_0 = 0.01$) are removed. At relatively higher V_c/V_0, the maximum displacement w_{max} of the bucket reaches 0.25L over 20–40 cycles, as Figure 13c,f show. At relatively lower V_c/V_0, w_{max} cannot reach 0.25L even under 600–800 cycles, as Figure 13a,d show. Although not measured in this study, the pore pressure may be accumulated much more slowly given that V_c/V_0 is below a threshold, then w_{max} becomes unchanged with cycles over long-term loading.

For the maximum displacement w_{max}, the accumulation rate is affected significantly by V_c and the number of cycles in the tests. At low V_c/V_0 as Figure 13a,d show, the accumulation rate of w_{max} over 10 cycles is relatively uniform. However, at high V_c/V_0 as Figure 13b,c show, the accumulation rate of w_{max} over 10 cycles is decreased with N. The phenomena may be related to the differences in the accumulation rates of pore pressure. At low V_c/V_0, the pore pressure is accumulated in a relatively uniform rate, while at high V_c/V_0, the accumulation rate of pore pressure is changed from high to low. Compared to w_{max}, the accumulation rate of the minimum displacement w_{min} is always lower. The above phenomena are due to the compressive capacity of the bucket being higher than the tensile one. Compared to the compression, the soil is disturbed more seriously by the tension, and then the strength softening becomes more significant on the tension side.

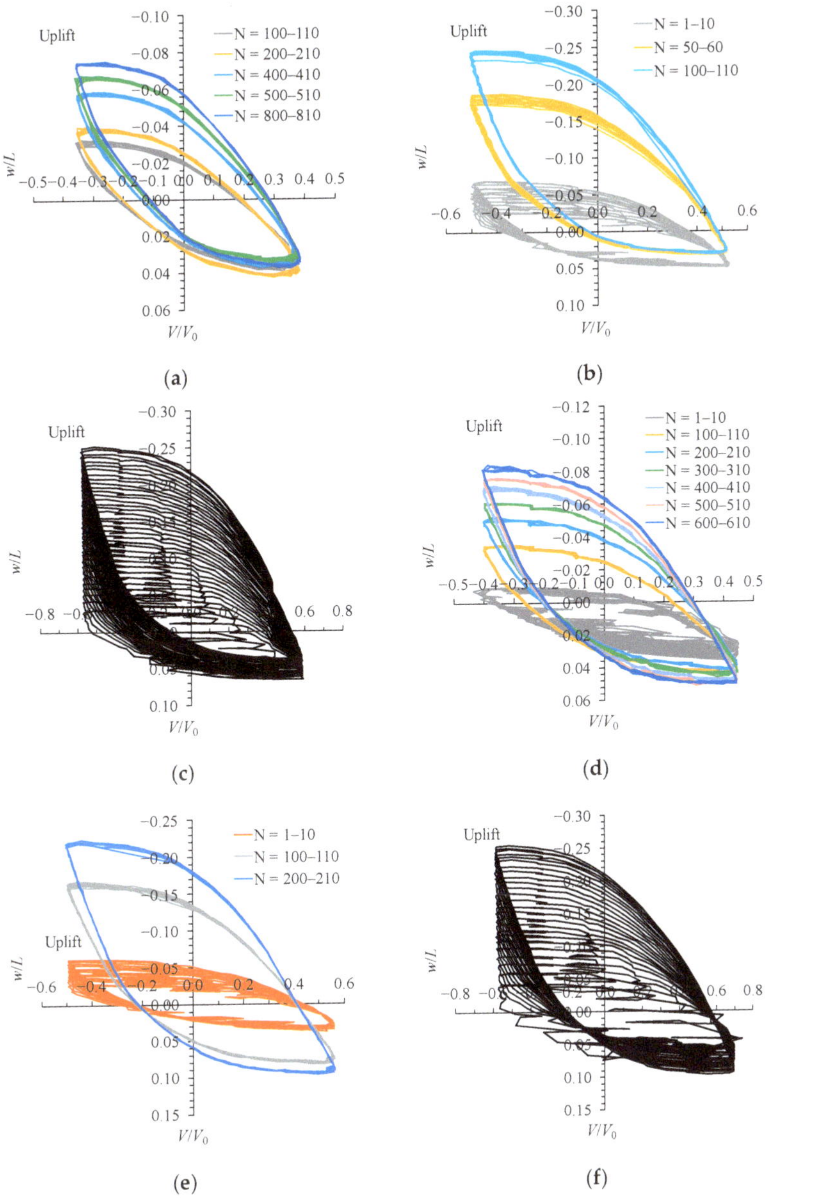

Figure 13. Normalized cyclic vertical loading and displacement relationship: (**a**) V_c/V_0 = 0.37, V_a/V_0 = 0.01; (**b**) V_c/V_0 = 0.51, V_a/V_0 = 0.01; (**c**) V_c/V_0 = 0.58, V_a/V_0 = 0.01; (**d**) V_c/V_0 = 0.42, V_a/V_0 = 0.02; (**e**) V_c/V_0 = 0.53, V_a/V_0 = 0.03; (**f**) V_c/V_0 = 0.64, V_a/V_0 = 0.05.

6.2. Evolution of Secant Stiffness

The secant stiffness K_N of the Nth hysteresis loop of bucket is defined as the slope of the line connecting the highest and lowest points of the hysteresis loop, as Figure 14 shows. It can be calculated as $K_N = (V_{max} - V_{min})/(w_{max} - w_{min})$ for the Nth cycle.

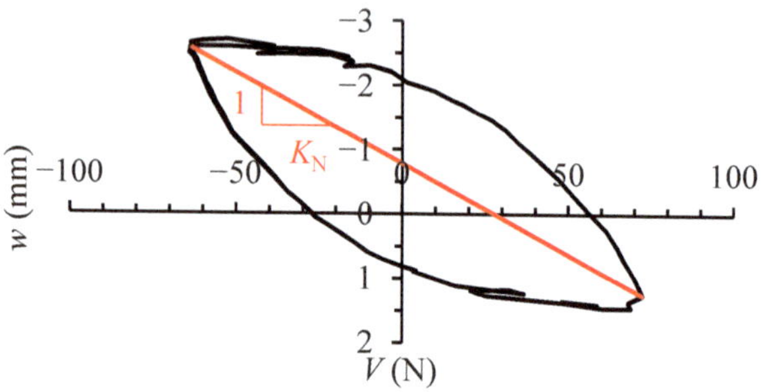

Figure 14. Secant stiffness of hysteresis loop.

Figure 15 shows the relationships between secant stiffness of hysteresis loops and the number of cycles. In general, K_N is decreased with the increase in V_c/V_0 and N. The reason may be that high V_c/V_0 or large N results in the accumulation of excess pore pressures in the soil. At V_c/V_0 = 0.51 and 0.53, K_N are close to each other since the loading amplitudes are similar. In Figure 15, the variations of K_N with N become gentle at V_c/V_0 = 0.37 and 0.42. It might be due to the fact that the soil around the bucket undergoes partial consolidation after a long-term loading. For example, the loading time is as large as 137 d in the prototype after 996 cycles with V_c/V_0 = 0.37. Also, the corresponding dimensionless time factor T after 996 cycles is at least 0.18, exceeding $T < 1.3 \times 10^{-3}$ for undrained clay as mentioned in Section 3. Therefore, partial drainage may occur in clay and cause an increase in shear strength.

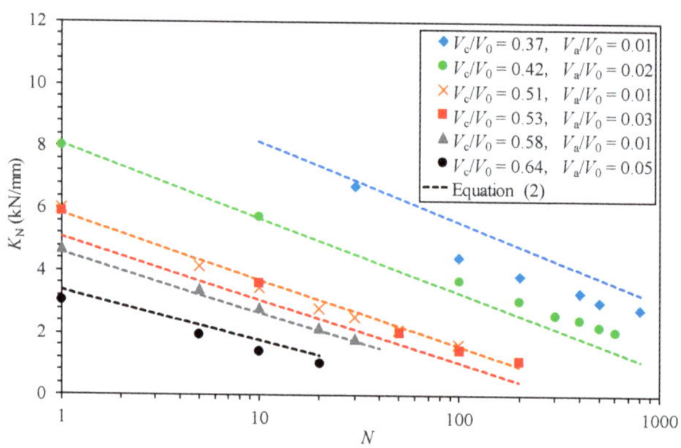

Figure 15. Relationship between secant stiffness of hysteresis loops and cyclic numbers (prototype scale).

By referring to the logarithmic function proposed by [20,54], the relationship between secant stiffness K_N and the number of cycles N is expressed as:

$$K_N = K_1 + A_k \ln N, \tag{2}$$

where K_1 is the secant stiffness of the first cycle and A_K is the fitting parameter. K_1 and A_K depend on ζ_b (=V_{max}/V_0), as Figure 16a,b show. Figure 15 also demonstrates the predictions of K_N by Equation (2).

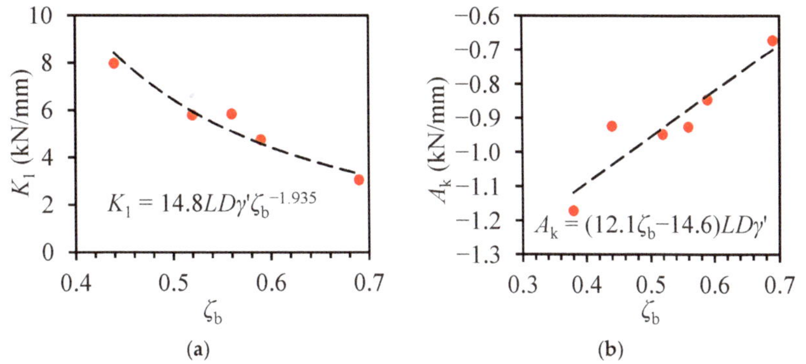

Figure 16. Fitting curves between K_1 and A_k with ζ_b (prototype scale): (**a**) K_1 with ζ_b; (**b**) A_k with ζ_b.

To describe the stiffness independent of ζ_b, the dimensionless variation in vertical displacement within one cycle $(w_{max} - w_{min})/L$ and the dimensionless secant stiffness $K^* = K_N/(V_0/L)$ are adopted by referring to [11]. As Figure 17 shows, K^* is gradually decreased with the increase in $(w_{max} - w_{min})/L$ and K^* is scattered in a relatively narrow range under the different loading amplitude ratio V_c/V_0. At the same $(w_{max} - w_{min})/L$, K^* is slightly increased with V_c/V_0. The evolution of K^* with $(w_{max} - w_{min})/L$ can be fitted as:

$$K^* = 1.5[(w_{max} - w_{min})/L]^{-0.8}, \quad (3)$$

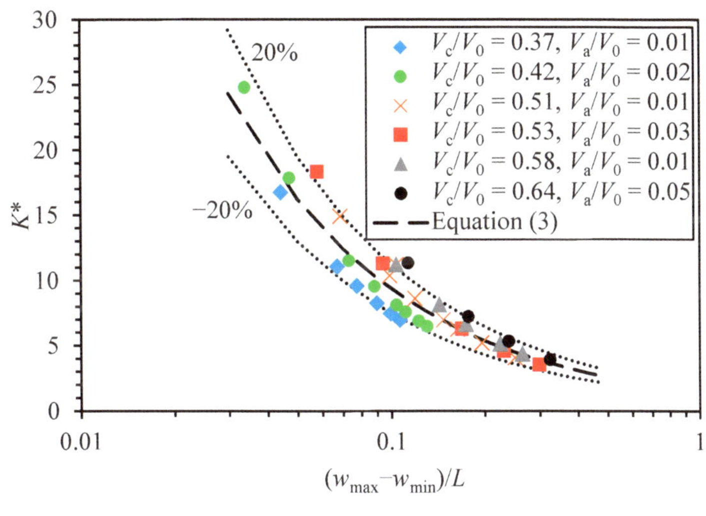

Figure 17. Relationship between dimensionless secant stiffness K^* and dimensionless variation in vertical displacement within one cycle (the error ranges between results by Equation (3) and tests are shown as dotted lines).

Figure 17 shows the fitting results of Equation (3) as the black dash line. The error between the fitting curve and the test results is within ±20%.

7. Conclusions

The monotonic vertical capacity and cyclic responses of a single bucket under symmetric vertical loading have been studied through centrifuge tests. The loading amplitude ratio V_c/V_0 is varied between 0.37 and 0.64 to investigate its effects on displacement and stiffness of the bucket. The main conclusions are as follows:

(1) The vertical capacities by centrifuge tests and numerical simulations are with errors less than 6%. The vertical capacity factors are in the range of 9.5–10.9.
(2) Under symmetric vertical loading, the normalized average displacements of the bucket, varied between −0.024 and −0.109 in six tests, are on the tensile side due to the tensile capacity being lower than the compressive. The accumulation rates of average displacement and displacement amplitude of the bucket are larger under higher V_c/V_0 due to the lower effective stress.
(3) The secant stiffness of the bucket is decreased with V_c/V_0 and N. Partial drainage may occur over 996 cycles with V_c/V_0 = 0.37, due to the loading time in the prototype of 137 d and the dimensionless time factor T > 0.18. Based on the experimental results, two simplified equations are proposed to describe the evolution of the secant stiffness of the bucket.

Author Contributions: Conceptualization, D.W., J.J. and D.F.; methodology, J.J.; software, D.W. and J.J.; validation, D.W., J.J. and D.F.; formal analysis, J.J.; investigation, D.W., J.J. and D.F.; data curation, J.J.; writing—original draft preparation, J.J.; writing—review and editing, D.W. and D.F.; funding acquisition, D.W. All authors have read and agreed to the published version of the manuscript.

Funding: This research was funded by the National Natural Science Foundation of China (grant Nos. 42025702 and 42177122).

Institutional Review Board Statement: Not applicable.

Informed Consent Statement: Not applicable.

Data Availability Statement: Not applicable.

Conflicts of Interest: The authors declare no conflict of interest.

References

1. Houlsby, G.; Byrne, B. Suction Caisson Foundations for Offshore Wind Turbines and Anemometer Masts. *Wind Eng.* **2000**, *24*, 249–255. [CrossRef]
2. Faizi, K.; Faramarzi, A.; Dirar, S.; Chapman, D. Investigating the Monotonic Behaviour of Hybrid Tripod Suction Bucket Foundations for Offshore Wind Towers in Sand. *Appl. Ocean Res.* **2019**, *89*, 176–187. [CrossRef]
3. Randolph, M.; Gourvenec, S. *Offshore Geotechnical Engineering*, 1st ed.; CRC Press: London, UK, 2011; ISBN 978-1-315-27247-4.
4. Vulpe, C. Design Method for the Undrained Capacity of Skirted Circular Foundations under Combined Loading: Effect of Deformable Soil Plug. *Géotechnique* **2015**, *65*, 669–683. [CrossRef]
5. Achmus, M.; Kuo, Y.S.; Abdel-Rahman, K. Behavior of Monopile Foundations under Cyclic Lateral Load. *Comput. Geotech.* **2009**, *36*, 725–735. [CrossRef]
6. Jeong, Y.; Kim, J.H.; Park, H.J.; Kim, D. Cyclic Behavior of Unit Bucket for Tripod Foundation System Supporting Offshore Wind Turbine via Model Tests. *Wind Energy* **2018**, *22*, 257–268. [CrossRef]
7. Houlsby, G.T.; Kelly, R.B.; Huxtable, J.; Byrne, B.W. Field Trials of Suction Caissons in Clay for Offshore Wind Turbine Foundations. *Géotechnique* **2005**, *55*, 287–296. [CrossRef]
8. He, B.; Jiang, J.; Cheng, J.; Zheng, J.; Wang, D. The Capacities of Tripod Bucket Foundation under Uniaxial and Combined Loading. *Ocean Eng.* **2021**, *220*, 108400. [CrossRef]
9. Kim, D.J.; Choo, Y.W.; Kim, J.H.; Kim, S.; Kim, D.S. Investigation of Monotonic and Cyclic Behavior of Tripod Suction Bucket Foundations for Offshore Wind Towers Using Centrifuge Modeling. *J. Geotech. Geoenvironmental Eng.* **2014**, *140*, 04014008. [CrossRef]
10. Carbon Trust. *Suction Installed Caisson Foundations for Offshore Wind: Design Guidelines*, 1st ed.; Carbon Trust: London, UK, 2019; pp. 1–92.
11. Villalobos, F.A.; Byrne, B.W.; Houlsby, G.T. Model Testing of Suction Caissons in Clay Subjected to Vertical Loading. *Appl. Ocean Res.* **2010**, *32*, 414–424. [CrossRef]
12. Jeong, Y.; Ko, K.W.; Kim, D.S.; Kim, J.H. Studies on Cyclic Behavior of Tripod Suction Bucket Foundation System Supporting Offshore Wind Turbine Using Centrifuge Model Test. *Wind Energy* **2020**, *24*, 515–529. [CrossRef]
13. Wang, Y.; Zhang, S.; Xu, H.; Zhang, Y.; Gaunt, P.; Ren, B.; Zhang, Y.; Yubin, R. Site Investigation and Soil Parameters for Offshore Suction Bucket Design: A Case Study of Houhu Wind Turbine. *Ocean Eng.* **2022**, *255*, 111458. [CrossRef]
14. Lau, B.H. Cyclic Behaviour of Monopile Foundations for Offshore Wind Turbines in Clay. Ph.D. Thesis, University of Cambridge, Cambridge, UK, 2015.

15. Bransby, M.F.; Yun, G.J. The Undrained Capacity of Skirted Strip Foundations under Combined Loading. *Géotechnique* **2009**, *59*, 115–125. [CrossRef]
16. Hung, L.C.; Kim, S.R. Evaluation of Vertical and Horizontal Bearing Capacities of Bucket Foundations in Clay. *Ocean Eng.* **2012**, *52*, 75–82. [CrossRef]
17. Cassidy, M.J.; Byrne, B.W.; Randolph, M.F. A Comparison of the Combined Load Behaviour of Spudcan and Caisson Foundations on Soft Normally Consolidated Clay. *Géotechnique* **2004**, *54*, 91–106. [CrossRef]
18. Gourvenec, S.; Acosta-Martinez, H.E.; Randolph, M.F. Experimental Study of Uplift Resistance of Shallow Skirted Foundations in Clay under Transient and Sustained Concentric Loading. *Géotechnique* **2009**, *59*, 525–537. [CrossRef]
19. Supachawarote, C. Inclined Load Capacity of Suction Caisson in Clay. Ph.D. Thesis, University of Western Australia, Perth, Australia, 2006.
20. Leblanc, C.; Houlsby, G.T.; Byrne, B.W. Response of Stiff Piles in Sand to Long-Term Cyclic Lateral Loading. *Géotechnique* **2010**, *60*, 79–90. [CrossRef]
21. Byrne, B.; Houlsby, G. Experimental Investigations of Response of Suction Caissons to Transient Vertical Loading. *J. Geotech. Geoenvironmental Eng.* **2002**, *128*, 926–939. [CrossRef]
22. Byrne, B.; Houlsby, G. Experimental Investigations of the Response of Suction Caissons to Transient Combined Loading. *J. Geotech. Geoenvironmental Eng.* **2004**, *130*, 240–253. [CrossRef]
23. Senders, M. Suction Caissons in Sand as Tripod Foundations for Offshore Wind Turbines. Ph.D. Thesis, University of Western Australia, Perth, Australia, 2009.
24. Bienen, B.; Klinkvort, R.T.; O'Loughlin, C.D.; Zhu, F.; Byrne, B.W. Suction Caissons in Dense Sand, Part II: Vertical Cyclic Loading into Tension. *Géotechnique* **2018**, *68*, 953–967. [CrossRef]
25. Stapelfeldt, M.; Bienen, B.; Grabe, J. Influence of Low-Permeability Layers on the Installation and the Response to Vertical Cyclic Loading of Suction Caissons. *J. Geotech. Geoenvironmental Eng.* **2021**, *147*, 04021076. [CrossRef]
26. Kelly, R.; Houlsby, G.; Byrne, B. A Comparison of Field and Laboratory Tests of Caisson Foundations in Sand and Clay. *Géotechnique* **2006**, *56*, 617–626. [CrossRef]
27. Zhang, Y.; Sudhakaran, K.; Askarinejad, A. Centrifuge Modelling of Suction Caissons Subjected to Cyclic Loading in Tension. In Proceedings of the 4th European Conference on Physical Modelling in Geotechnics, Lulea, Sweden, 15 March 2020.
28. Gütz, P.; Achmus, M. Suction Bucket Foundations under Cyclic Tensile Loading–Physical and Numerical Modeling. *Geotech. Test. J.* **2021**, *44*, 20200056. [CrossRef]
29. Kou, H.; Fang, W.; Zhou, N.; Huang, J.; Zhang, X. Dynamic Response of Single-Bucket Foundation in Clay under Vertical Variable Amplitude Cyclic Loadings. *Ocean Eng.* **2023**, *273*, 113973. [CrossRef]
30. Chen, W.; Randolph, M. Uplift Capacity of Suction Caissons under Sustained and Cyclic Loading in Soft Clay. *J. Geotech. Geoenvironmental Eng.* **2007**, *133*, 1352–1363. [CrossRef]
31. Zografou, D.; Gourvenec, S.; O'Loughlin, C. Vertical Cyclic Loading Response of Shallow Skirted Foundation in Soft Normally Consolidated Clay. *Can. Geotech. J.* **2018**, *56*, 473–483. [CrossRef]
32. Iskander, M.; El-Gharbawy, S.; Olson, R. Performance of Suction Caissons in Sand and Clay. *Can. Geotech. J.* **2002**, *39*, 576–584. [CrossRef]
33. Al-Janabi, H.; Aubeny, C. Experimental and Numerical Investigation of the Performance of Piles and Suction Caissons Subjected to Inclined Cyclic Loading in Cohesive Soils. *J. Geotech. Geoenvironmental Eng.* **2022**, *148*, 04022036. [CrossRef]
34. Wang, T.; Yu, S.; Liu, W.; Bao, X.; Liu, J. Cyclic Bearing Mechanism of Suction Caissons Supporting Offshore Wind Turbines in Clay. *China Ocean Eng.* **2021**, *35*, 135–144. [CrossRef]
35. Zhou, N.; Kou, H.; Chen, Q. Horizontal Cyclic Response of Bucket Caisson for Offshore Wind Turbines in Over-Consolidated Clay. *Appl. Ocean Res.* **2021**, *118*, 102973. [CrossRef]
36. Zhao, X.L.; Wang, X.; Ding, P.C.; Sui, S.H.; Deng, W.N. Development and Influence of Pore Pressure around a Bucket Foundation in Silty Soil. *J. Mar. Sci. Eng.* **2022**, *10*, 2020. [CrossRef]
37. Zhu, W.; Dai, G.; Wang, B.; Gong, W.; Sun, J.; Hu, H. Study on cyclic characteristics and equivalent cyclic creep model of the soft clay at the bottom of suction caisson foundation. *ROCK SOIL Mech.* **2022**, *43*, 466–478. [CrossRef]
38. Vicent, S.; Kim, S.R.; Van Tung, D.; Bong, T. Effect of Loading Rate on the Pullout Capacity of Offshore Bucket Foundations in Sand. *Ocean Eng.* **2020**, *210*, 107427. [CrossRef]
39. Chung, S.F.; Randolph, M.F.; Schneider, J.A. Effect of Penetration Rate on Penetrometer Resistance in Clay. *J. Geotech. Geoenvironmental Eng.* **2006**, *132*, 1188–1196. [CrossRef]
40. Lehane, B.M.; O'loughlin, C.D.; Gaudin, C.; Randolph, M.F. Rate Effects on Penetrometer Resistance in Kaolin. *Géotechnique* **2009**, *59*, 41–52. [CrossRef]
41. Zhang, W.; Liu, K.; Wang, D.; Zheng, J. Coefficient of Consolidation Measured by Cone Penetration Tests in Overconsolidated Cohesive Soils. *Ocean Eng.* **2023**, *276*, 114301. [CrossRef]
42. Lunne, T.; Robertson, P.; Powell, J. Cone Penetration Testing in Geotechnical Practice. *Soil Mech. Found. Eng.* **1997**, *46*, 237. [CrossRef]
43. Low, H.E.; Lunne, T.; Andersen, K.H.; Sjursen, M.A.; Li, X.; Randolph, M.F. Estimation of Intact and Remoulded Undrained Shear Strengths from Penetration Tests in Soft Clays. *Géotechnique* **2010**, *60*, 843–859. [CrossRef]

44. Liu, B.; Zhang, Y.; Ma, Z.; Andersen, K.; Jostad, H.; Liu, D.; Pei, A. Design Considerations of Suction Caisson Foundations for Offshore Wind Turbines in Southern China. *Appl. Ocean Res.* **2020**, *104*, 102358. [CrossRef]
45. Dong, Y.K.; Wang, D.; Randolph, M. Investigation of Impact Forces on Pipeline by Submarine Landslide with Material Point Method. *Ocean Eng.* **2017**, *146*, 21–28. [CrossRef]
46. Sun, Q.L.; Wang, Q.; Shi, F.Y.; Alves, T.; Gao, S.; Xie, X.N.; Wu, S.G.; Li, J.B. Runup of Landslide-generated Tsunamis Controlled by Paleogeography and Sea-level Change. *Commun. Earth Environ.* **2022**, *3*, 244. [CrossRef]
47. Randolph, M.F. Offshore Design Approaches and Model Tests for Sub-Failure Cyclic Loading of Foundations. In *Mechanical Behaviour of Soils under Environmentally Induced Cyclic Loads*; Di Prisco, C., Wood, D.M., Eds.; CISM Courses and Lectures; Springer: Vienna, Austria, 2012; pp. 441–480. ISBN 978-3-7091-1068-3.
48. Zhu, F.Y.; O'Loughlin, C.D.; Bienen, B.; Cassidy, M.J.; Morgan, N. The Response of Suction Caissons to Long-Term Lateral Cyclic Loading in Single-Layer and Layered Seabeds. *Géotechnique* **2018**, *68*, 729–741. [CrossRef]
49. Gourvenec, S.; Randolph, M. Consolidation beneath Circular Skirted Foundations. *Int. J. Geomech.* **2010**, *10*, 22–29. [CrossRef]
50. Feng, X.; Gourvenec, S. Consolidated Undrained Load-Carrying Capacity of Subsea Mudmats under Combined Loading in Six Degrees of Freedom. *Géotechnique* **2015**, *65*, 563–575. [CrossRef]
51. DNV-RP-E303 Geotechnical Design and Installation of Suction Anchors in Clay. Available online: https://www.dnv.com/Default (accessed on 22 October 2023).
52. Manuals-PLAXIS-GeoStudio | PLAXIS Wiki-GeoStudio | PLAXIS-Bentley Communities. Available online: https://communities.bentley.com/products/geotech-analysis/w/wiki/46137/manuals{-}{-}-plaxis (accessed on 25 September 2023).
53. Kanmin, S.; Jiang, J.; Wang, D.; Zheng, J. Capacities of Tripod Bucket Foundation under Uniaxial and Combined Loading Considering Adhesion Factor. *Mar. Georesources Geotechnol.* **2021**, *40*, 1520–1528. [CrossRef]
54. Hung, L.C.; Lee, S.H.; Vicent, S.; Kim, S.R. An Experimental Investigation of the Cyclic Response of Bucket Foundations in Soft Clay under One-Way Cyclic Horizontal Loads. *Appl. Ocean Res.* **2018**, *71*, 59–68. [CrossRef]

Journal of
Marine Science and Engineering

MDPI

Article

Dynamic Performance of Suspended Pipelines with Permeable Wrappers under Solitary Waves

Youkou Dong [1], Enjin Zhao [1,*], Lan Cui [2], Yizhe Li [1] and Yang Wang [3]

1 College of Marine Science and Technology, China University of Geosciences, 388 Lumo Road, Wuhan 430074, China; dongyk@cug.edu.cn (Y.D.); yizheli@cug.edu.cn (Y.L.)
2 Institute of Rock and Soil Mechanics, Chinese Academy of Sciences, Wuhan 430071, China; lcui@whrsm.ac.cn
3 Haikou Marine Geological Survey Center, China Geological Survey, Haikou 570100, China; wangyang01@mail.cgs.gov.cn
* Correspondence: zhaoej@cug.edu.cn

Abstract: Submarine pipelines are widely adopted around the world for transporting oil and gas from offshore fields. They tend to be severely ruined by the extreme waves induced by the natural disaster, such as hurricanes and tsunamis. To maintain the safety and function integrity of the pipelines, porous media have been used to wrap them from the external loads by the submarine environment. The functions of the porous wrappers under the hydrodynamic impact remain to be uncovered before they are widely accepted by the industry. In this study, a numerical wave tank is established with the immersed boundary method as one of the computational fluid dynamics. The submarine pipelines and their porous wrappers are two-way-coupled in terms of displacement and pressure at their interfaces. The impact from the solitary waves, which approximately represent the extreme waves in the reality, on the pipelines with different configurations of the porous wrapper is investigated. The results present significant protective functions of the wrappers on the internal pipelines, transferring the impact forces from the pipelines to the wrappers. The protective effects tend to be enhanced by the porosity and thickness of the wrappers. The influence of the pipeline configurations and the marine environment are then analysed. As for the front pipeline, an increase in the gap leads to a slight increase in the horizontal forces on both the wrapper and the pipeline, but a significant increase in the vertical forces. As for the rear pipeline, because of the shield function of the front pipeline, the velocity within the gap space and the forces on the pipes are decreased with the decrease in the gap size. The complex flow fields around the pipelines with wrappers are also illuminated, implying that the protection function of the wrapper is enhanced by the wave height reduction.

Keywords: extreme wave; submarine pipeline; external wrapper; coupling analysis; computational fluid dynamics

Citation: Dong, Y.; Zhao, E.; Cui, L.; Li, Y.; Wang, Y. Dynamic Performance of Suspended Pipelines with Permeable Wrappers under Solitary Waves. *J. Mar. Sci. Eng.* **2023**, *11*, 1872. https://doi.org/10.3390/jmse11101872

Academic Editors: José António Correia, José-Santos López-Gutiérrez and Abdellatif Ouahsine

Received: 14 June 2023
Revised: 15 August 2023
Accepted: 5 September 2023
Published: 26 September 2023

1. Introduction

Pipelines that are laid on or below the seabed and continuously transport large amounts of oil (or gas) are collectively referred to as submarine pipelines. They constitute the main transporting structures and currently they are the most economical and reliable selections in the design of transportation tools. Pipelines are usually installed within the seabed sediments under the protection of rock berms [1]. However, the sediments around the pipelines may be scoured by contour currents and internal waves, which expose the pipelines to the threat of complex marine environments [2]. The scour mechanism and its evolution process around the in-position pipelines were investigated by many scholars, such as Reference [3]. Occasionally, segments of a pipeline may be suspended between high points through continental slopes due to an uneven seabed profile. For example, suspended pipelines were widely used in the Ormen Lange projects, with massive depressions and landslide blocks scattered along the 120-km-long route [4].

Natural disaster, such as hurricanes and tsunamis, may induce extreme waves that generate enormous impact loads on the pipelines and may cause serious ruins to the whole production and transportation system [5–7]. Tsunamis, one of the major marine disasters caused by earthquakes and submarine landslides [8,9], send surges of water with extremely long waves that are not especially steep [10]. The tsunami triggered by a 9.0-Mw earthquake in 2011 extensively destroyed 70% of the total 200,000 structures along the Miyagi coastline, including submarine pipelines, seawalls, and coastal bridges. A tsunami is typically composed of several transient waves with varying amplitudes, wavelengths, and wave periods during propagation. Solitary waves were proposed to simulate the tsunami waves by decomposing them into N-waves through the Korteweg-de Vries equation [11–14]. Since then, the run-up process of the tsunami waves along the shoreline was investigated with the depth-averaged smooth particle hydrodynamics method [15,16]. References [17,18] quantified the impact loads over cylinders from a tsunami wave.

To protect the marine structures from potential damages due to extreme marine conditions, engineers have developed outer protections in terms of wrappers made of porous media. A porous medium enhances the buffering performance of the structures and dissipates part of the incoming wave energy [19]. For example, the turbulent intensities on a permeable breakwater were significantly attenuated in the numerical analysis by References [20–22]. Naturally, porous media are expected to be protective to submarine pipelines under extreme marine conditions, although thermal insulation and erosion prevention were mainly considered in designing pipeline coatings in the industry [23,24]. Reference [25] quantified the wave forces on pipelines buried in an impermeable bed with coverings of porous media. References [26,27] evaluated the protective performance of a porous polymer coating on subsea pipelines under sudden impacts. The drag reduction function of the porous coatings over cylinders were then quantified by Reference [28]. Two factors were considered to influence the stabilization effect of the porous coatings on pipelines: the production of an entrainment layer through the coating and the triggering of turbulent transition of the detaching shear layers. In engineering practice, applications of porous coatings on submarine pipelines are limited. Concrete wrappers, mainly designed to counteract the buoyancy forces of pipelines, can be considered as one kind of porous wrapper with medium permeability. In addition, porous wrappers made with woven carbon-fiber materials or polyurethane foam may be designed in future for pipeline protection.

The above literature review revealed that few studies were performed to examine the protective effect by the porous media on submarine pipelines, which is the main aim of this study. The porous wrapper and the submarine pipeline modules are simulated in a numerical wave tank (NWT) with the immersed boundary (IB) method. The numerical methods and equations will be provided in Section 2. Verification of the numerical model is provided in Section 3. The parametric simulations are in Section 4, in which the effects of different waves on various pipelines with porous wrappers are analysed. The conclusions are given in Section 5.

2. Numerical Methods

For simulating the interactions between pipelines and waves, the finite volume methods have been widely used. In this study, the commercial finite volume package FLOW-3D® (version 11.1.0; 2014; https://www.flow3d.com (accessed on 10 December 2022); Flow Science, Inc., Santa Fe, NM, USA). Flow-3D aims to solve the transient response of fluids under interactions with structures, internal and external loads and multi-physical processes. It features some advantages in terms of a high level of accuracy in solving the Navier-Stokes equation with the volume of fluid (VOF) method, efficient meshing techniques for complex geometries, and high efficiency level for large-scale problems. Also, Flow-3D provides the flexibility and utility for flowing through porous media. A two-dimensional numerical wave tank was constructed by using the immersed boundary (IB) method and an in-house subroutine termed as IFS_IB. A submarine pipeline and porous medium were two-way coupled at the interface described by the individual volume fractions [29]. The pipeline was wrapped with a layer of a porous medium. A solitary wave was generated at the inlet boundary of the tank to simulate an approaching tsunami. Non-slip wall conditions were assigned at the bottom of the tank and the pipe surface, which was also specified with a roughness coefficient. The top boundary was defined as a free boundary and configured with the atmospheric pressure. A Neumann-type absorbing boundary condition, a stable, local, and absorbing numerical boundary condition for discretized transport equations [30], was imposed on the outlet boundary to attenuate the reflections of the outgoing waves. A transition zone is set within a certain range from the boundary to reduce the horizontal gradient force of the elements near the boundary and suppress the calculation wave caused by this boundary condition. Through the relaxation coefficient, the predicted value on the inner boundary of the transition zone and the initial value on the outer boundary are continuously transitioned to achieve the purpose of reducing the reflection of propagating waves. The CUSTOMIZATION function of the software FLOW-3D was utilised to impose the Neumann-type absorbing boundary condition. The FLOW-3D distribution includes a variety of FORTRAN source subroutines that allow the user to customize FLOW-3D to meet their requirements. The FORTRAN subroutines provided allow the user to customize boundary conditions, include their own material property correlations, specify custom fluid forces (i.e., electromagnetic forces), add physical models to FLOW-3D, and have additional benefits. Several "dummy" variables have been provided in the input file namelists that users may use for custom options. A user definable namelist has also been provided for customization. Makefiles are provided for Linux and Windows distributions and Visual Studio solution files are provided for Windows distributions to allow users to recompile the FLOW-3D code with their customizations.

2.1. Governing Equations

The governing equations involved include the continuity equations and Reynolds-averaged Navier-Stokes equations. The mass and momentum are conserved in a two-dimensional zone [31]:

$$\frac{\partial \rho}{\partial t} + \nabla \cdot (\rho \mathbf{U}) = 0 \tag{1}$$

$$\frac{\partial (\rho \mathbf{U})}{\partial t} + \nabla \cdot (\rho \mathbf{U}\mathbf{U}) = -\nabla P + \mathbf{g} \cdot \mathbf{X} \nabla \rho + \mu \nabla^2 \cdot \mathbf{U} + \sigma \kappa \nabla \alpha \tag{2}$$

where $\mathbf{U}$ is the velocity vector, $\mathbf{X}$ is the Cartesian position vector, $\mathbf{g}$ denotes the gravitational acceleration vector, and ρ represents the weighted averaged density. The term μ is the viscosity. $\sigma\kappa\nabla\alpha$ identifies the surface tension effects with σ as the surface tension and α as the fluid volume fraction. Each cell in the fluid domain has a water volume fraction (α) ranging between 0 and 1, where 1 represents cells that are fully occupied with water, while 0 represents cells that fully occupied with air. Values between 1 and 0 represent free surface between air and water. The free surface elevation is defined by using the volume of fluid (VOF) function:

$$\frac{\partial F}{\partial t} + \frac{1}{V_F}\left[\frac{\partial}{\partial x}(\alpha A_x u) + R\frac{\partial}{\partial y}(\alpha A_y v) + \frac{\partial}{\partial z}(\alpha A_z w) + \varepsilon \frac{F A_x u}{x}\right] = F_{DIF} + F_{SOR} \tag{3}$$

where V_F is the volume of fluid fraction, F_{SOR} is the source function, F_{DIF} is the diffusion function; A_x, A_y, and A_z represent the fractional areas; and u, v, and w are the velocity components in the x, y, and z directions.

2.2. Porous Media Module

In FLOW-3D, the porous medium's flow resistance is modelled by the inclusion of a drag term in the momentum equations (Equation (2)). Coarse granular material is used in most coastal engineering applications, in which case the Forchheimer model is suitable. Using this model, a drag term $\mathrm{F_d u_i}$ is added to the righthand-side of Equation (2):

$$F_d\mathbf{U} = -g\left(an\mathbf{U} + bn^2|\mathbf{U}|\mathbf{U}\right) \quad (4)$$

where $|\mathbf{U}|$ is the norm of the velocity vector, n the porosity, and a and b are the factors.

2.3. Solitary Wave Boundary

The solitary wave is generated in terms of variations of the surface elevation η and velocities u and v by following McCowan's theory [32]:

$$\eta = Qh \quad (5)$$

$$u = c_0 E \frac{1+\cos\left(M\frac{z}{h}\right)\cosh\left(M\frac{x}{h}\right)}{\left[\cos\left(M\frac{z}{h}\right)+\cosh\left(M\frac{x}{h}\right)\right]^2}\ v = c_0 E \frac{\sin\left(M\frac{z}{h}\right)\sinh\left(M\frac{x}{h}\right)}{\left[\cos\left(M\frac{z}{h}\right)+\cosh\left(M\frac{x}{h}\right)\right]^2} \quad (6)$$

where h is the still water depth; Q is the reference value

$$Q = \frac{E}{M}\frac{\sin[M(1+Q)]}{\cos[M(1+Q)]+\cosh(+X/h)} \quad (7)$$

$$E = \frac{2}{3}\sin^2[M(1+\frac{2H}{3h})]\ M = E\frac{h}{H}\tan[\frac{1}{2}M(1+\frac{H}{h})] \quad (8)$$

where $X = x - c_0 t$; $c_0 = g\sqrt{H+h}$; H is the wave height; and t is the elapsed time.

3. Validation

3.1. Propagation over a Porous Breakwater

An experimental test on the propagation process of a solitary wave over a permeable breakwater was performed by Reference [20], which was simulated in this study to validate the adopted two-way coupling model (Figure 1a). The length, width, and depth of the flume tank were 25, 0.5, and 0.6 m, respectively. A permeable breakwater was mounted at the bottom of the flume, which had dimensions of 13 cm and 6.5 cm in the length and height, respectively. The porous breakwater with an average porosity of 0.52 was configured by glass beads with a constant diameter of 1.5 cm. Two wave gauges were fixed before (WG_1) and behind (WG_2) the breakwater, respectively. The initial still water depth h was assumed to be 10.6 cm. Height of the solitary wave H was considered to be 4.77 cm. In the numerical model, the calculation zone had dimensions of 5 m in length and 0.25 m in height. The second order quadrilateral mesh elements were adopted. The grid around the breakwater was the finest of 0.001 m. The adopted time step size was 0.05 s. The numerical predictions of the water elevations at the locations WG_1 and WG_2 by the adopted numerical tool FLOW-3D are close to both the experimental measurements and the numerical predictions from another CFD FLUENT version 14.0.1 [33] (Figure 1). Figure 1b,c show the comparison of monitored water levels at the two water level monitoring points in Figure 1a. It can be seen that the experimental results of the two monitoring points are consistent with the numerical simulation results, indicating that the propagating solitary wave energy is basically completely dissipated and then flows out. If the propagating wave energy is not dissipated, the phenomenon of wave reflection will occur. The waves monitored at the two monitoring points will appear superposition of propagating waves and reflected waves. The numerical simulation results do not agree with the physical experiment results. The fluctuations of the water surface elevation after the bypass of the incoming wave are due to its residual reflection at the right absorbing boundary condition, which arrives at WG_2 at an earlier time than WG_1. Evolution of the wave surfaces was also compared between the experimental and the numerical models (Figure 2), which demonstrates that the numerical tool is sufficiently reliable. The velocity of the wave is reduced by the porous medium as it partially infiltrates into the breakwater, which is shown as in Figure 3 by comparing the horizontal velocity distributions between the experimental and numerical results at times of 1.5 s and 2 s. The numerical predictions of the flow velocities have slight discrepancies with the experimental measurements, which are attributed to the material assumptions made in the numerical model for the glass beads in the experimental setup.

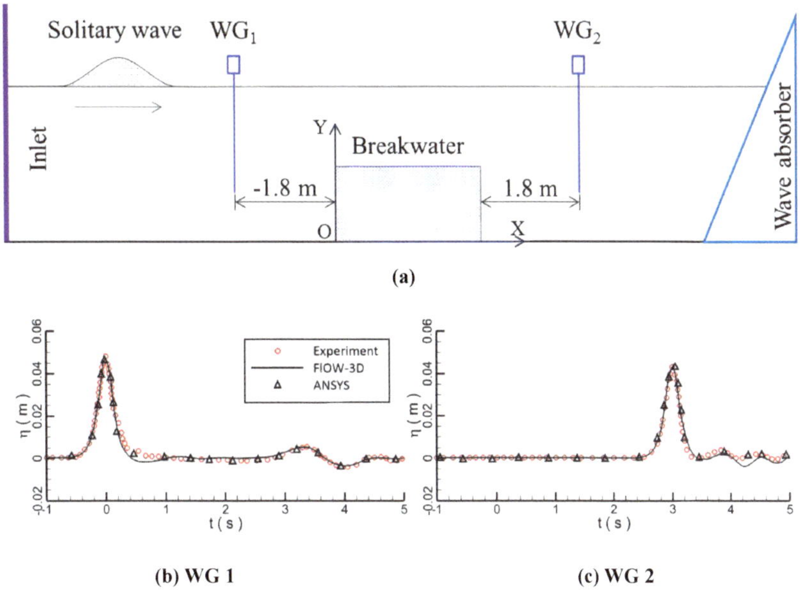

Figure 1. The diagrammatic sketch of the numerical setup (non–scaled) (**a**) and the temporal evolution comparison of water surface between experimental and numerical results (**b**,**c**).

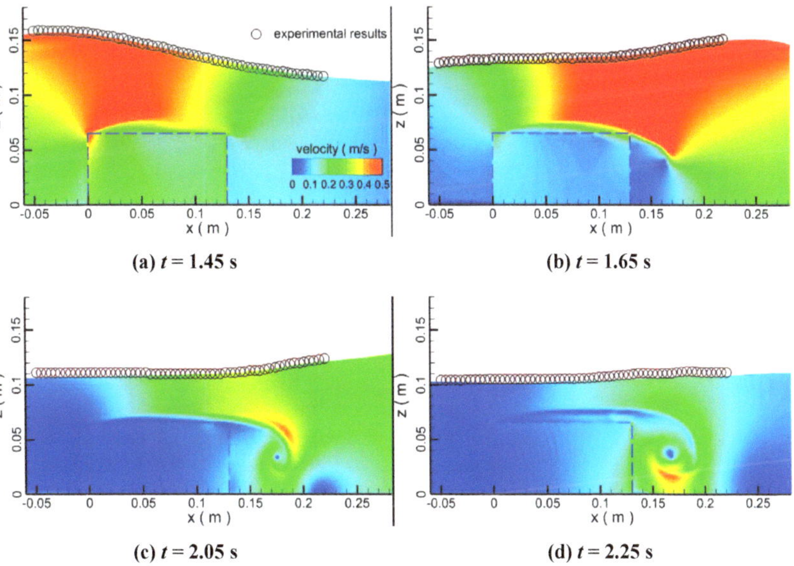

Figure 2. Water surface comparison between experimental and numerical results.

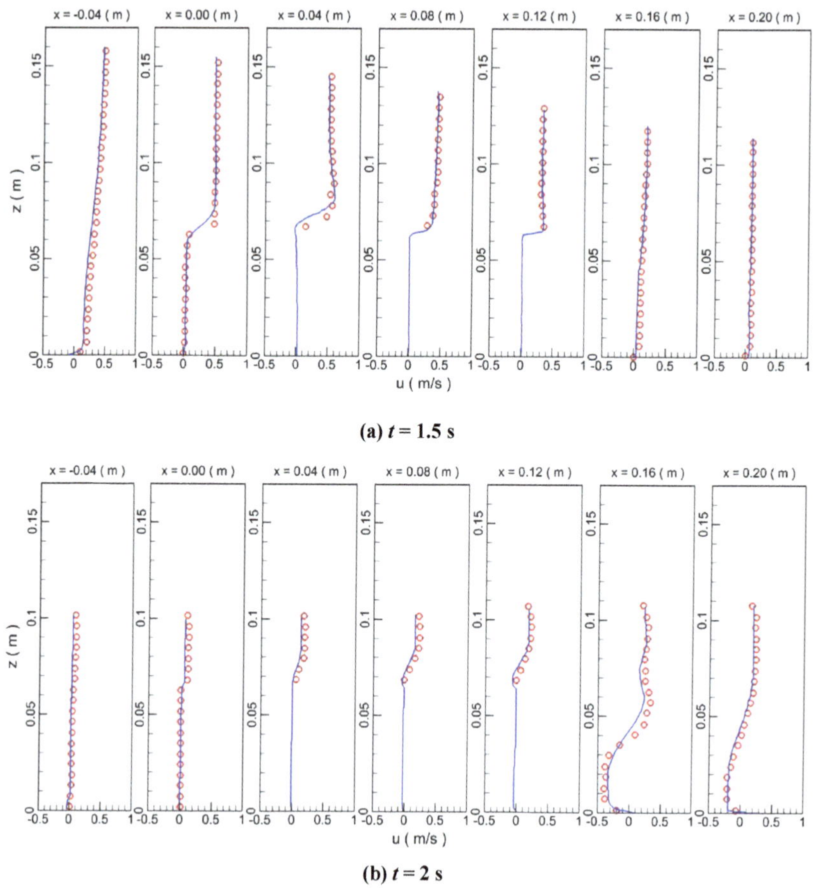

Figure 3. Comparison of horizontal velocity distribution between experimental and numerical results.

3.2. Forces on Pipeline

Another experimental test of a solitary wave impacting a pipeline was performed by Reference [34], which was also reproduced in this study for validation purposes. The calculation zone had dimensions of 40 m in length and 0.6 m in height. The solitary wave had a height of 0.0555 m with the initial water depth of 0.192 m. The pipe had a diameter of 0.048 m, which had a distance of 0.136 m over the bottom boundary of the model. A dense mesh consisting of 413,411 cells was employed with a mesh size of 0.1 mm around the pipe, which proved to be sufficiently fine through convergence studies. History of the horizontal and vertical forces, normalized by $\rho g L(\pi D^2/4)$ with L as the unit length of 1 m, is compared between the experimental and numerical results (Figure 4). Both the peak values and the transient variations of the forces predicted by the numerical analysis converge to the measured values in the experimental test. The slight discrepancy between the numerical and experimental results at 2.5 s and 3.1 s, which may be induced by the error of the numerical model simulating the complicated turbulence behaviour, is acceptable in relation to the requirements of this study as our concern is mainly the peak values of forces.

Therefore, the adopted numerical tool is sufficiently reliable to simulate the interactions between solitary waves and the permeable structure through the above validation cases.

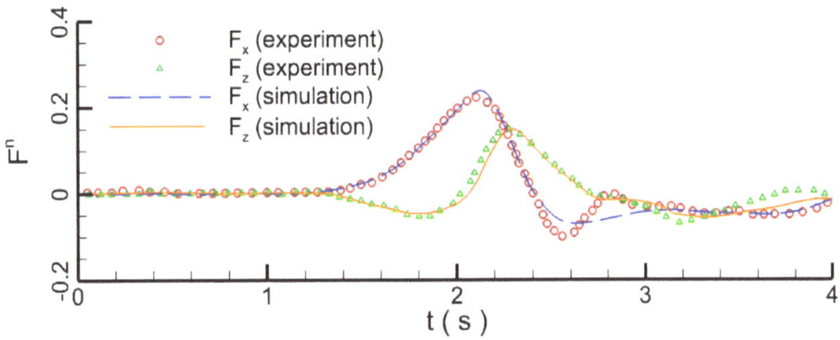

Figure 4. Force comparison between the experimental and numerical results.

4. Results and Discussion

Influence of the solitary waves on the performance of wrapped pipelines was investigated by considering different wave heights (H) and thicknesses (T) and wrapper porosities (n). The still water depth (h) was taken to be 6 m (Figure 5). The diameter of the porous medium was assumed to be 0.05 m. The pipeline diameter D was set at 1 m. In Figure 5 the variable G represents the gap between the permeable wrapper and the seabed. The scouring process had been completed before the simulation; therefore, the seabed boundary was taken as a rigid wall. The tandem pipelines had a distance of S between each other. The whole model had dimensions of 400 m in length and 12 m in height. The finest mesh around the pipeline was configured as 0.0025 m, which was verified to be sufficiently fine through trial calculations with finer meshes.

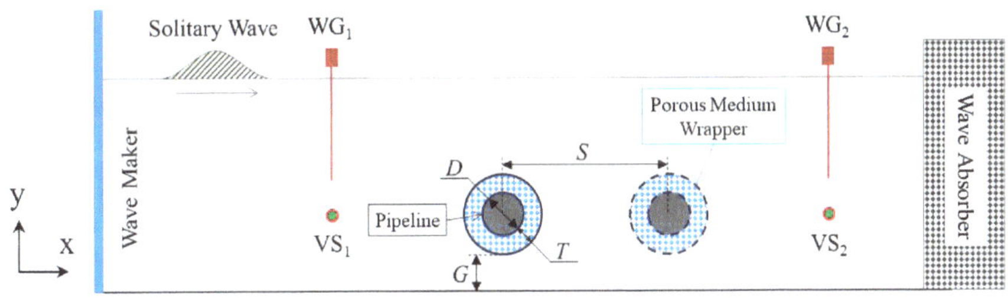

Figure 5. Layout for solitary wave impinging on the submarine pipeline encased in porous media.

4.1. *Effect of Porous Wrapper*

4.1.1. Wrapper Porosity

The pipeline was put on the seabed. The gap (G) between the wrapper and the seabed was considered to be zero. The height (H) of the solitary wave was considered to be 2 m. The porosity (n) was taken to be 0.0, 0.4, 0.6, and 1.0. Note that n = 0.0 indicates the impervious condition, while n = 1.0 corresponds to the non-wrapping condition. The thickness of the permeable wrapper remained at 0.5 m. In calculation, the wave approaches the pipe at around 6.3 s and departs from it at 10.2 s. When the wave approaches, the kinematic performance over the pipe is enhanced (Figure 6). Due to the wave disturbance, a number of small vortices are generated around the pipe (Figure 7). At the departure of the wave, the disturbance to the flow field seems to be more intense than that at its arrival, which further generates vortices around the pipeline. Without a wrapper, the pipe is fully exposed to the disturbance of the incoming wave, which maximises the velocity and vorticity values around the pipe. When the pipeline is wrapped by a porous medium, some water seeps into the wrapper, and the velocity in the wrapper is reduced to a very low value, which implies that the porous medium is capable of absorbing the dynamic energy of the flowing fluid. With an external coverage ($n < 1.0$), the disturbance is generated mainly at the outer surface of the wrapper. As the wrapper porosity increases, the domain of the low-speed flow underneath the pipeline expands.

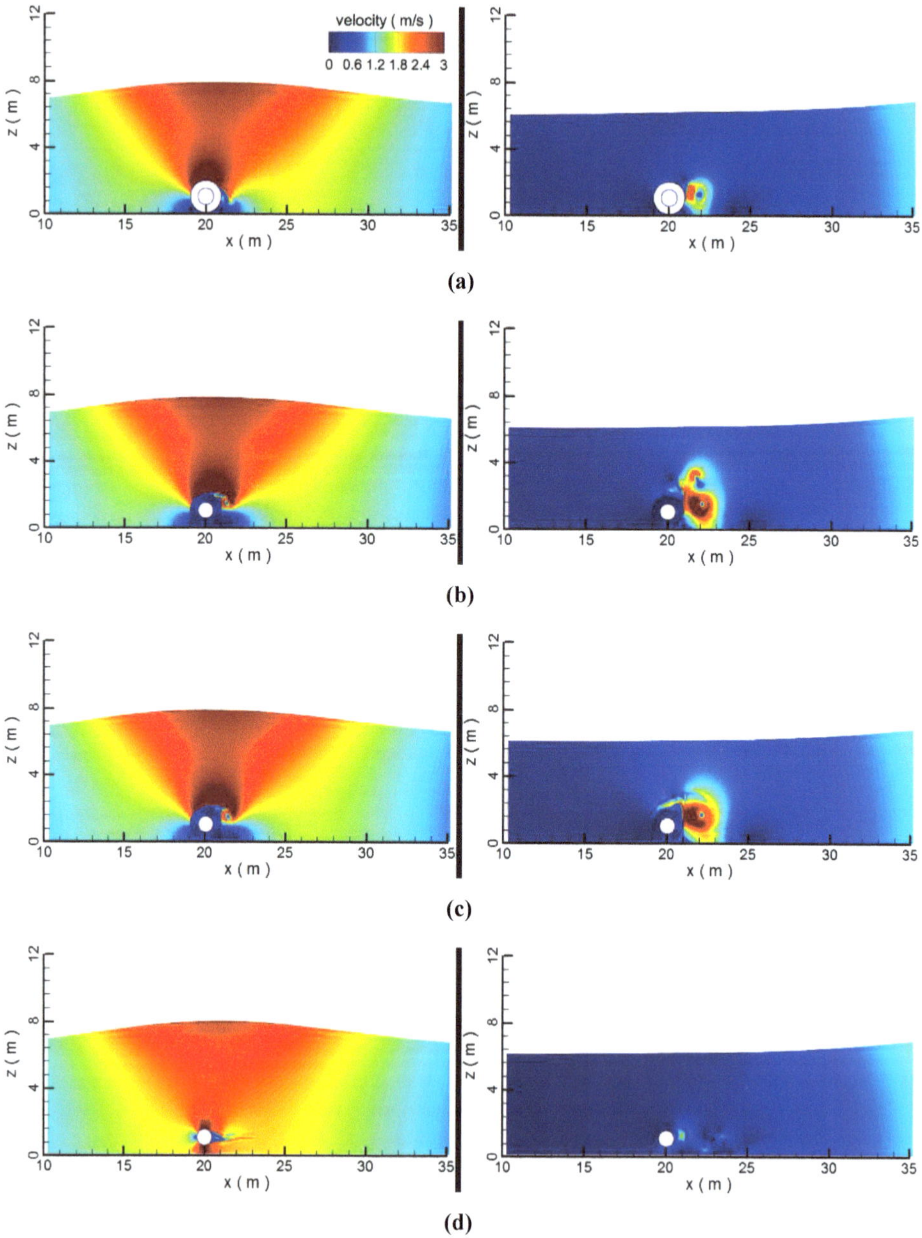

Figure 6. The velocity contours of the flow fields under different porosities; (**a**) $n = 0.0$; (**b**) $n = 0.4$; (**c**) $n = 0.6$; (**d**) $n = 1.0$; left to right: arrival, departure.

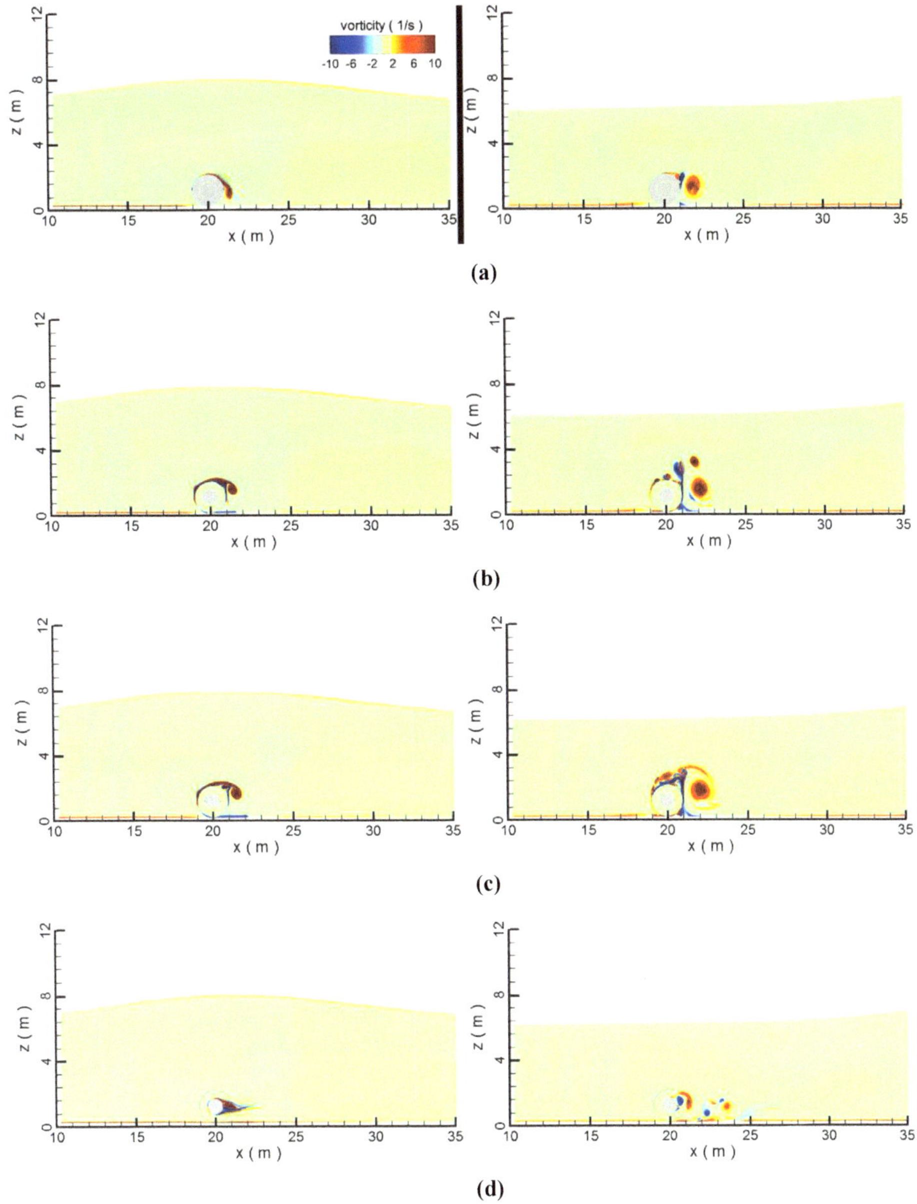

Figure 7. The vorticity contours of the flow fields under different porosities; (**a**) n = 0.0; (**b**) n = 0.4; (**c**) n = 0.6; (**d**) n = 1.0; left to right: arrival, departure.

The peak velocity around the pipeline without a wrapper (1.9 m/s) is larger than that with a wrapper (1.6 m/s) (Figure 8). For pipes with wrappers, the peak velocities around them are similar to one another. In contrast, the velocity profiles at x = 23 m are quite different. When the pipeline has no wrapper (i.e., n = 1.0), the change in velocity is fairly moderate. When the pipeline has a wrapper, the porous wrapper causes a secondary fluctuation in the rear water body after the primary fluctuation due to the peak of the wave passing through the pipeline. This generates a series of velocity peaks. The secondary velocity peaks for a porosity coefficient of 0.4 are higher than those for a porosity coefficient of 0.6. Accordingly, the turbulent kinetic energy (TKE) also changes with the porosities, as shown in Figure 9. The TKE is expressed as

$$\mathrm{TKE} = \int \frac{1}{2}\rho(|u|^2 + |v|^2)dV_f \tag{9}$$

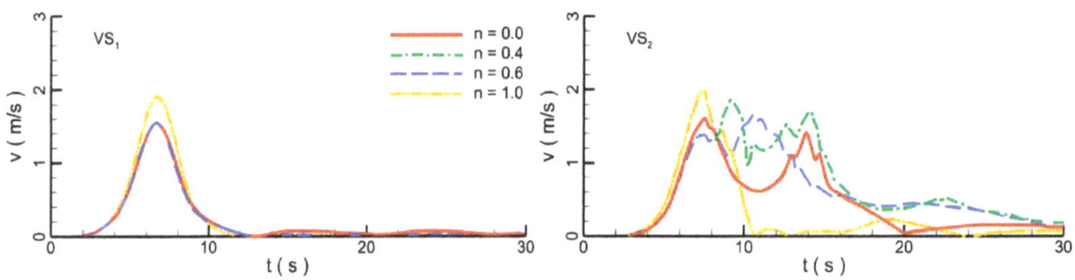

Figure 8. Comparison of horizontal and vertical velocities at front and rear of pipeline under different porosities.

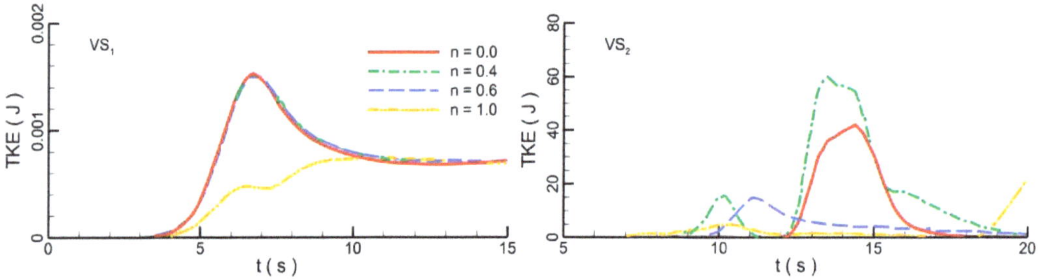

Figure 9. Comparison of turbulent kinetic energy at front and rear of wrapper under different porosities.

With the propagation of the wave, the TKE increases gradually in front of the pipeline. The TKE value under the pipeline without a wrapper (n = 1.0) (0.0008 kJ) is nearly half of that with a wrapper (0.0015 kJ). In comparison, the TKE values for the wrapped pipelines (n < 1.0) are very close to each other. After the wave leaves the pipeline, the TKE in front of the pipeline decreases for around 50%. Then, the TKE in the rear of the pipeline with a porous wrapper increases intensively because the porous media perturb the flow field. Compared with the pipeline without the wrapper, the interaction between the wrapped pipeline with the flow field is more severe. Furthermore, the solid wrapper can cause a strong disturbance to the flow, but the interference of the solid wrapper (n = 0.0) in the rear flow is still weaker than the wrapper with the porosity of 0.4.

The hydrodynamic forces (F), including the pressure and shear stress, are normalized by $\rho g L(\pi D^2/4)$ (Figure 10). With a fully solid (i.e., n = 0.0) wrapper, the pipeline tends to be unaffected by the external flow. Hence, the hydrodynamic forces are zero while the forces on the wrapper reach their maximum. With porous wrappers, water seeps into the wrapper, buffering the impact of the incoming waves on the pipe. As the porosity coefficient increases, the induced forces on the pipeline increase while those on the wrapper decrease. When the porosity coefficient is 0.4, the forces on the external wrapper become higher than that on the internal pipeline. Therefore, the porous wrapper is capable of protecting the pipeline. The smaller the porosity coefficient the better protection the wrapper provides to the pipeline. The pressure gradient and shear stress forces are also shown in Figure 11.

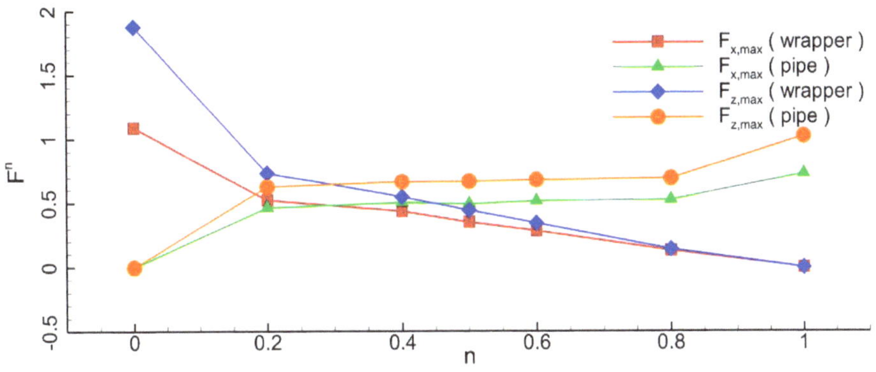

Figure 10. Comparisons of the maximum hydrodynamic forces on the pipeline and wrapper.

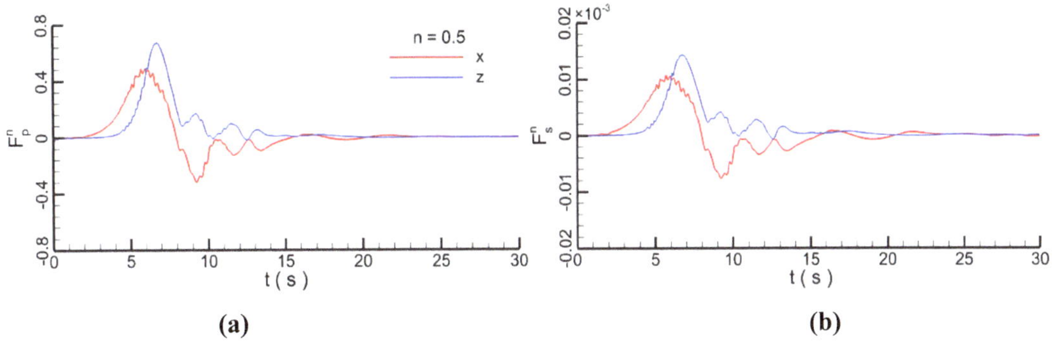

Figure 11. Decomposed pressure gradient force (**a**) and shear stress (**b**) force on the pipeline.

4.1.2. Thickness of Wrapper

Seven wrapper thicknesses are considered: T = 0.2, 0.25, 0.3, 0.35, 0.4, 0.45, and 0.5 m. The porosity coefficient is taken to be 0.6. At the moment that the wave goes through the pipe, the transient evolution of the vorticity contours around the pipeline with a wrapper thickness of 0.25 m is depicted in Figure 12. A couple of vortices emerge on the upper and lower vertices of the pipeline as the wave approaches the pipeline. As the wave propagates, many vortices flow along the wrapper and then shed off. Compared with the top vortices, the bottom vortices are shed off faster for two reasons. Firstly, as the friction at the seabed is small, the bottom flow velocity is higher than that on the top. Secondly, when the wave peak departs from the pipeline, a strong disturbance by the water body occurs behind the pipeline, followed by the irregular swing and fall off of the vortices. After the wave travels far away, the water flow near the pipeline becomes weak, and the vortices are scattered around the pipeline.

Figure 13 shows a comparison of flow field stream traces and the velocity contours. When the fluid penetrates the wrapper, the streamline starts to diverge, which indicates that the free flow is hindered. Therefore, the flow becomes slower and the flow direction becomes non-uniform. For the fluid flows out of the wrapper, the stream traces are quite complex and chaotic. The reason is that the seeping fluid mixes with the bypass flows and causes strong interference in the water body behind the pipeline. The streamlines passing through the wrapper indicates frequent water exchange at the wrapper surface. Along with the small-attached vortices on the wrapper surface, more fluid passes over the wrapper and causes a large vortex behind the wrapper.

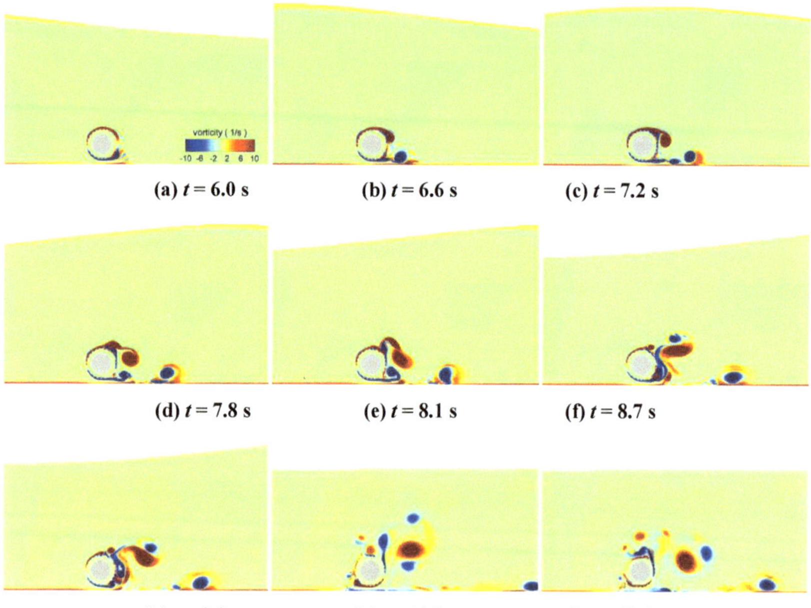

Figure 12. Temporal evolutions of vorticity contours around pipeline with wrapper thickness of 0.25 m (**a**) 6.0 s (**b**) 6.6 s (**c**) 7.2 s (**d**) 7.8 s (**e**) 8.1 s (**f**) 8.7 s (**g**) 9.0 s (**h**) 10.2 s (**i**) 12.6 s.

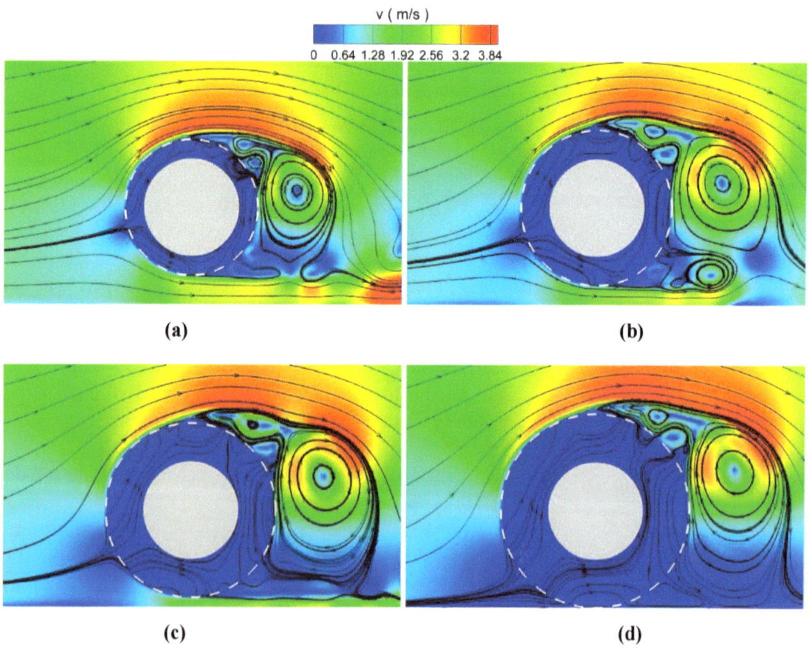

Figure 13. Comparisons of flow field streamtraces and velocity contours under different wrapper thicknesses; (**a**) T = 0.2 m; (**b**) T = 0.3 m; (**c**) T = 0.4 m; (**d**) T = 0.5 m.

The highest free surface elevations and velocities at the front and at the rear of the pipeline with different wrapper thicknesses are depicted in Figure 14. As the wrapper thickness increases, the highest elevation at the front of the pipeline seems to be quite stable, although the peak velocity increases by around 6%. At the moment that the wave bypasses the pipeline, the maximum elevation reduces with an increase in the wrapper thickness. This is because the pipeline blocks the wave propagation. However, due to the strong mixing effect of the seepage and bypass water, the maximum velocity rises to be higher than that in front of the pipe. The maximum forces on the wrapper and the pipeline for different wrapper thicknesses are shown in Figure 15. With an increase in the wrapper thickness from 0.2 to 0.5 m, the normalized forces on the wrapper are doubled as a larger interaction area is involved. In contrast, the vertical forces on the pipeline decrease by 12.5%. Therefore, the larger the thickness of the wrapper the safer the pipeline.

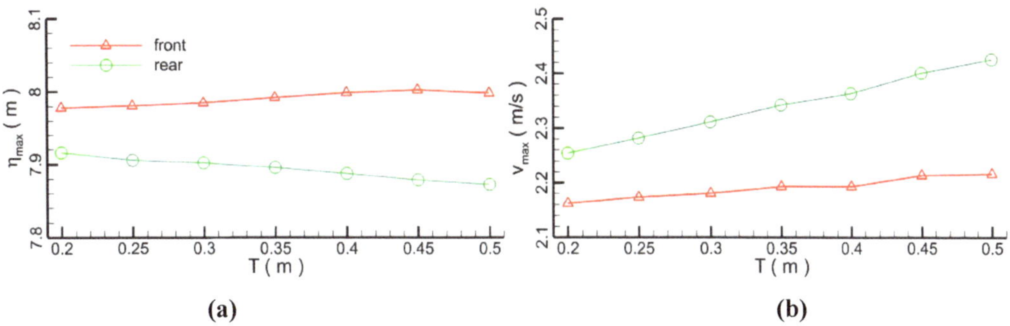

Figure 14. Comparisons of the maximum elevations and velocities in front and rear of the pipelines with different wrapper thicknesses; (**a**) free surface elevation (note: original water depth is 6 m); (**b**) velocity.

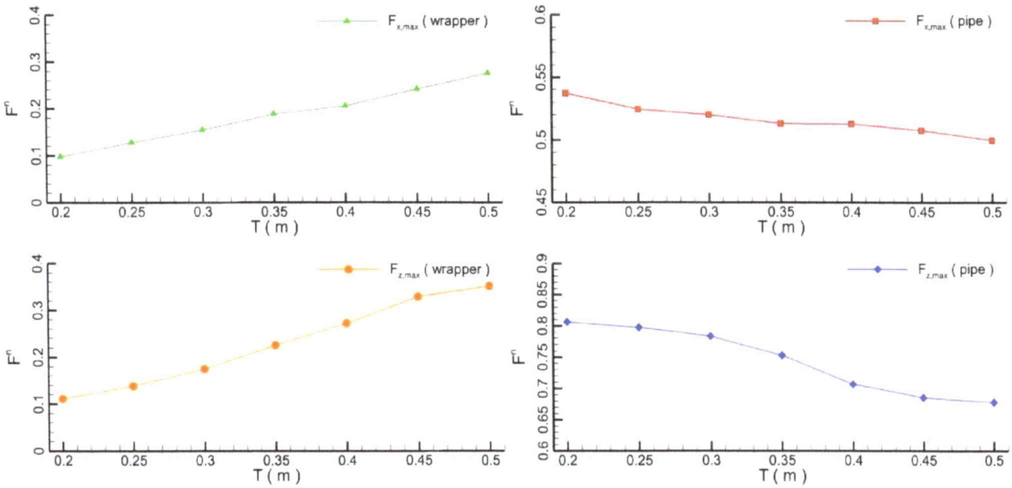

Figure 15. Hydrodynamic forces on the pipeline and wrapper.

4.2. *Effect of Pipeline Structure*

The in-situ pipelines may be under various suspended conditions since the seabed topography is often uneven. Some pipelines are also laid in tandem for the sake of the transportation efficiency. In order to examine the effects of porous wrappers on pipelines under different conditions, a study was carried out considering two scenarios, namely, suspended pipelines and pipelines in tandem. In the numerical models, the porosity coefficient (n) remained at 0.6, the thickness (T) of the wrapper was kept at 0.5 m, and the wave height (H) was assumed to be 2.0 m.

4.2.1. Suspended Pipelines

Six gaps (G) between the wrapper and the seabed (0.0, 0.2, 0.4, 0.6, 0.8, and 1.0 m) were considered [35–37]. The representative flow field at three points in time (6.3, 7.2, and 10.2 s) are shown in Figure 16. At the arrival of the wave at the pipeline (at 6.3 s), the flow is accelerated and the velocities over and beneath the pipe reach the maximum values due to the bypass effect of the fluid. At the moment that the wave peak is above the pipe (at 7.2 s), all the velocities around the pipe reach their highest values. After the wave passes over the pipe (at 10.2 s), the velocity decreases and several vortices are formed behind the pipeline. With a tiny wrapper-seabed gap, the velocity within the gap is very high while the flux is relatively small. An increase in the gap will result in an increase in the flux and a decrease in the velocity. A symmetric velocity distribution similar to a fisheye is observed behind the pipeline, which becomes more obvious when the gap increases (Figure 16c).

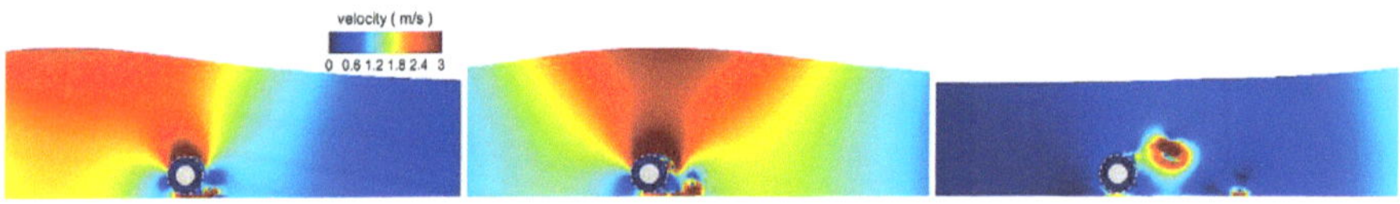

(a)

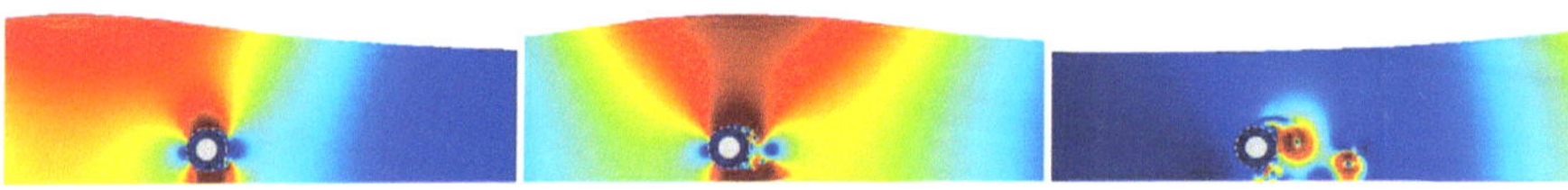

(b)

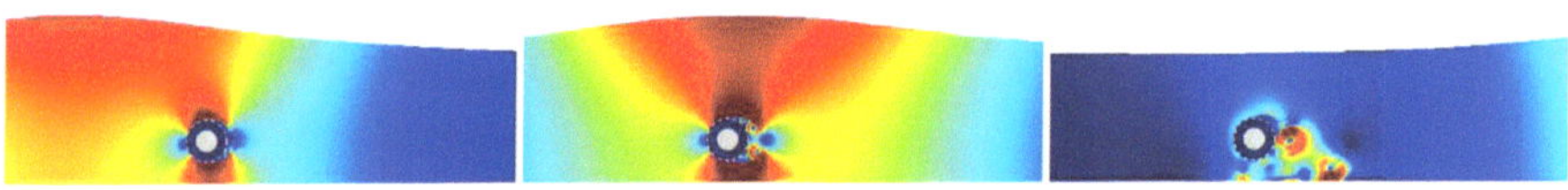

(c)

Figure 16. The velocity contours of the flow fields under different gaps; (**a**) G = 0.2 m; (**b**) G = 0.6 m; (**c**) G = 1.0 m. Left to right: 6.3 s, 7.2 s, and 10.2 s. Left to right: arrival, stay, departure.

With the bypass of the wave, the vortices generated around the pipeline become larger. The vorticity contours and the streamlines of the flow field are shown in Figure 17. As the solitary wave approaches, a pair of whirlpools shed off from the wrapper with a gap of 0.2 m. With an increase in the gap, the two whirlpools gradually disappear and are replaced with two smaller vortices. Due to the internal pores within the wrapper, the streamlines in the wrapper are dispersed, and it is hard for a vortex to be generated. With an increase in the gap, two anti-symmetric vortices shed off from the wrapper. Besides, some tiny vortices remain adhered to the wrapper due to the interaction by the seepage and the external flow. When the gap is very small, a few small vortices are generated between the wrapper and the seabed. In contrast to the interface of vortex from the flow around a solid cylinder, the vortex interface at the wrapper is not fully smooth. Because of the strong interactions of fluid over the wrapper surface, several small vortices mingle with the large shedding vortices. The flow direction also varies greatly according to the streamline mobilisation.

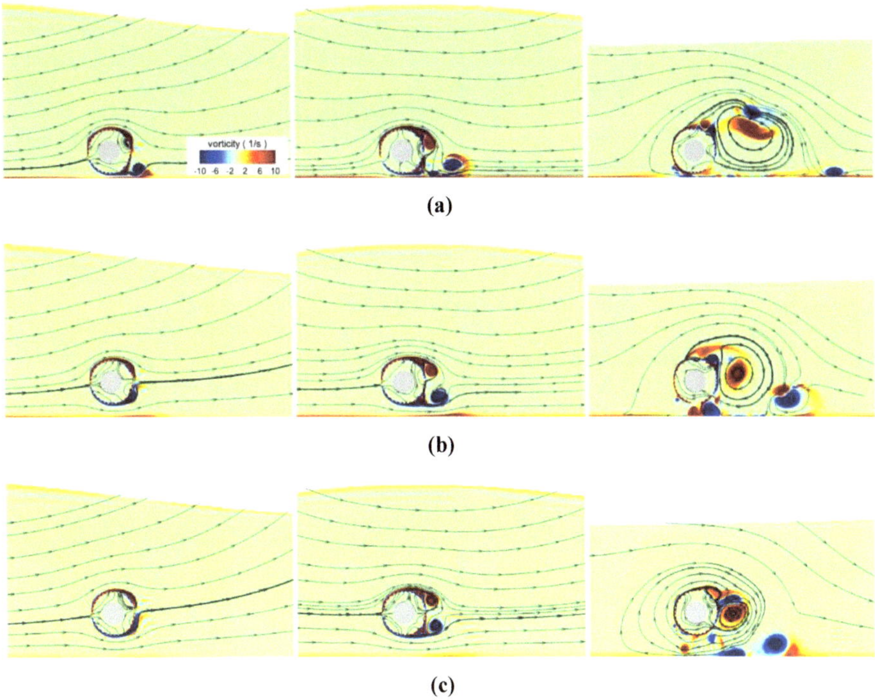

Figure 17. The vorticity contours of the flow fields under different gaps; (**a**) G = 0.2 m; (**b**) G = 0.6 m; (**c**) G = 1.0 m. Left to right: 6.3 s, 7.2 s and 10.2 s.

The gap is normalized by the pipeline diameter as $\beta = G/D$. With a small gap ($\beta < 0.2$), the horizontal forces on both the wrapper and the pipeline are slightly smaller than those on the wrapper and pipeline without a gap (Figure 18). With a further rise of the gap width, the horizontal forces are accordingly enlarged due to higher velocity around the pipeline as shown in Figure 16.

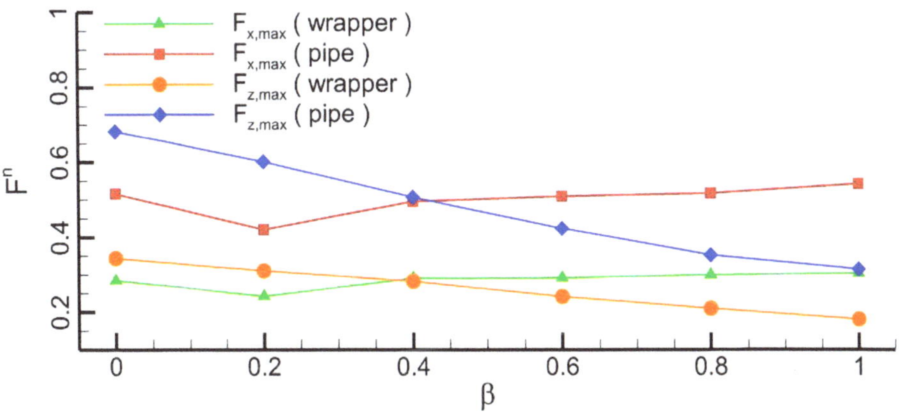

Figure 18. Comparisons of the maximum horizontal and vertical hydrodynamic forces on the pipeline and wrapper under different gaps.

In contrast, an increase in the gap width may inversely cause the reduction of vertical forces on both the wrapper and the pipeline. The vertical forces can be considered to consist of two parts. One is caused by the weight of the water body at the bypass of the wave from the pipeline, while the other can be caused by the velocity difference between the flow above and below the pipeline after the flow passes over. In summary, as the gap increases, the flow velocity within the gap initially increases when $\beta < 0.2$ and then decreases when $\beta > 0.2$. In contrast, the vertical forces caused by the wave's weight always decrease with an increase in the gap.

4.2.2. Pipelines in Tandem

The hydrodynamic forces on pipelines in tandem are investigated considering five different distances (S) between the two pipeline centres (2.5, 3.0, 3.5, 4.0, and 4.5 m). The velocity and vorticity fields at 6.3, 7.2, and 10.2 s around the tandem pipelines with distances of 2.5, 3.5, and 4.5 m are depicted (Figures 19 and 20). As the wave approaches the pipeline, the velocity within the pipeline gap is very small due to the blockage effect of the pipeline in front. As the distance increases, the velocity field within gap space is enhanced as more water flow is allowed. The velocity above the pipeline has its maximum value, and part of the high-speed fluid flows into the gap through the space underneath the pipeline. With a small distance, the vortices shedding off from the front pipeline impinge directly on the rear pipeline without any stretching. When the distance is increased, noticeable vortex shedding emerges in the middle space (Figure 20c). Similar vortex shedding behind the rear pipeline is observed for different distances. After the wave bypasses the pipeline, the increase in the distance between the pipelines will cause an increase in the velocity magnitudes in the space among the pipelines. As the distance increases, the flow becomes more chaotic due to the seepage from the wrapper and the limited flow space. In summary, influence of the distance between the pipelines over the whole kinematic field is not significant, although the local flow field around the pipelines is severely affected. When the wave bypasses the tandem pipelines, the largest forces on structures (i.e., the pipelines and wrappers) are shown in Figure 21, in which the distance ratio (θ) is calculated as $\theta = S/D$.

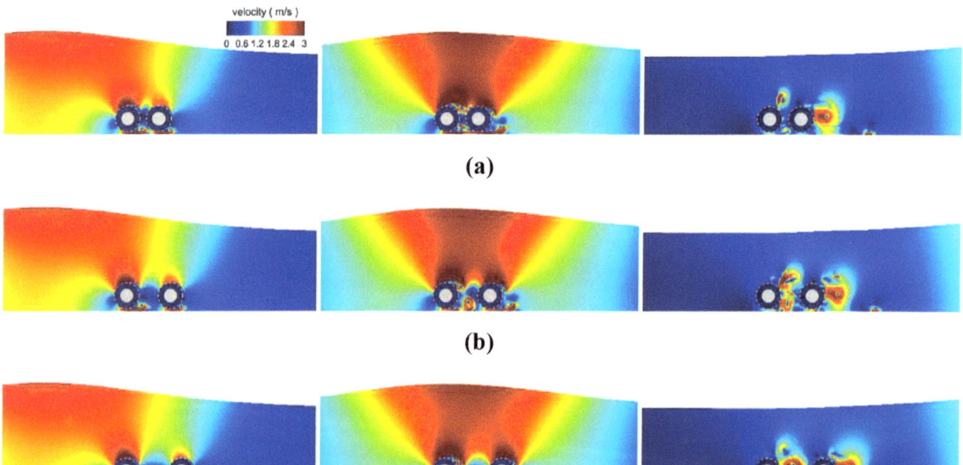

Figure 19. The velocity contours of the flow fields under different spacings; (**a**) S = 2.5 m; (**b**) S = 3.5 m; (**c**) S = 4.5 m. Left to right: 6.3 s, 7.2 s and 10.2 s.

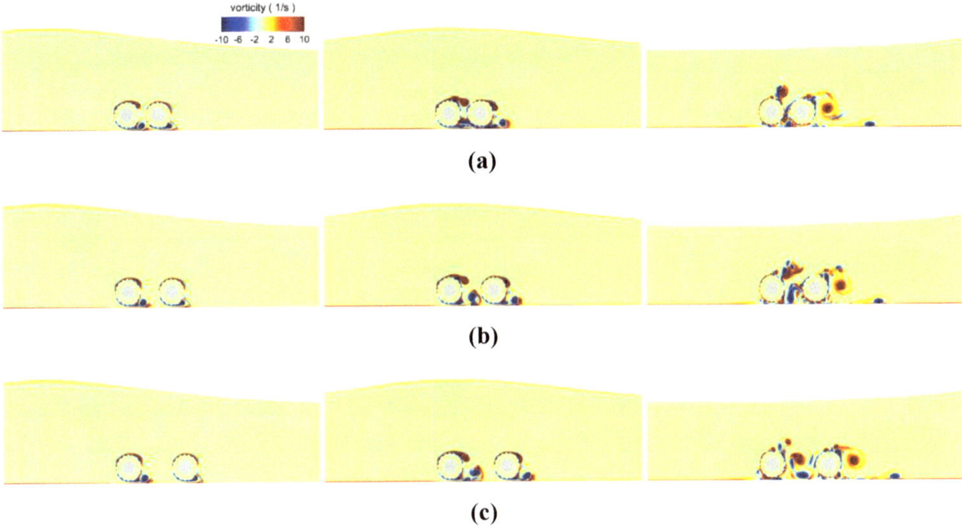

Figure 20. The vorticity contours of the flow fields under different porosities; (**a**) S = 2.5 m; (**b**) S = 3.5 m; (**c**) S = 4.5 m. Left to right: 6.3 s, 7.2 s and 10.2 s.

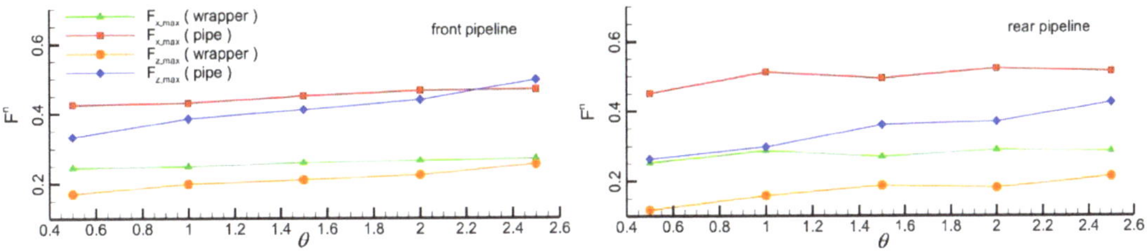

Figure 21. The maximum forces on the pipeline and wrapper under different distances.

As for the pipeline in front, as the distance ratio increases, the horizontal loads on the wrapper and pipeline increase slightly, while the vertical forces are almost doubled. As for the rear pipeline, as the distance reduces, the velocity in the gap becomes smaller and the forces on the pipelines and wrappers are also reduced, which is mainly attributed to the shield effect from the front pipeline. With an increase in the distance, the forces increase due to the increase in the turbulence energy in the gap.

Different ratios of the forces on the front and rear pipelines are depicted in Figure 22. The difference ratio is defined as $\Delta F^n = (f_{f,max} - f_{r,max})/f_{f,max}$, where $f_{f,max}$ and $f_{r,max}$ are the maximum forces on the pipeline or wrapper. It is found that the horizontal loads on the rear pipe and wrapper tend to be always higher than their counterparts on the front pipe. This means that a turbulent flow in the horizontal direction on rear pipe is more intense than that on the front pipe. For different distances, deviations for the forces on the pipelines and wrappers are also different. The deviation is found to be maximized at a distance of 1 m and this indicates that the pipeline is not well protected and needs to be avoided in engineering practice.

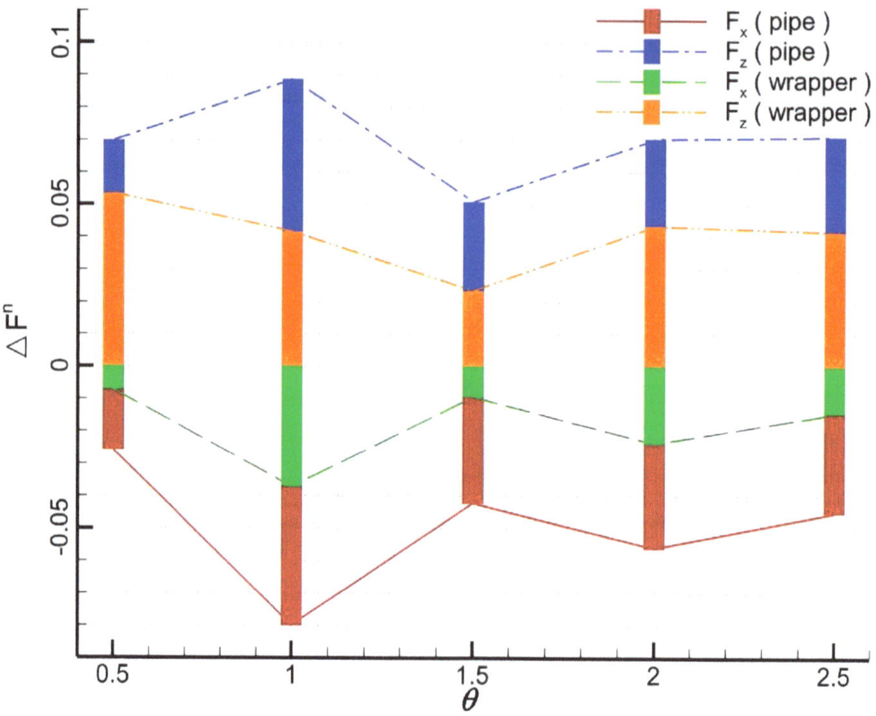

Figure 22. The deviation of the forces on the front and rear pipelines and wrappers under different distances.

4.3. Effect of Wave Height

Six groups of wave heights (H), i.e., 1.6, 1.8, 2.0, 2.2, 2.4, and 2.6 m, are selected to consider different marine environment. After bypassing the pipeline, the height of the wave decreases because of the blockage effect of the pipeline and the dissipation of the flow energy (Figure 23a). The deviation ratio of the wave heights before and after the wave passes over the pipeline is shown in Figure 23b and is defined as $\delta = (H_{f,max} - H_{r,max})/H_{f,max}$. The wave height attenuation becomes more significant as the wave height increases. This means that waves with larger heights are more easily affected by the pipelines.

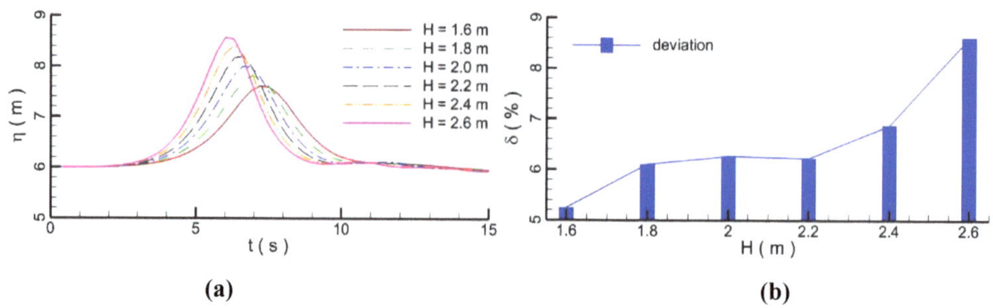

Figure 23. Waves with different wave heights; (**a**) temporal evolutions; (**b**) attenuation deviation.

At the bypass of the wave through the pipe, the loads are increased until they reach the maximum values at the moment that the wave peak appears above pipeline (Figure 24). The forces gradually decrease as the wave propagates. Because of some reflux after the wave bypasses the pipeline, the flow is in the opposite direction to that of the wave propagation, resulting in a negative force. The vibration of the water body by the wave propagation induces oscillations of the forces on the pipeline and wrapper. When the wave height is larger, the force oscillation becomes fiercer and the maximum loads on the pipeline and the wrapper increase (Figure 25). The vertical forces on the pipeline are the largest compared with other forces under the same conditions. Besides, as the wave height increases, the increased amplitude of vertical forces on the pipeline is the most significant change since the weight of the water above the pipeline increases. Therefore, given that the wave height is very high, the protective function of wrapper on the pipeline tends to be weakened compared with that of the wrapper for a low wave height.

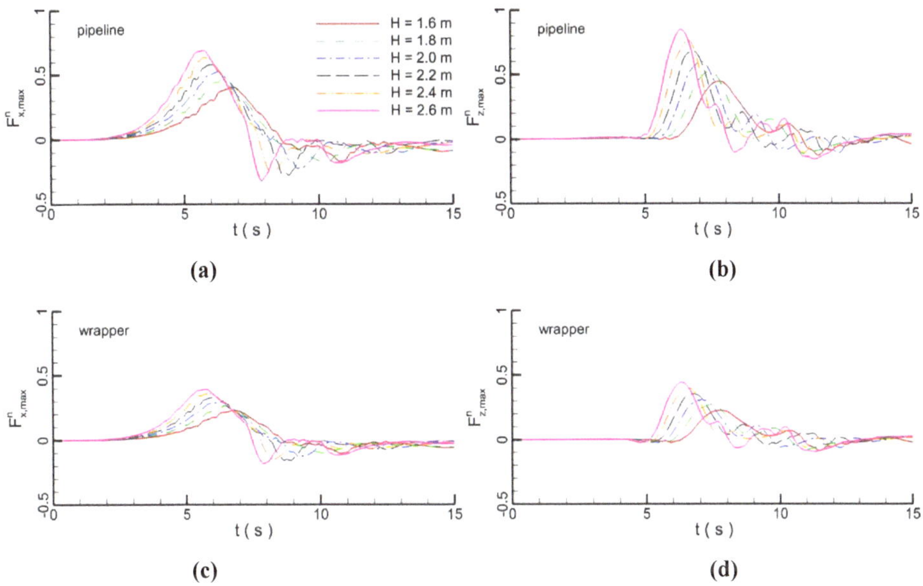

Figure 24. The temporal evolutions of forces on the pipeline and wrapper; (**a**) Horizontal maximum force on pipeline; (**b**) Vertical maximum force on pipeline; (**c**) Horizontal maximum force on wrapper; (**d**) Vertical maximum force on wrapper.

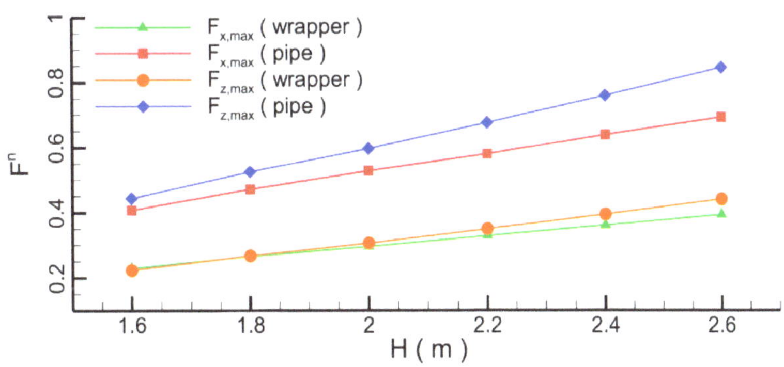

Figure 25. Variation of hydrodynamic forces on the pipeline and wrapper under different distances.

5. Conclusions

The effect of porous media on the dynamic performance of submarine pipelines under solitary waves was investigated. The porosity of the wrapper, the seabed topography, the structure of the pipeline, and the marine environment were considered. The study had a limitation of the model sizes due to the limited computational resource and the simplification of the solitary wave due to its mathematical complication, which will be tackled in future works. The following main conclusions have been made.

(1) When a pipe is wrapped by a porous medium, the velocity in the wrapper is relatively small because the porous medium can consume the water energy and weaken the flow. With an increase in the porosity, the range of the low-speed flow at the bottom of the pipeline expands. This indicates that the porous wrapper can slow down the flow and affect a wider region of the surrounding water. After the bypass of the wave through the pipe, the number and volume of the vortices behind the porous wrapper are larger than those for a pipeline with a solid wrapper or without a wrapper. As the porosity coefficient increases, the impact forces on the pipe increase, while those on the wrapper decrease. This implies that the porous wrapper is capable of protecting the pipeline.

With an increase in the wrapper's thickness, the hydrodynamic forces on the wrapper tend to increase. In particular, the horizontal forces on the pipeline decrease with an increase in the thickness due to the protection of the wrapper, while the vertical forces are increased because of variations in the fluid's stagnation point.

(2) For a wave bypassing a pipe with different heights, a symmetric speed change similar to a fisheye appears behind the pipeline, along with two antisymmetric vortices shedding off from the wrapper.

As the internal seepage interacts with the external fluid flow, several small vortices are still attached to the wrapper. The hydrodynamic vertical forces on both the wrapper and the pipeline decrease with the pipeline distance. With an increase in the suspension of the pipe, the velocity and TKE within the gap space increase and both the vortex intensity and the number of vortices increase. Therefore, the flow pattern appears to be chaotic. As for the front pipeline, an increase in the gap leads to a slight increase in the horizontal forces on both the wrapper and the pipeline, but a significant increase in the vertical forces. As for the rear pipeline, because of the shield function of the front pipeline, the velocity within the gap space and the forces on the pipes decrease with a decrease in the gap size.

(3) When the waves with different heights pass over the pipeline, the height of the wave is reduced because of the blockage function from the pipeline and the dissipation characteristic of the flow energy. When the wave height is increased, the velocity around the pipeline increases, inducing an increase in the TKE. As the wave height increases, all the maximum forces on the pipeline and wrapper also increase. Note that an increase in the vertical forces on the pipeline is the most significant change because the weight of the water above the pipeline increases, which implies that the protection function of the wrapper is enhanced by the reduction in the wave height.

From the above investigation, the mechanism of load transfer from the pipeline to the external wrapper has been presented. This encourages industrial experts and academic scholars to arrange more investigations of the functions and cost-efficiency of porous wrappers, which could form a new branch of the pipeline design practice.

Author Contributions: Contributor Roles Taxonomy: E.Z.: Conceptualization, Methodology, Validation, Investigation and Writing—Original Draft; Y.D.: Data Curation, Formal analysis; Y.D.: Visualization, Project administration; Y.D., L.C., Y.W. and Y.L.: Writing—Review & Editing. All authors have read and agreed to the published version of the manuscript.

Funding: The paper was supported by the National Natural Science Foundations of China (Grants No. 52001286 and No. 42272328), GuangDong Basic and Applied Basic Research Foundation (Grant No. 2022A1515240002) and Comprehensive Survey of Natural Resources in Huizhou-Shanwei Coastal Zone (Grant No. DD20230415).

Data Availability Statement: The data that support the findings of this study are available from the corresponding author upon request.

Conflicts of Interest: The authors declare that they have no conflict of interest.

References

1. Jeanjean, P.; Liedtke, E.; Clukey, E.C.; Hampson, K.; Evans, T. An operator's perspective on offshore risk assessment and geotechnical design in geohazard-prone areas. In Proceedings of the International Symposium on Frontiers in Offshore Geotechnics, Perth, Australia, 19–21 September 2005; CRC Press: London, UK, 2005; pp. 115–143.
2. Fuhrman, D.R.; Baykal, C.; Sumer, B.M.; Jacobsen, N.; Fredsøe, J. Numerical simulation of wave-induced scour and backfilling processes beneath submarine pipelines. *Coast. Eng.* **2014**, *94*, 10–22. [CrossRef]
3. Zhao, M.; Cheng, L. Numerical modeling of local scour below a piggyback pipeline in currents. *J. Hydraul. Eng.* **2008**, *134*, 1452–1463. [CrossRef]
4. Eklund, T.; Høgmoen, K.; Paulsen, G. Ormen Lange Pipelines Installation and Seabed Preparation. In Proceedings of the Offshore Technology Conference, Houston, TX, USA, 30 April–3 May 2007; OTC-18967-MS. [CrossRef]
5. Dong, Y.; Wang, D.; Randolph, M. Investigation of impact forces on pipeline by submarine landslide with material point method. *Ocean Eng.* **2017**, *146*, 21–28. [CrossRef]
6. Dong, Y.; Wang, D.; Randolph, M. Quantification of impact forces on fixed mudmats from submarine landslides using the material point method. *Appl. Ocean Res.* **2020**, *102*, 102227. [CrossRef]
7. Dong, Y.; Liao, Z.; Liu, Q.; Cui, L. Potential failure patterns of a large landslide complex in the Three Gorges Reservoir area. *Bull. Eng. Geol. Environ.* **2023**, *82*, 41–52. [CrossRef]
8. Zhao, E.J.; Dong, Y.; Tang, Y.Z.; Sun, J.K. Numerical investigation of hydrodynamic characteristics and local scour mechanism around submarine pipelines under joint effect of solitary waves and currents. *Ocean Eng.* **2021**, *222*, 108553. [CrossRef]
9. Sun, Q.L.; Wang, Q.; Shi, F.Y.; Alves, T.; Gao, S.; Xie, X.N.; Wu, S.G.; Li, J.B. Runup of landslide-generated tsunamis controlled by paleogeography and sea-level change. *Commun. Earth Environ.* **2022**, *3*, 244. [CrossRef]
10. Rabinovich, A.B.; Titov, V.V.; Moore, C.W.; Eblé, M.C. The 2004 Sumatra tsunami in the southeastern Pacific Ocean: New global insight from observations and modeling. *J. Geophys. Res. Ocean.* **2017**, *122*, 7992–8019. [CrossRef]
11. Madsen, P.A.; Fuhrman, D.R. Run-up of tsunamis and long waves in terms of surf-similarity. *Coast. Eng.* **2008**, *55*, 209–223. [CrossRef]
12. Madsen, P.A.; Fuhrman, D.R.; Schäffer, H.A. On the solitary wave paradigm for tsunamis. *J. Geophys. Res. Ocean.* **2008**, *113*, C12012. [CrossRef]
13. Fan, N.; Jiang, J.; Dong, Y.; Guo, L.; Song, L. Approach for evaluating instantaneous impact forces during submarine slide-pipeline interaction considering the inertial action. *Ocean Eng.* **2022**, *245*, 110466. [CrossRef]
14. Dong, Y.; Cui, L.; Zhang, X. Multiple-GPU for three dimensional MPM based on single-root complex. *Int. J. Numer. Methods Eng.* **2022**, *123*, 1481–1504. [CrossRef]
15. Xie, P.; Chu, V.H. The forces of tsunami waves on a vertical wall and on a structure of finite width. *Coast. Eng.* **2019**, *149*, 65–80. [CrossRef]
16. Smith, L.; Kolaas, J.; Jensen, A.; Sveen, K. X-ray measurements of plunging breaking solitary waves. *Eur. J. Mech. B Fluids* **2019**, *73*, 112–121. [CrossRef]
17. Lin, M.Y.; Liao, G.Z. Vortex shedding around a near-wall circular cylinder induced by a solitary wave. *J. Fluids Struct.* **2015**, *58*, 127–151. [CrossRef]
18. Aristodemo, F.; Tripepi, G.; Meringolo, D.D.; Veltri, P. Solitary wave-induced forces on horizontal circular cylinders: Laboratory experiments and SPH simulations. *Coast. Eng.* **2017**, *129*, 17–35. [CrossRef]
19. Van Gent, M.R.A. Wave Interaction with permeable coastal structures. *Int. J. Rock Mech. Min. Sci. Geomech.* **1996**, *33*, 227A. [CrossRef]
20. Wu, Y.T.; Hsiao, S.C. Propagation of solitary waves over a submerged permeable breakwater. *Coast. Eng.* **2013**, *81*, 1–18. [CrossRef]
21. Qu, K.; Sun, W.Y.; Deng, B.; Kraatz, S.; Jiang, C.B.; Chen, J.; Wu, Z.Y. Numerical investigation of breaking solitary wave runup on permeable sloped beach using a nonhydrostatic model. *Ocean Eng.* **2019**, *194*, 106625. [CrossRef]
22. Jiménez, J.; Uhlmann, M.; Pinelli, A.; Kawahara, G. Turbulent shear flow over active and passive porous surfaces. *J. Fluid Mech.* **2001**, *442*, 89–117. [CrossRef]
23. Liu, W.; Li, X.; Hu, J. Research on flow assurance of deepwater submarine natural gas pipelines: Hydrate prediction and prevention. *J. Loss Prev. Process Ind.* **2019**, *61*, 130–146. [CrossRef]
24. Bruneau, C.; Mortazavi, I. Passive control of the flow around a square cylinder using porous media. *Int. J. Numer. Methods Fluids* **2004**, *46*, 415–433. [CrossRef]

25. Akbari, H.; Pooyarad, A. Wave force on protected submarine pipelines over porous and impermeable beds using SPH numerical model. *Appl. Ocean Res.* **2020**, *98*, 102118. [CrossRef]
26. Vestrum, O.; Kristoffersen, M.; Polanco-Loria, M.A.; Ilstad, H.; Langseth, M.; Børvik, T. Quasi-static and dynamic indentation of offshore pipelines with and without multi-layer polymeric coating. *Mar. Struct.* **2018**, *62*, 60–76. [CrossRef]
27. Vestrum, O.; Langseth, M.; Børvik, T. Finite element analysis of porous polymer coated pipelines subjected to impact. *Int. J. Impact Eng.* **2021**, *152*, 103825. [CrossRef]
28. Klausmann, K.; Ruck, B. Drag reduction of circular cylinders by porous coating on the leeward side. *J. Fluid Mech.* **2017**, *813*, 382–411. [CrossRef]
29. Hirt, C.W.; Sicilian, J.M. A porosity technique for the definition of obstacles in rectangular cell meshes. In Proceedings of the 4th International Conference on Numerical Ship Hydrodynamics, Washington, DC, USA, 24–27 September 1985.
30. Coulombel, J.F.; Lagoutière, F. The Neumann numerical boundary condition for transport equations. *arXiv* **2018**, arXiv:1811.02229. [CrossRef]
31. Ding, D.; Ouahsine, A.; Xiao, W.; Du, P. CFD/DEM coupled approach for the stability of caisson-type breakwater subjected to violent wave impact. *Ocean Eng.* **2021**, *223*, 108651. [CrossRef]
32. Wu, T. A Numerical Study of Three Dimensional Breaking Waves and Turbulence Effects. Ph.D. Thesis, Cornell University, Ithaca, NY, USA, 2004.
33. ANSYS. *ANSYS FLUENT 14.0 Theory Guide. v.14.0.1*; ANSYS, Inc.: Canonsburg, PA, USA, 2011.
34. Sibley, P.O. The Solitary Wave and the Forces It Imposes on a Submerged Horizontal Circular Cylinder: An Analytical and Experimental Study. Ph.D. Thesis, City University London, London, UK, 1991.
35. Zhao, E.J.; Shi, B.; Qu, K.; Dong, W.B.; Zhang, J. Experimental and numerical investigation of local scour around submarine piggyback pipeline under steady current. *J. Ocean Univ. China* **2018**, *17*, 244–256. [CrossRef]
36. Zhao, E.J.; Qu, K.; Mu, L. Numerical study of morphological response of the sandy bed after tsunami-like wave overtopping an impermeable seawall. *Ocean Eng.* **2019**, *186*, 106076. [CrossRef]
37. Zhao, E.J.; Sun, J.K.; Tang, Y.Z.; Mu, L.; Jiang, H.Y. Numerical investigation of tsunami wave impacts on different coastal bridge decks using immersed boundary method. *Ocean Eng.* **2020**, *201*, 107132. [CrossRef]

Journal of
Marine Science
and Engineering

MDPI

Article

Experimental Investigation on Particle Breakage Behavior of Marine Silica Sand under High-Stress Triaxial Shear

Rong Chen [1,2,*], Tong Zhao [1,2], Zhiyong Wu [3], Dongxue Hao [1,2,*], Nan Xue [4] and Chi Yuan [5]

1 Key Lab of Electric Power Infrastructure Safety Assessment and Disaster Prevention of Jilin Province, Northeast Electric Power University, Jilin 132012, China; 13384329601@163.com
2 School of Civil Engineering and Architecture, Northeast Electric Power University, Jilin 132012, China
3 State Grid Jibei Beijing Power Transmission and Transformation Co., Ltd., Beijing 102401, China; 15049335961@163.com
4 Zhejiang Electric Transmission and Transformation Co., Ltd., Hangzhou 310016, China; zettxn@163.com
5 College of Architecture and Civil Engineering, Beijing University of Technology, Beijing 110124, China; yuanc@emails.bjut.edu.cn
* Correspondence: 20112384@neepu.edu.cn (R.C.); 20102291@neepu.edu.cn (D.H.); Tel.: +86-432-6480-6481 (D.H.)

Abstract: There is often obvious particle breakage for silica sand under high-stress, which will lead to the bearing capacity reduction and excessive settlement of the foundation. This paper focuses on the particle breakage characteristics of marine silica sand from the East China Sea under high-stress conditions. A series of conventional triaxial tests for silica sand, including consolidated drained (CD) and consolidated undrained (CU) shear tests, were conducted under the confining pressures in the range of 2–8 MPa to investigate the breakage rule during the shearing process. The developments of particle breakage index B_r with axial strain ε_1 and volumetric strain ε_v present hyperbolic and linear trends, respectively. A hyperbolic model was adopted to describe the relationship of B_r and ε_1 and the corresponding model parameters were obtained. The particle breakage index also has a good correlation with the input work per unit volume under various average stresses, regardless of the stress history. Furthermore, the relationship between the fractal dimension and the particle breakage was studied based on the particle size distribution curve. It is concluded that the fractal dimension increases in an up–convex hyperbolic trend with the increase of particle breakage index. The dividing radius for whether the silica sand particles exhibit the fractal features is determined as approximately 0.4 mm. This is anticipated to provide reference and supplementary test data for analyzing sand constitutive models/environments regarding particle crushing.

Keywords: marine silica sand; mechanical characteristics; particle breakage; fractal distribution; evolution

check for updates

Citation: Chen, R.; Zhao, T.; Wu, Z.; Hao, D.; Xue, N.; Yuan, C. Experimental Investigation on Particle Breakage Behavior of Marine Silica Sand under High-Stress Triaxial Shear. *J. Mar. Sci. Eng.* **2023**, *11*, 1825. https://doi.org/10.3390/jmse11091825

Academic Editor: Carl T. Friedrichs

Received: 12 August 2023
Revised: 15 September 2023
Accepted: 17 September 2023
Published: 19 September 2023

1. Introduction

Silica sand is widely distributed in the offshore area of the East China Sea and is the main engineering material for offshore foundations. It is commonly believed that particle breakage often occurs when silica sands are subjected to high stress [1–8]. The soil at the tip of a deep pile foundation in an offshore oil drilling platform or at the bottom of rockfill dams may bear a high load, with the stress level sometimes reaching approximately 10 MPa [9]. Due to changes in the initial stress state or the groundwater pressure, the earth pressure and water pressure generated by these large loads will cause more particle breakage in silica sand [10,11]. Moreover, continuous particle breakage causes the gradation of silica sand in offshore foundations to change continuously [12,13], affecting its compressibility [14], dilatancy [15], critical state [16], and other mechanical properties, thus significantly changing the engineering characteristics of foundations, and ultimately directly affecting the design scheme of offshore foundation engineering and the safety of the project [17,18].

In the past few decades, various laboratory tests have been conducted to study the mechanical properties of silica sand and the evolution process of particle breakage, including ring shear tests [19,20], direct shear tests [21,22], one-dimensional compression tests [23,24], isotropic compression tests [25–27], conventional triaxial tests [28], true triaxial tests [29], cyclic tests [30–32], and creep tests [33]. Several studies have investigated the effect of particle breakage on the mechanical properties of sand. Kikumoto et al. [34] found that in the e–lgp' plane, the critical-state line moved downward with an increase in the particle breakage index. Through triaxial tests, Carrera et al. [35] found that an increase in particle breakage may lead to an upward movement of the critical-state line. Bandini et al. [36] found that particle breakage in the triaxial shear process led to translation and rotation of the critical-state line; however, the critical-state friction angle did not change. Yu et al. [37] conducted a triaxial test on pre-crushed silica sand and found that particle breakage caused the critical-state line on the e–lgp' plane to shift downward and rotate counterclockwise, while on the q–p' plane, all the critical-state points were on the initial critical-state line; however, the critical-state points varied with particle breakage under different drainage conditions. Silica sand particle breakage is related to factors such as the particle mineral composition, particle shape, effective stress path, and pore water pressure [38,39], and the characteristics of particle breakage can be typically explained by the fractal theory [40,41] and energy theory [42]. Afshar et al. [43] conducted several compression tests using X-ray and scanning electron microscopy and found that in the process of particle breakage, the sphericity and aspect ratio of fine silica sand decreased continuously, and the fractal condition of large granular silica sand ceased. Zhao et al. [44] found that when the particle size of Leighton Buzzard sand is lower than a certain value, regardless of the assumed shape, the characteristic dimension is proportional to the corresponding particle size, and the fractal dimension of the sand remains constant. In offshore foundation engineering applications, stress is a key parameter when considering particle breakage. However, most of the above studies analyzed the amount of particle breakage when the sample was loaded to the failure or critical state and rarely discussed the intermediate development process of particle breakage along specific stress paths [45]; that is, the influence of particle breakage on the strength parameters of the loading process (such as the peak strength of the softened material) could not be truly reflected. Hence, it is necessary to study the evolution of the particle-crushing process and its correlation with the mechanical properties of the silica sand.

Currently, research on particle breakage during the shear process of silica sand is limited. Existing studies have generally focused on the evolution process of the particle breakage of calcareous sand [46] and rockfill [47] under general stress conditions. Whether the crushing theory based on the above materials under general stress conditions can be applied to describe the particle-crushing evolution process of silica sand under high-stress conditions requires further verification. Owing to the strict conditions of high-stress triaxial tests, experiments on the evolution of the particle breakage of silica sand under high-stress conditions are lacking, thus, limited theoretical research on the evolution model of the particle breakage of quartz sand under high-stress conditions. In addition, the fractal characteristics of silica sand under different drainage conditions, particularly high stresses, have not been thoroughly studied.

In this study, a GDS high-stress triaxial test system was used to conduct shear tests on consolidated drained (CD) and consolidated undrained (CU) silica sand specimens at high stresses to investigate the effects of stress level, axial strain, and drainage conditions on the degree of particle breakage. The evolution law of the particle breakage of silica sand during shearing was analyzed, and the parameters of the particle breakage model of silica sand were obtained. The evolution of the fractal dimension in the particle-breaking process was discussed, and the relationship between the volume deformation, fractal dimension, input work, and particle breaking was studied.

2. Materials and Methods

2.1. Triaxial Test Equipment

A British GDS high-stress environment triaxial test system (HPETTS) was used, as shown in Figure 1. The test system was mainly composed of GDSLAB data acquisition software (GDSLab v2.5.4.42), a channel data acquisition board, a pressure/volume controller, a 500 kN VIS load frame, and a three-axis pressure chamber. The three-axis pressure chamber comprised an upper shaft pressure chamber and a lower confining pressure chamber, and the maximum bearing capacity can reach 32 MPa. In addition to the test chamber, the high-stress triaxial test system was equipped with control and data acquisition systems. The control system includes an axial pressure controller, a confining pressure controller, and a reverse pressure controller. Both systems can apply pressures of up to 32 MPa with an accuracy of 0.001%. The stress and volume changes were recorded every second using the data acquisition system. The measuring range of the confining pressure and reverse pressure volume controller was 0–16 MPa, and the measuring range of the axial pressure volume controller was 0–32 MPa. The sensor and data conversion device of the high-stress triaxial test system automatically collected the test data for the deviatoric stress q, effective axial pressure, effective confining pressure, pore pressure u, volume change, and axial strain ε_1.

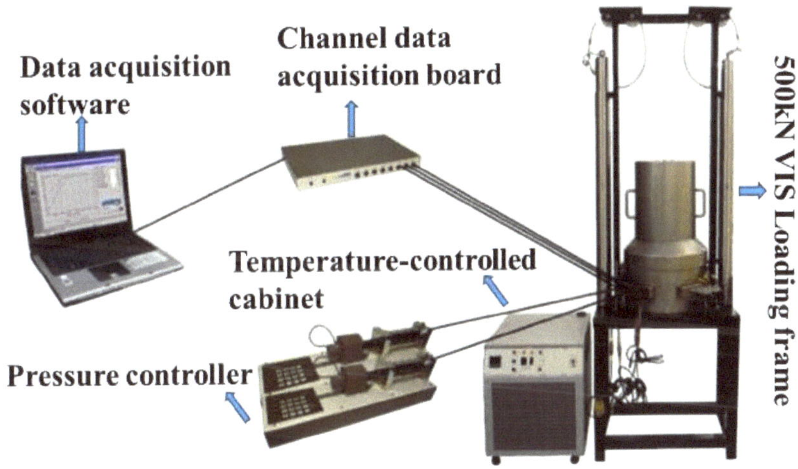

Figure 1. GDS high-stress triaxial test system.

2.2. Test Material and Condition

Sand samples were obtained from Pingtan, Fujian province, near the East China Sea, and the geographic location is shown in Figure 2. The main component of the Fujian sand is quartz. An image of the silica sand sample is shown in Figure 3a. Silica sand has a uniform shape and smooth surface, similar to the characteristics of Leighton Buzzard sand [38,44], Ottawa sand, and Toyoura sand. Figure 3b shows the initial particle size distribution (PSD) curve of the silica sand. Fujian sand was sieved into five-grain groups: d_1 (0.5–0.6 mm), d_2 (0.6–0.7 mm), d_3 (0.7–0.8 mm), d_4 (0.8–0.9 mm), and d_5 (0.9–1 mm). Silica sand was then prepared according to the principle of equal proportion. This ensured that the initial particle composition of each sample was identical. Table 1 lists the physical characteristics of the silica sand.

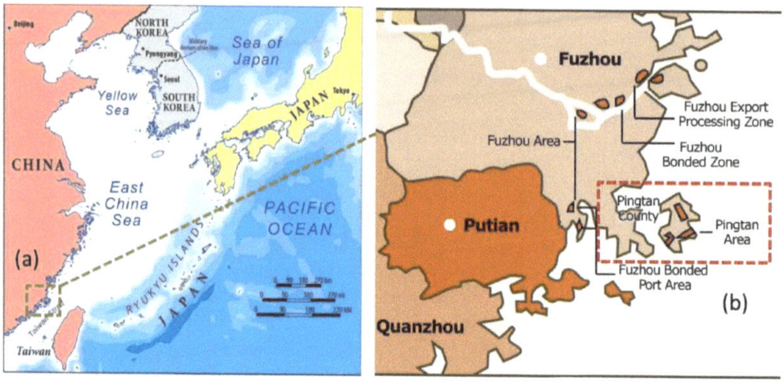

Figure 2. The collection point of sand samples is located on Pingtan Island, Pingtan County, Fuzhou City, Fujian Province, adjacent to the East China Sea. (**a**) East China Sea, (**b**) Pingtan Area.

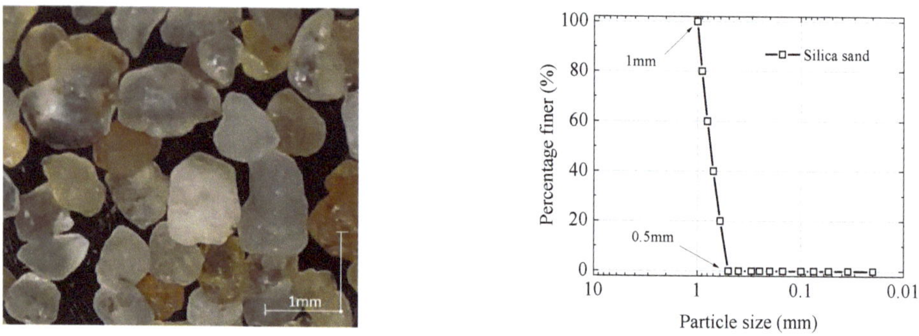

(**a**) (**b**)

Figure 3. (**a**) Image of silica sand sample. (**b**) Particle size distribution curve of silica sand under initial condition.

Table 1. Properties of silica sand.

Property	Value
Specific gravity, G_s	2.65
Average grain size d_{50}: mm	0.75
Coefficient of uniformity, C_u	1.45
Curvature coefficient, C_c	0.96
Maximum void ratio, e_{max}	0.78
Minimum void ratio, e_{min}	0.55
Particle size range, mm	0.5–1.0

2.3. Test Program

Triaxial compression tests under two different drainage conditions (CD and CU tests) were conducted to study the influence of drainage conditions on the mechanical properties and particle breakage of silica sand. A sample with an initial relative density of 75% (dense sand) was prepared using the drop sand method (a reconstruction method that simulates the free-falling behavior of natural objects). The mass of the sand was predetermined, and all the samples prepared had a diameter of 50 mm and a height of 100 mm. Before the triaxial compression test, to ensure that the sand sample was fully saturated, it was subjected to ventilation saturation for 3 h and water head saturation for 10 h. A reverse pressure was applied step-by-step to 500 kPa in increments of 50 kPa for reverse pressure saturation, ensuring that Skempton's B value after the saturation test was greater than 0.97. To prevent the rubber film from being punctured by the corners of silica sand samples under high-pressure conditions, a rubber film with a thickness of 1 mm was adopted. The application of this method to a structural sample can be referred to in [8]. A total of 32 high-pressure triaxial shear tests were conducted, as listed in Table 2 (ε_v is the volumetric strain, D is the fractal dimension, and W is the input work per unit volume), R^2 is the correlation coefficient, and the larger the R^2, the higher the degree of coincidence. To quantify particle breakage after compression, all the tests were performed in increments of 2, 4, 6, and 8 MPa when the specified average effective stress range was 2–8 MPa, and the strain shear rate in the test was 0.05%/min. To analyze the change in the particle breakage during the shearing process, parallel tests with axial strains of 5%, 10%, 15%, and 20% were conducted in each confining pressure test, and the corresponding PSD curves were obtained by performing a sieving analysis.

Table 2. Test program and results of high-stress triaxial tests.

Test (No.)	Type of Shear	Termination Axial Strain (%)	σ_3 (MPa)	B_r	W (MJ/m^3)	D	R^2
TCD-1	CD	5	2	0.040	0.160	0.170	0.940
TCD-2	CD	10	2	0.065	0.319	0.390	0.975
TCD-3	CD	15	2	0.079	0.480	0.550	0.981
TCD-4	CD	20	2	0.097	0.584	0.667	0.970
TCD-5	CD	5	4	0.070	0.367	0.302	0.983
TCD-6	CD	10	4	0.109	0.707	0.691	0.988
TCD-7	CD	15	4	0.130	1.074	0.895	0.981
TCD-8	CD	20	4	0.160	1.323	1.142	0.976
TCD-9	CD	5	6	0.077	0.577	0.519	0.989
TCD-10	CD	10	6	0.013	1.116	0.916	0.988
TCD-11	CD	15	6	0.175	1.713	1.244	0.986
TCD-12	CD	20	6	0.208	2.129	1.387	0.976
TCD-13	CD	5	8	0.099	0.776	0.763	0.975
TCD-14	CD	10	8	0.015	1.538	1.239	0.994
TCD-15	CD	15	8	0.206	2.384	1.514	0.990
TCD-16	CD	20	8	0.248	2.987	1.928	0.980
TCU-1	CU	5	2	0.057	0.176	0.197	0.923
TCU-2	CU	10	2	0.062	0.234	0.319	0.964
TCU-3	CU	15	2	0.067	0.290	0.611	0.978
TCU-4	CU	20	2	0.071	0.341	0.917	0.987
TCU-5	CU	5	4	0.073	0.413	0.143	0.959
TCU-6	CU	10	4	0.084	0.518	0.517	0.973
TCU-7	CU	15	4	0.091	0.617	0.839	0.982
TCU-8	CU	20	4	0.099	0.734	0.967	0.983
TCU-9	CU	5	6	0.093	0.651	0.287	0.960
TCU-10	CU	10	6	0.105	0.798	0.727	0.960
TCU-11	CU	15	6	0.114	0.935	0.964	0.973
TCU-12	CU	20	6	0.119	1.121	1.144	0.971
TCU-13	CU	5	8	0.103	0.879	0.477	0.962
TCU-14	CU	10	8	0.119	1.061	0.865	0.971
TCU-15	CU	15	8	0.123	1.234	1.048	0.972
TCU-16	CU	20	8	0.129	1.484	1.158	0.976

3. Results and Analysis

To study the influence of the stress level, drainage condition, strain termination point, and other factors on the particle crushing characteristics of silica sand, several high-pressure triaxial tests were conducted under CD and CU conditions. After the tests, the corresponding particle size distribution curves were obtained through the sieving analysis. Considering the high workload involved in the tests, the initial relative compactness was set to D_r = 75%, with four pressure levels of 2, 4, 6, and 8 MPa. Two drainage conditions were set for each stress level. To measure the amount of particle breakage during shearing, corresponding parallel tests were performed under different stress levels. The test was stopped at the termination strain point ε_1 = 5%, 10%, 15%, and 20%, and the particle breakage index at the termination strain points was calculated. Table 2 presents the detailed test data.

3.1. Examining Similarity of Parallel Specimens

The degree of particle breakage can be reflected by the change in the particle scale. However, due to the lack of a method for monitoring the change in the particle scale during the test process, the method of testing the parallel sample was used to study the evolution of particle crushing with loading. Several samples with the same initial state were loaded to different axial strains along the same stress path, and then the grading curves under the corresponding axial strains were obtained through indoor screening. By comparing the changes in the grading, the specific particle breakage was obtained under the current axial strain. Figure 4 shows the ratio of the deviatoric stress to the average effective stress under confining pressure q/p'–axial strain ε_1 relationship curve. The graph shows that the difference of q/p'–ε_1 curves under a CD test for different termination strains is small, except when σ_3 = 2 MPa and 8 MPa, axial strain of ε_1 = 5% corresponding to the extreme difference for q/p' of 13.4% and 14.7%, respectively, and the extreme differences for other confining pressures and strains are between 3% and 9%. The reason for this difference is that in the triaxial drained tests, a small difference in the volumetric strain can lead to a large difference in the effective confining pressure, resulting in a large deviation in the stress–strain curves. Therefore, the stress–strain difference of the parallel specimens in the CD test is acceptable. Overall, the extreme differences of q/p' for the parallel tests are all less than 15%, indicating that the repeatability of the CD test is good at different axial strain termination points under different confining pressures. For the CU test, the four curves under σ_3 = 4 MPa have the largest extreme difference of 9.2% when ε_1 = 5%, and the range of the other conditions is 1.2–7.2%, indicating that the CU test with different strain termination points under different confining pressures has good repeatability.

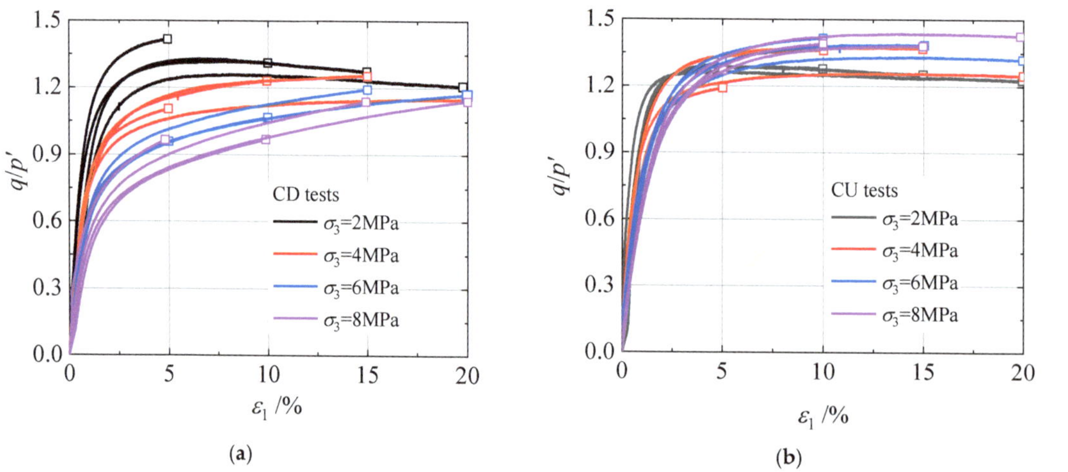

Figure 4. Relationships between q/p' and ε_1 with various confining pressures. (**a**) CD tests, (**b**) CU tests.

Figure 5 shows the effective stress paths under different confining pressures during the CD and CU tests. The hollow circles in the figure represent the points at which the axial strains were 5%, 10%, 15%, and 20%. The peak strength points under different confining pressures are between the particle breakage measurement points and close to the axial strain when the test was stopped.

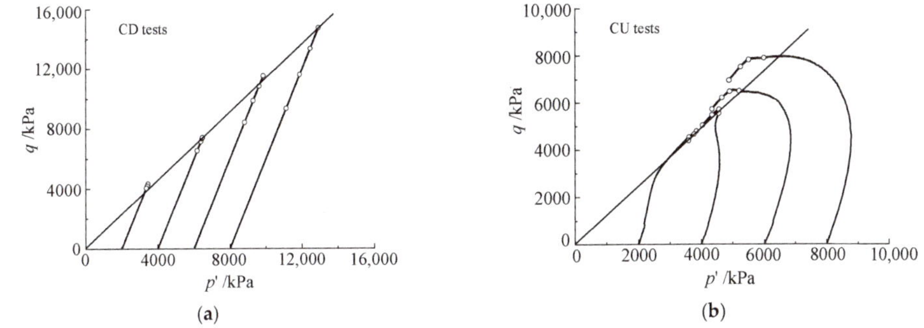

Figure 5. Stress paths under various confining pressures. (**a**) CD tests, (**b**) CU tests.

3.2. Mechanical Properties of Silica Sand

Figure 6a shows the deviatoric stress–axial strain relationship curves of silica sand. The deviatoric stress and initial modulus increase with an increase in the confining pressure. When σ_3 = 2 MPa, the $q - \varepsilon_1$ curve shows a softening characteristic, and the deviatoric stress reaches a peak value at ε_1 of 6%, then gradually decreases, but still does not reach a stable critical state even at ε_1 = 20%. When $\sigma_3 \geq 4$ MPa, the $q - \varepsilon_1$ curves of the sand showed different degrees of strain hardening; that is, the deviatoric stresses increase with the increase in the axial strain. The greater the confining pressure, the more evident the hardening feature for the stress–strain curve. Figure 6b shows the deviatoric stress–axial strain relationship curves obtained from the CU tests. The deviatoric stress increases with the increase in the confining pressure, and the strain at the peak point gradually advances. With the increase in the axial strain, the deviatoric stress curves under different confining pressures all show a strain-softening behavior.

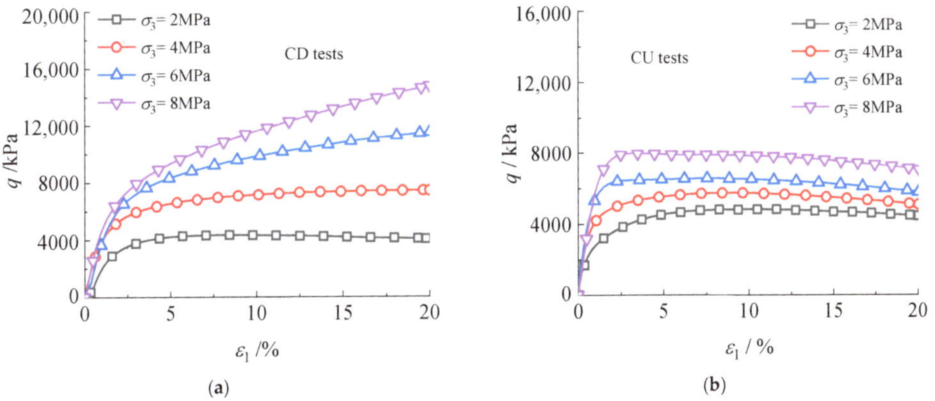

Figure 6. Relationships between q and ε_1 under various confining pressures. (**a**) CD tests, (**b**) CU tests.

Figure 7a shows the volumetric strain–axial strain ($\varepsilon_v - \varepsilon_1$) relationship curves for different confining pressures; shear contraction is positive, and shear expansion is negative. As shown, the sample undergoes a second phase change at σ_3 = 2 MPa, in which the sample presents reduced volume first, followed by shear expansion. With the increase in the axial strain, the sample again exhibits shear contraction. The corresponding axial strains of the two phase-change points in the diagram are 2.5% and 15.9%, respectively. When σ_3 > 2 Mpa, the sand sample was subjected to shear contraction with an increase in the axial strain; the greater the confining pressure, the more significant the shear contraction. Figure 7b shows the pore pressure in the CU test and axial strain relationship curve. From the graph, when σ_3 = 2 Mpa, the pore pressure is positive initially, and it reached a maximum at ε_1 = 1%, and then began to decrease, gradually decreasing to a negative value, reaching a minimum at ε_1 = 12.8% and then increasing. Similar to the CD test, two-phase transitions can be observed in the test process, indicated by black circles. When σ_3 > 2 Mpa, with an increase in the axial strain, the pore pressure increases rapidly in the early stage and is always positive in the shear process; the greater the confining pressure, the higher the pore pressure. Under different confining pressures, the pore pressure, after reaching the peak deviatoric stress, gradually tended to be flat but continued to increase gradually.

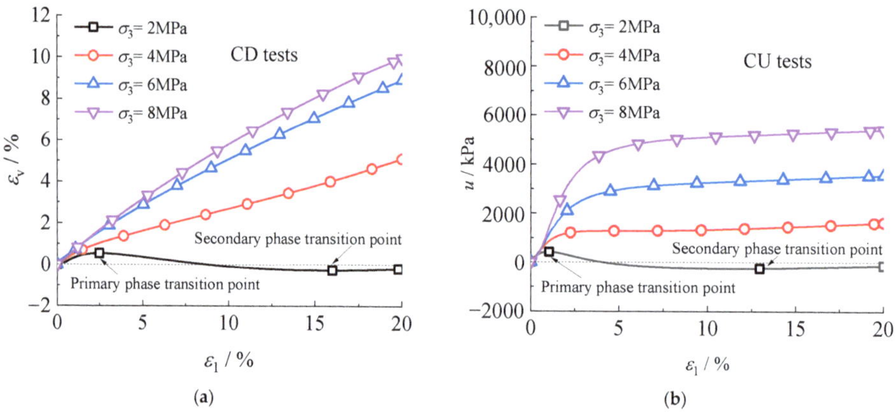

Figure 7. Relationships of $\varepsilon_v - \varepsilon_1$ and $u - \varepsilon_1$ with various confining pressures. (**a**) CD tests, (**b**) CU tests.

Figure 8 shows the relationship between the void ratio *e* of silica sand samples and the effective mean stress p' under different confining pressures. It can be seen that the void ratio gradually decreases with the increase of mean effective stress during the consolidation compression process, and a steep drop does not occur. The decrease in void ratio is commonly due to compaction and particle breakage, and particle breakage is the main factor leading to a sharp drop in the e–lgp' curve [37,48]. Therefore, it is inferred from the curve's characteristics that the particle breakage during consolidation is extremely small and can be ignored.

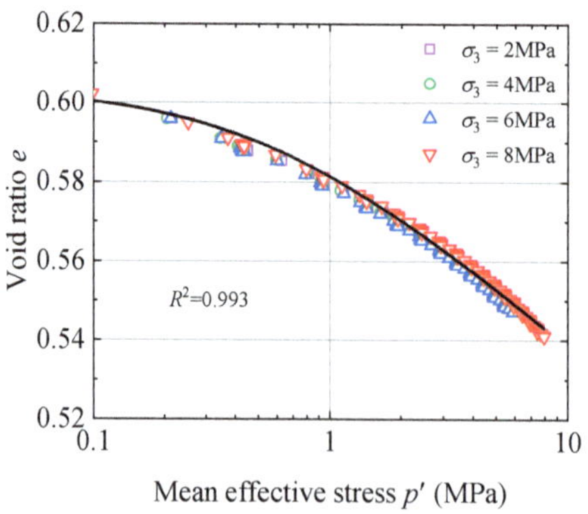

Figure 8. Void ratio versus mean effective stress during consolidation.

The relationship between the void ratio e and the mean effective stress p' is expressed in Equation (1):

$$e = 0.579 - 0.015 \times \ln\left(\frac{p' + 0.187}{1 + 0.187}\right) \tag{1}$$

3.3. Evolution of PSD Curves with Axial Strain

Figure 9 shows the distribution of the particle size before and after the test under different confining pressures. Under the high-pressure conditions, the fine particles produced by the crushing of each sample increased significantly. With the increase in the axial strain, the content of large particles decreased while that of the small particles increased, and the greater the axial strain or confining pressure, the more significant this phenomenon was. The grading of the samples changed to a wide particle size distribution.

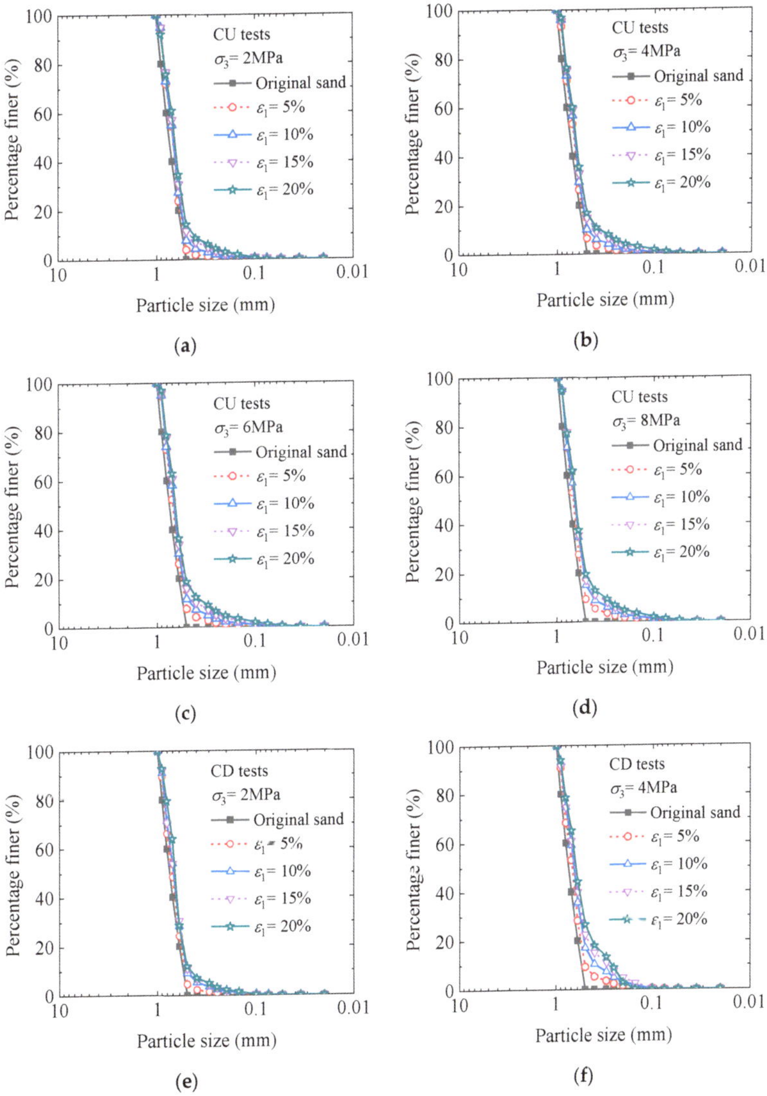

Figure 9. *Cont.*

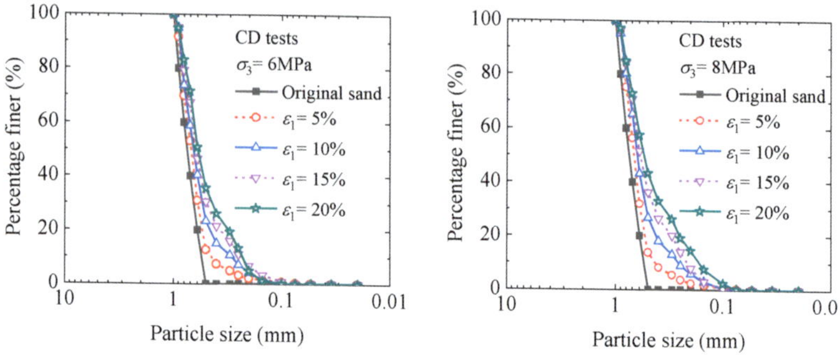

(g) (h)

Figure 9. Particle size distribution curves after triaxial tests. (**a**) CU test under σ_3 = 2 MPa, (**b**) CU test under σ_3 = 4 MPa, (**c**) CU test under σ_3 = 6 MPa, (**d**) CU test under σ_3 = 8 MPa, (**e**) CD test under σ_3 = 2 MPa, (**f**) CD test under σ_3 = 4 MPa, (**g**) CD test under σ_3 = 6 MPa, and (**h**) CD test under σ_3 = 8 MPa.

To quantify the particle breakage degree of silica sand under high stresses, this study adopted the particle breakage index theory proposed by Hardin [42], as shown in Figure 10. The particle breakage index, B_r is given by Equation (2):

$$B_r = \frac{B_t}{B_p} = \frac{S_{<ABCD>}}{S_{<ABCF>}} \tag{2}$$

where B_t is the area enclosed by the initial and shear grading curves, and B_p is the area enclosed by the initial grading curve and a straight line with a particle size of 0.074 mm.

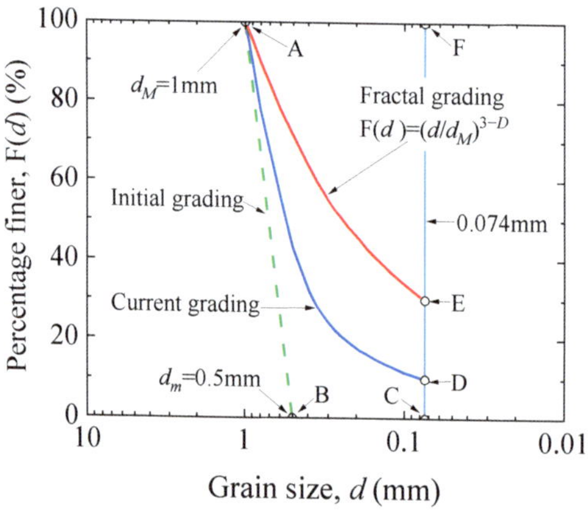

Figure 10. Particle breakage index definition proposed by Hardin [42].

Based on the definition of the particle breakage index B_r proposed by Hardin, that is, Equation (2), the relationship curves between the B_r and the axial strain ε_1 of silica sand in the CD and CU tests were plotted, as shown in Figure 11. The particle breakage index B_r increases with increasing axial strain. The particle breakage index B_r of the CU test under each confining pressure is between the range of the values of B_r for σ_3 = 2–4 MPa under the CD test. When σ_3 = 2 MPa, the difference in particle breakage index in the shear process of the CD and CU tests is small. The difference in the particle breakage index between the CD and CU tests increases with the axial strain and confining pressure when σ_3 > 2 MPa, that is, the particle breakage index of the CD test is significantly higher than that of the CU test. This is mainly because the positive pore pressure produced in the CU test reduces the effective stress, thereby inhibiting particle breakage. The higher the confining pressure, the greater the positive pore pressure, and the greater the difference in the particle breakage under different drainage conditions.

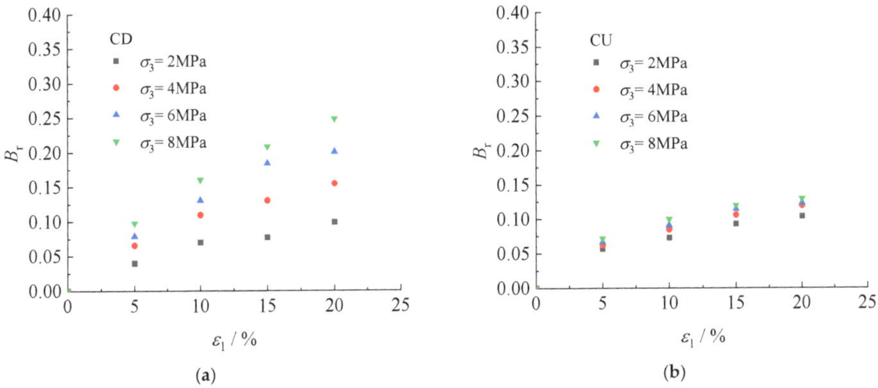

Figure 11. Relationship between B_r and ε_1 under various confining pressures applied in the CD and CU tests. (**a**) CD test (**b**) CU test.

3.4. Correlations between Relative Breakage and Input Work and Volumetric Strain

Particle breakage in the shearing process of silica sand is due to the continuous work of the external forces on the material, and this process is irreversible; therefore, there is a good correspondence between the particle breakage index and input work. The total input work was mainly converted into the work consumed by particle breakage, frictional, and particle rearrangement. In this study, the input work W per unit volume is defined as follows [49]:

$$W = \int (p\prime d\varepsilon_v + q d\varepsilon_1) \tag{3}$$

where p' is the effective mean stress, ε_v is the volumetric strain increment, q is deviatoric stress, and ε_1 is the axial strain.

Figure 12 shows the relationship between the particle breakage index of all the samples and the input work per unit volume. Without considering the influences of the stress level, stress path, and end-strain point, the particle breakage index shows a hyperbolic nonlinear increase with increasing input work. The relationship between the particle breakage index B_r and input work W per unit volume is given by Equation (4):

$$B_r = \frac{W}{\chi_w + \zeta_w W} \tag{4}$$

where W is the unit volume input work (MJ/m^3 or MPa). The method of input work is like Equation (3).

The particle breakage index B_r is the derivative of the unit volume of input work, which is the initial tangent modulus of W = 0, that is $k_{B_{r0}}$.

$$\frac{1}{\chi_W} = \frac{\mathrm{d}B_r}{\mathrm{d}W} = \lim_{W \to 0} \frac{1}{\chi_W + \zeta_W W} \triangleq k_{B_{r0}} \tag{5}$$

When W approaches infinity, the maximum particle crushing rate can be obtained using the limit of B_r, that is $k_{B_{ru}}$:

$$\frac{1}{\zeta_W} = \lim_{W \to \infty} \frac{W}{\chi_W + \zeta_W W} \triangleq k_{B_{ru}} \tag{6}$$

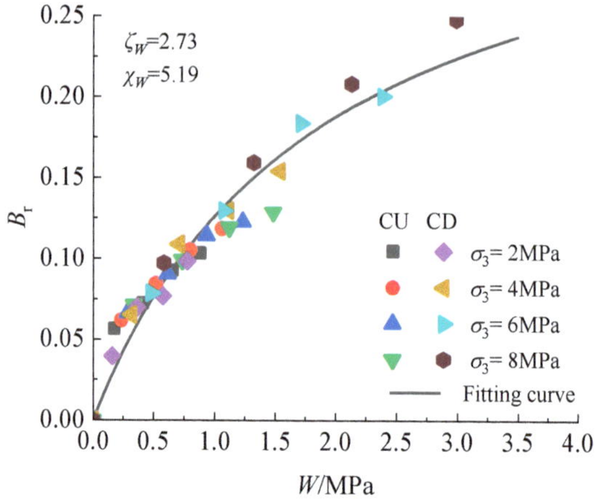

Figure 12. Relationship between the particle breakage index and input energy.

In the equation, the material parameters χ_W and ζ_W are 5.19 and 2.73, respectively, $k_{B_{r0}}$ is the reciprocal of χ_W, and $k_{B_{ru}}$ is the reciprocal of ζ_W.

Figure 13 shows the relationship between the particle breakage index and the volumetric strain of silica sand. Regardless of the stress level, stress path, and end-strain point, the particle breakage index increased linearly with the increase in the volumetric strain. This relationship has also been observed in triaxial tests on rockfills under drainage conditions [46], calcareous sand drainage triaxial tests [45], and silica sand compression tests with different stress paths [38]. The slope of the fitted curve in this study was greater than the value (0.018) adopted by Wu et al. [8] for dense specimens. This is because the particle coordination number (the number of accessible particles around the particle) increases with relative compactness, and this effect reduces the average contact force on the silica sand particles. The relationship between the particle breakage index and the volumetric strain of silica sand can be expressed as follows:

$$B_r = 0.038 + 2.0\varepsilon_v \tag{7}$$

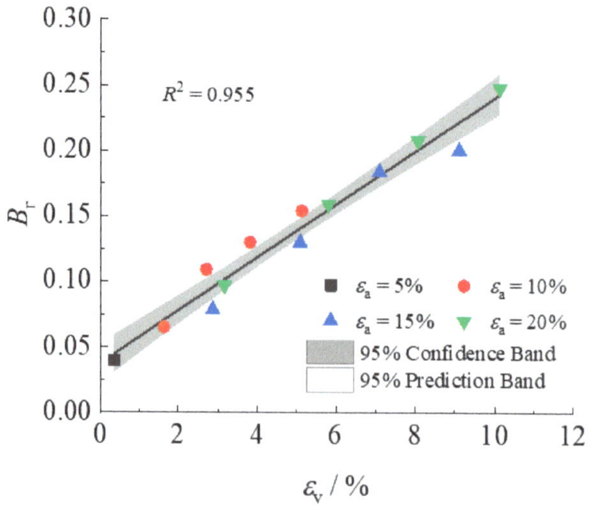

Figure 13. Relationship between the particle breakage index and the volumetric strain.

3.5. Evolution Law of Particle Breakage of Silica Sand

The particle breakage evolution law of geotechnical materials is nonlinear, with an increase in the axial strain during soil shearing. The particle breakage evolution law proposed by several researchers can be applied to the analysis of rockfills [47] and calcareous sands [46]. However, whether it can be applied to silica sand, how to select the relevant model parameters, and whether the fracture evolution process of silica sand under high pressures can be reasonably described remains to be verified. Therefore, based on the test results of all the samples in this study, we found that the particle breakage and the axial strain of the silica sand soil exhibited a good hyperbolic relationship. The relationship between the particle breakage index and the axial strain of silica sand soil can be expressed as follows:

$$B_r = \frac{\varepsilon_a}{1/R_{B0} + \varepsilon_a/\overline{B_r}} \tag{8}$$

Here, ε_α is the axial strain, $\overline{B_r}$ is the maximum particle breakage index produced under the current confining pressure, and R_{B0} represents the initial growth rate of particle breakage when $\varepsilon_\alpha = 0$. The expressions for these two physical quantities are as follows:

$$R_{B0} = c_1 \left(p'/p_a\right)^{c_2} \tag{9}$$

$$\overline{B_r} = \frac{p'}{Ap_a + p'}\overline{\overline{B_r}} \tag{10}$$

Here, c_1 and c_2 are the material parameters; P_a is the standard atmospheric pressure; A is a dimensionless parameter; and $\overline{\overline{B_r}}$ is the final particle breakage index, representing the final degree of particle breakage that can be produced under extremely high confining pressure and continuous loading conditions. For the Hardin particle breakage index, it is believed that when sand is subjected to sufficient pressure, particles of any size will eventually break into particles with a size lower than 0.075 mm, which means that $\overline{\overline{B_r}} = 1$.

Figure 14 shows the fitting effect of the particle breakage model. In both the CD and CU tests, the particle breakage model results are in good agreement with the test results, indicating that the model can accurately describe the particle breakage change process of silica sand at high stresses under CD and CU conditions. The model parameters c_1, c_2, and A of silica sand soil were 0.43, 0.38, and 96.8, respectively, as listed in Table 3.

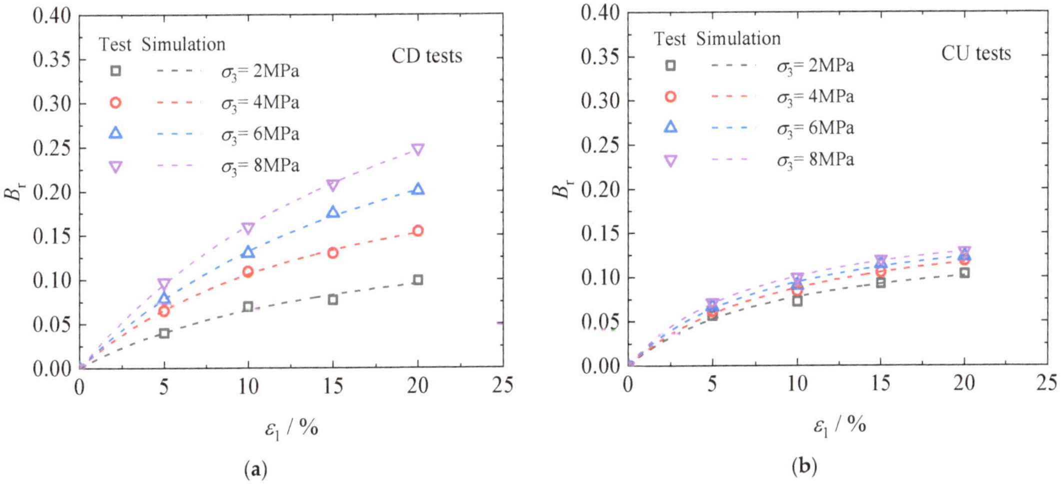

Figure 14. Performance of breakage model. (**a**) CD tests, (**b**) CU tests.

Table 3. Parameters of the particle breakage model.

c_1	c_2	A	$\overline{\overline{B_r}}$	R^2
0.43	0.38	96.8	1	0.997

3.6. Effect of Particle Breakage on Fractal Dimension

The compaction and crushing of the silica sand particles can be considered an energy-dissipation process with self-similar characteristics [50]. Therefore, the fractal theory was applied to describe the particle size distribution of the silica sand after compaction. Based on the associated particle number and the characteristic scale [51], the basic definition of a fractal can be obtained as follows:

$$\frac{M_d(x < d)}{M_T} = (\frac{d}{d_{\max}})^{3-D} \tag{11}$$

where D is the fractal dimension, M_d is the particle mass with a radius less than r, M_T is the total mass, and $d_{\max}$ is the dimension of the largest particle.

Figure 15 shows the fractal distribution of the silica sand samples under different stress levels. Table 2 presents the fractal dimensions. The fractal dimension in each graph increases with an increase in the fine particle content, and the fractal characteristics become more evident with an increase in the termination strain. For silica sand grains, the PSD curves show self-similarity at high stresses owing to particle breakage.

Figure 16 shows the fractal distribution of silica sand samples under different effective mean stresses. Under the same effective confining pressure conditions, the fractal degree of the CD tests is obviously greater than that of the CU tests for the cases of $\sigma_3 > 2$ MPa. This is because the confining pressures in the CD tests for $\sigma_3 > 2$ MPa are higher, causing the fractal dimension to extend in a larger direction.

Figure 17 shows the curve of the particle breakage index and fractal dimensions of silica sand samples under different stress ratios. As shown in the figure, the fractal dimension gradually increases with an increase in the relative particle crushing amount, exhibiting hyperbolic characteristics consistent with the results of the one-dimensional compression test of the silica sand. The crushing strength of particles depends on their size and coordination number, and the dominance of the coordination number on the particle size makes silica sand particles exhibit fractal characteristics in essence and have real fractal dimensions [44]. The force distribution of the large particles with high coordination numbers is uniform, and the probability of breakage is significantly lower than that of small particles with low coordination numbers. Therefore, particle breakage is dominant in smaller particles, satisfying the fractal condition and continuously protecting larger particles [43]. In the shearing process, with an increase in the axial strain, the increase in the relative fractal dimension of the particles decreases to zero (i.e., the limiting fractal dimension). The relationship between the particle breakage index and the fractal dimension is expressed as follows:

$$D = \frac{B_r}{0.192B_r + 0.056} - 0.606 \tag{12}$$

Figure 18 shows the particle size distribution of silica sand under different stress levels. As shown, with an increase in the stress level, the particle size distribution of the silica sand changes from a uniform distribution to a uniform gradient distribution. The higher the stress level, the greater the shift in the particle size distribution curve toward the direction of a greater uniform gradient.

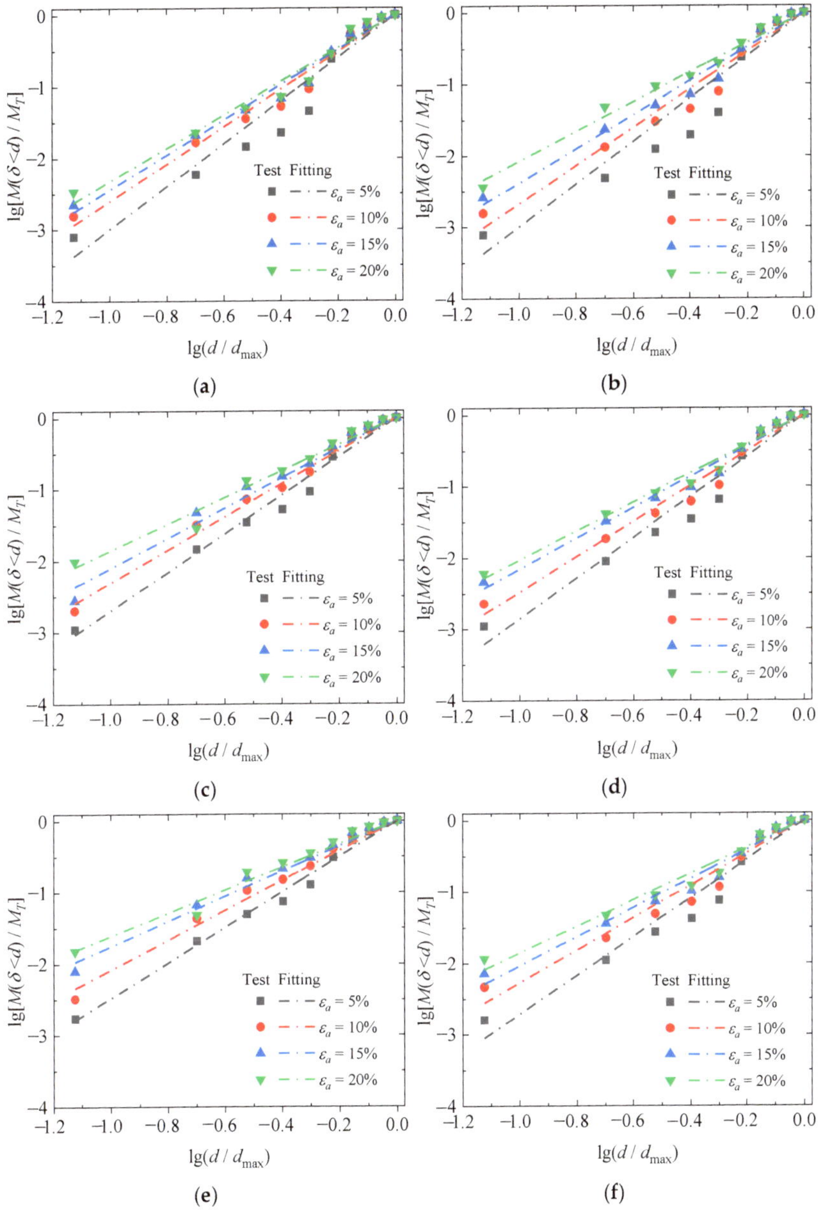

Figure 15. *Cont.*

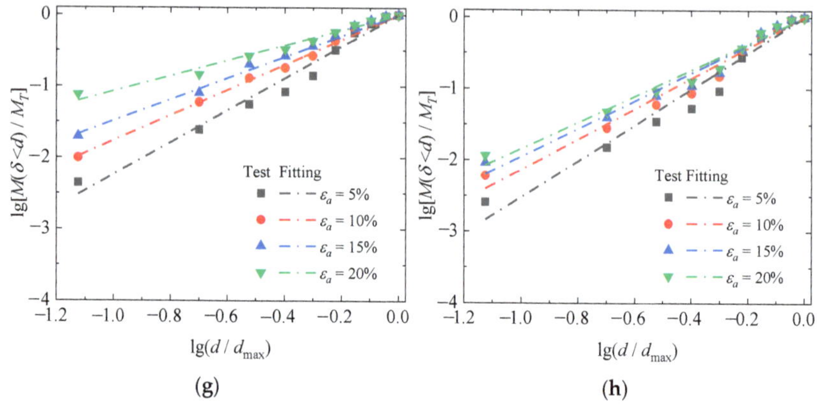

Figure 15. Relationship between $\lg[M(\delta < d)/M_T]$ and $\lg(d/d_{max})$ at different axial strains. (**a**) CD test under σ_3 = 2 MPa, (**b**) CU test under σ_3 = 2 MPa, (**c**) CD test under σ_3 = 4 MPa, (**d**) CU test under σ_3 = 4 MPa, (**e**) CD test under σ_3 = 6 MPa, (**f**) CU test under σ_3 = 6 MPa, (**g**) CD test under σ_3 = 8 MPa, and (**h**) CU test under σ_3 = 8 MPa.

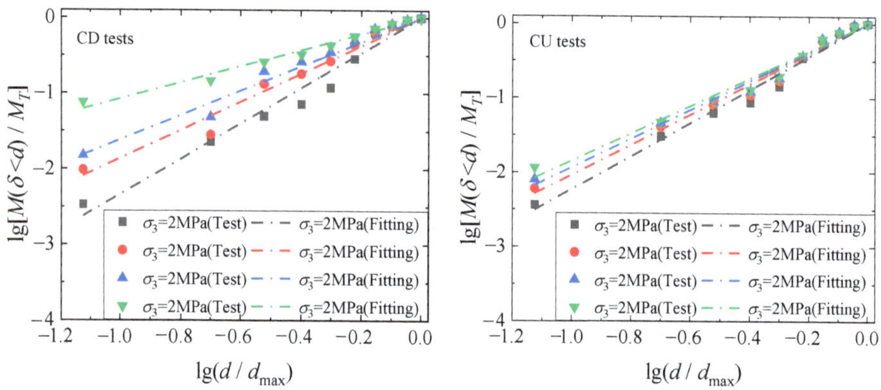

Figure 16. Relationship curves of $\lg[M(\delta < d)/M_T]$ and $\lg(d/d_{max})$ with various confining pressures.

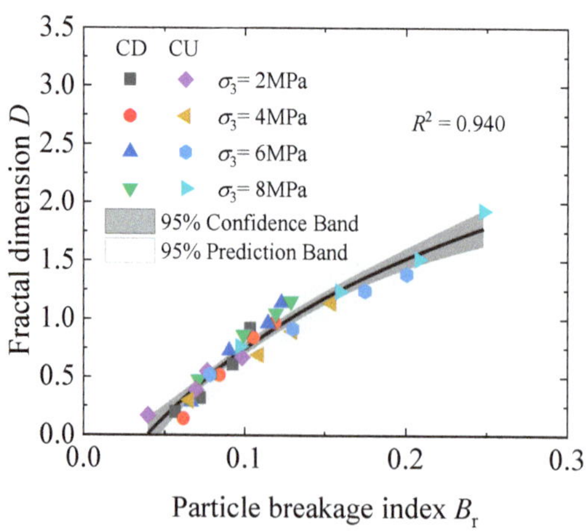

Figure 17. Relationship between fractal dimension and the particle breakage index.

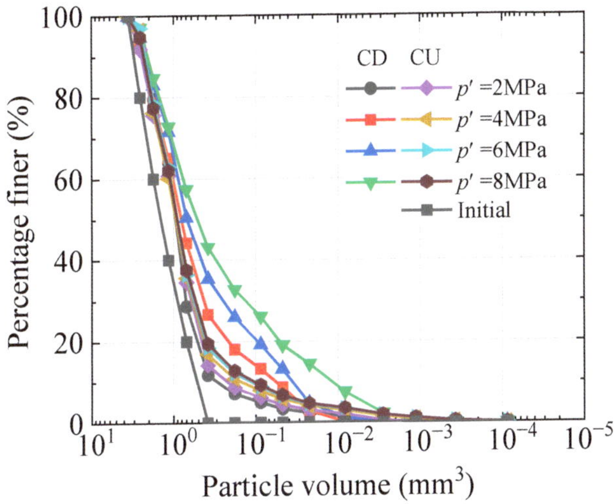

Figure 18. Particle size distributions of silica sand under different stress levels.

The fractal dimension was measured using the relationship between the number of particles and their sizes. The particle radius r is equivalent to the radius of a sphere with the same particle volume, and N is the number of particles. The relationship in logarithmic coordinates is expressed as follows:

$$3\lg(r) + \lg N(> r) \propto (3 - D)\lg(r) \quad (13)$$

Figure 19 shows the fractal distribution of the particle sizes of silica sand at different stress levels. The line with a 3D slope in the figure represents the fractal case of a particle. The silica sand particles exhibited self-similarity under different drainage conditions, and the fractal dimension increased with an increase in the particle breakage index. Some large particles with limited fragmentation terminated the particle fractal condition, and it was determined that the dividing line between silica sand particles with fractal characteristics and those without fractal characteristics was approximately 0.4 mm, similar to the results of the 1D compression test of Leighton Buzzard sand (LBS) reported by Zhao et al. [38].

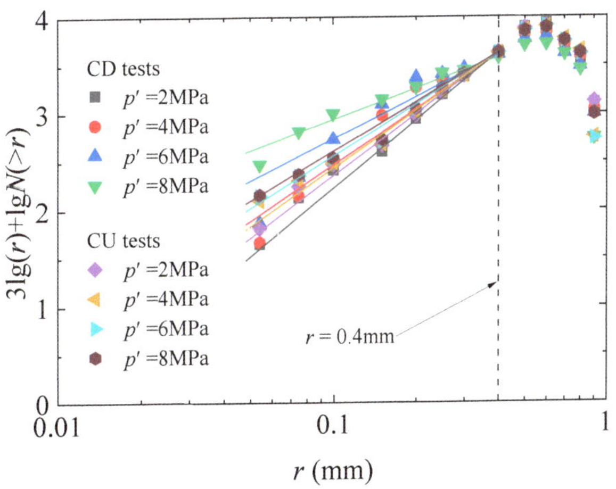

Figure 19. Fractal distributions for silica sand under different stress levels.

Figure 20 shows a histogram of the particle content of silica sand particles with a radius of no more than 0.4 mm ($r \leq 0.4$ mm) obtained from the high-stress triaxial test of CD and CU. Clearly, the content w of a particle with a radius $r \leq 0.4$ mm increases with the increase of the confining pressure σ_3 when the drainage conditions are the same. When the confining pressure is $\sigma_3 = 2$ MPa, the difference of sand particle content w with $r \leq 0.4$ mm between the CD test and the CU test is small; they are 11.95% and 14.52%, respectively. With the increase of confining pressure, the sand particle content w with $r \leq 0.4$ mm in the CD test obviously increases, while the content w in the CU test gradually increases. This is attributed to the different effective stresses generated under different drainage conditions.

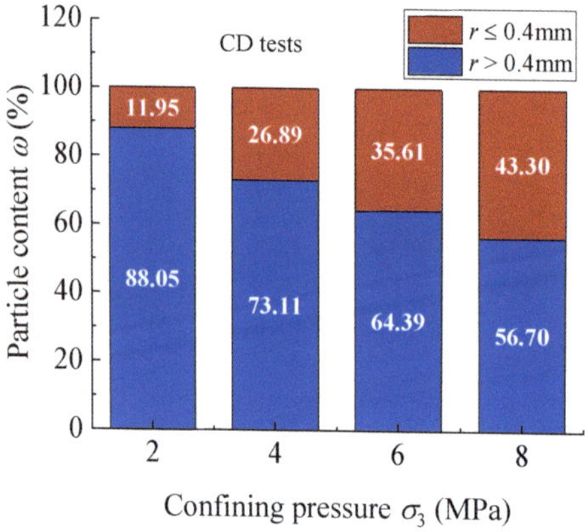

(a)

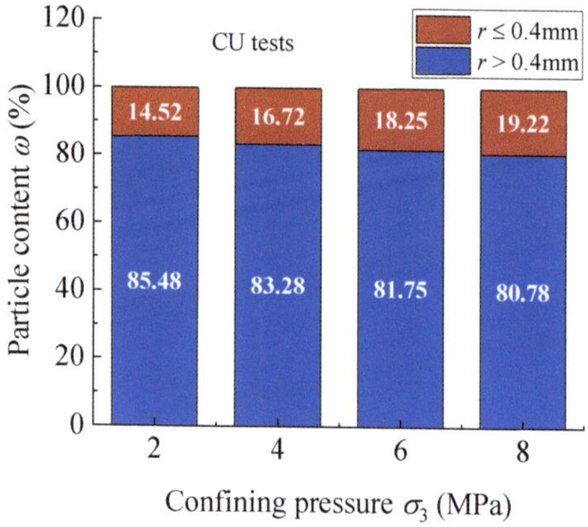

(b)

Figure 20. Bar graphs of the evolution of the particle content (**a**) CD tests, (**b**) CU tests.

Figure 21 shows the relationship between the fractal dimensions of the particles with a radius less than 0.4 mm and the fine particle content. Clearly, the fractal dimension increases with the increase in the silica sand particle content ($\omega \leq 0.4$), the slope of the curve is linear, and the linear growth form is independent of the stress path and stress level. The fitting line is expressed as follows:

$$D = 0.477 + 0.041\omega_{\leq 0.4} \tag{14}$$

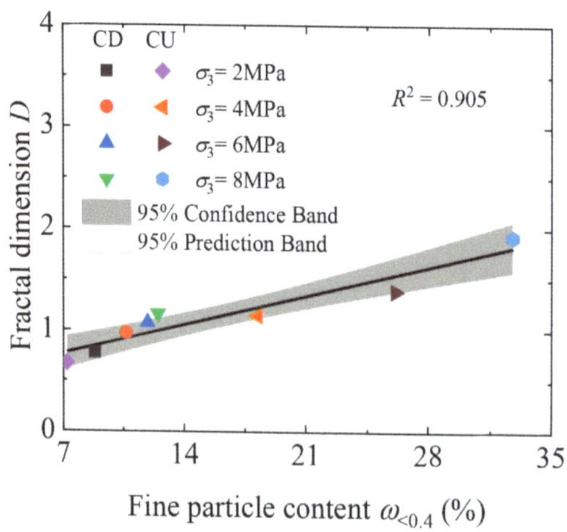

Figure 21. Relationship between the fractal dimension and the fine particle content ($r < 0.4$ mm).

4. Conclusions

In this study, high-stress triaxial shear tests were conducted on silica sand, and the evolution law of particle breakage during shearing under different drainage conditions was obtained. The relationship between the particle breakage and fractal dimension during shearing was studied. The conclusions obtained from the tests are as follows:

- In the CD and CU tests, the particle breakage index B_r of silica sand increased with increasing confining pressure and axial strain. However, the particle breakage index in the CD test was more evidently affected by the confining pressure and shearing process, whereas that in the CU test exhibited relatively small changes and was generally lower than that in the CD test, mainly because the pore water pressure generated in the CU test reduced the effective stress, thus significantly inhibiting the influence of the confining pressure on particle breakage.
- In the high-stress shearing process, the particle breakage of silica sand increased with an increase in the axial strain in a hyperbolic form, and a mathematical model was developed to describe the change in the particle breakage index of silica sand under CD and CU conditions. A hyperbolic model was proposed to describe the relationship between the particle breakage index amount and the input work per unit volume under different drainage conditions.
- An up-convex hyperbolic model was proposed to correlate all the test results of the fractal dimension and relative fragmentation. The fractal feature terminated at the radius of the particles that were broken to a certain extent. The dividing line between silica sand particles with the fractal features and those without the fractal features was approximately 0.4 mm. For particles with radii less than or equal to 0.4 mm, the fractal dimension increased linearly with increasing particle content.

Author Contributions: Conceptualization, formal analysis, and writing—review and editing, R.C. and D.H.; investigation and writing—original draft preparation, T.Z. and Z.W.; investigation and data curation, T.Z., Z.W., C.Y. and N.X. All authors have read and agreed to the published version of the manuscript.

Funding: This research was funded by the National Natural Science Foundation of China (grant number 52078108); Jilin Province Youth Science and Technology Innovation Leader, Team Project of Provincial Department of Science and Technology (grant number 20210509058RQ); and the Scientific Research Project of Jilin Provincial Department of Education (grant number JJKH20210103KJ).

Institutional Review Board Statement: Not applicable.

Informed Consent Statement: Not applicable.

Data Availability Statement: Authors have confirmed the upload data.

Conflicts of Interest: The authors declare no conflict of interest.

References

1. Hagerty, M.M.; Hite, D.R.; Ullrich, C.R.; Hagerty, D.J. One-dimensional high-pressure compression of granular media. *J. Geotech. Eng.* **1993**, *119*, 1–18. [CrossRef]
2. Wang, Y.; Zhang, S.; Ao, D.H.; Yu, Y.; Sun, X. Particle breakage characteristics of rockfills under complex stress paths. *Chin. J. Geotech. Eng.* **2018**, *40*, 698–706.
3. Dong, Y.; Wang, D.; Randolph, M.F. Investigation of impact forces on pipeline by submarine landslide with material point method. *Ocean. Eng.* **2017**, *146*, 21–28. [CrossRef]
4. Sun, Q.L.; Wang, Q.; Shi, F.Y.; Alves, T.; Gao, S.; Xie, X.N.; Wu, S.G.; Li, J.B. Runup of landslide-generated tsunamis controlled by paleogeography and sea-level change. *Commun. Earth Environ.* **2022**, *3*, 244. [CrossRef]
5. Fan, N.; Jiang, J.; Nian, T.; Dong, Y.; Guo, L.; Fu, C.; Tian, Z.; Guo, X. Impact action of submarine slides on pipelines: A review of the state-of-the-art since 2008. *Ocean. Eng.* **2023**, *286*, 115532. [CrossRef]
6. Xiao, Y.; Desai, C.S.; Daouadji, A.; Stuedlein, A.W.; Liu, H.; Abuel-Naga, H. Grain crushing in geoscience materials–Key issues on crushing response, measurement and modeling: Review and preface. *Geosci. Front.* **2020**, *11*, 363–374. [CrossRef]
7. Yamamuro, J.A.; Lade, P.V. Drained sand behavior in axisymmetric tests at high pressures. *J. Geotech. Eng.* **1996**, *122*, 109–119. [CrossRef]

8. Wu, Y.; Yamamoto, H.; Cui, J.; Cheng, H. Influence of load mode on particle crushing characteristics of silica sand at high stresses. *Int. J. Geomech.* **2020**, *20*, 04019194.
9. Yu, F. Characteristics of particle breakage of sand in triaxial shear. *Powder Technol.* **2017**, *320*, 656–667. [CrossRef]
10. Yu, F. Particle breakage and the drained shear behavior of sands. *Int. J. Geomech.* **2017**, *17*, 04017041. [CrossRef]
11. Yu, F. Particle breakage and the undrained shear behavior of sands. *Int. J. Geomech.* **2018**, *18*, 04018079. [CrossRef]
12. Nakata, Y.; Hyodo, M.; Hyde, A.F.L.; Kato, Y.; Murata, H. Microscopic particle crushing of sand subjected to high pressure one-dimensional compression. *Soils Found.* **2001**, *41*, 69–82. [CrossRef]
13. Lade, P.V.; Yamamuro, J.A.; Bopp, P.A. Significance of particle crushing in granular materials. *J. Geotech. Eng.* **1996**, *122*, 309–316. [CrossRef]
14. Mehta, A.A.; Patel, A. An investigation on the particle breakage of Indian River sands. *Eng. Geol.* **2018**, *233*, 23–37. [CrossRef]
15. Xiao, Y.; Liu, H.; Ding, X.; Chen, Y.; Jiang, J.; Zhang, W. Influence of particle breakage on critical state line of rockfill material. *Int. J. Geomech.* **2016**, *16*, 04015031. [CrossRef]
16. Xiao, Y.; Liu, H. Elastoplastic constitutive model for rockfill materials considering particle breakage. *Int. J. Geomech.* **2017**, *17*, 04016041. [CrossRef]
17. Yu, F. Influence of particle breakage on behavior of coral sands in triaxial tests. *Int. J. Geomech.* **2019**, *19*, 04019131. [CrossRef]
18. Wu, Y.; Li, N.; Wang, X.; Cui, J.; Chen, Y.; Wu, Y.; Yamamoto, H. Experimental investigation on mechanical behavior and particle crushing of calcareous sand retrieved from South China Sea. *Eng. Geol.* **2021**, *280*, 105932. [CrossRef]
19. Miao, G.; Airey, D. Breakage and ultimate states for a carbonate sand. *Géotechnique* **2013**, *63*, 1221–1229. [CrossRef]
20. Liu, L.; Cai, G.; Liu, S. Compression properties and micro-mechanisms of rubber-sand particle mixtures considering grain breakage. *Constr. Build. Mater.* **2018**, *187*, 1061–1072. [CrossRef]
21. Luzzani, L.; MR, C. On the relationship between particle breakage and the critical state of sands. *Soils Found.* **2002**, *42*, 71–82. [CrossRef]
22. Coop, M.R.; Sorensen, K.K.; Bodas Freitas, T.; Georgoutsos, G. Particle breakage during shearing of a carbonate sand. *Géotechnique* **2004**, *54*, 157–163. [CrossRef]
23. Nakata, Y.; Kato, Y.; Hyodo, M.; Hyde, A.F.; Murata, H. One-dimensional compression behaviour of uniformly graded sand related to single particle crushing strength. *Soils Found.* **2001**, *41*, 39–51. [CrossRef]
24. Mesri, G.; Vardhanabhuti, B. Compression of granular materials. *Can. Geotech. J.* **2009**, *46*, 369–392. [CrossRef]
25. Mun, W.; McCartney, J.S. Roles of particle breakage and drainage in the isotropic compression of sand to high pressures. *J. Geotech. Geoenviron. Eng.* **2017**, *143*, 04017071. [CrossRef]
26. Altuhafi, F.N.; Coop, M.R. Changes to particle characteristics associated with the compression of sands. *Géotechnique* **2011**, *61*, 459–471. [CrossRef]
27. Xiao, Y.; Liu, H.; Chen, Q.; Ma, Q.; Xiang, Y.; Zheng, Y. Particle breakage and deformation of carbonate sands with wide range of densities during compression loading process. *Acta Geotech.* **2017**, *12*, 1177–1184. [CrossRef]
28. Yu, F.; Su, L. Particle breakage and the mobilized drained shear strengths of sand. *J. Mt. Sci.* **2016**, *13*, 1481–1488. [CrossRef]
29. Xiao, Y.; Liu, H.; Chen, Y.; Chu, J. Influence of intermediate principal stress on the strength and dilatancy behavior of rockfill material. *J. Geotech. Geoenviron. Eng.* **2014**, *140*, 04014064. [CrossRef]
30. Wang, G.; Zha, J. Particle breakage evolution during cyclic triaxial shearing of a carbonate sand. *Soil Dyn. Earthq. Eng.* **2020**, *138*, 106326. [CrossRef]
31. Hyodo, M.; Wu, Y.; Aramaki, N.; Nakata, Y. Undrained monotonic and cyclic shear response and particle crushing of silica sand at low and high pressures. *Can. Geotech. J.* **2017**, *54*, 207–218. [CrossRef]
32. Ding, Z.; He, S.H.; Sun, Y.F.; Xia, T.-D.; Zhang, Q.-F. Comparative study on cyclic behavior of calcareous sand and terrigenous siliceous sand for transportation infrastructure applications. *Constr. Build. Mater.* **2021**, *283*, 122740. [CrossRef]
33. Xiao, Y.; Yuan, Z.; Desai, C.S.; Zaman, M.; Ma, Q.; Chen, Q.; Liu, H. Effects of load duration and stress level on deformation and particle breakage of carbonate sands. *Int. J. Geomech.* **2020**, *20*, 06020014. [CrossRef]
34. Kikumoto, M.; Wood, D.M.; Russell, A. Particle crushing and deformation behaviour. *Soils Found.* **2010**, *50*, 547–563. [CrossRef]
35. Carrera, A.; Coop, M.R.; Lancellotta, R. Influence of grading on the mechanical behaviour of Stava tailings. *Géotechnique* **2011**, *61*, 935–946. [CrossRef]
36. Bandini, V.; Coop, M.R. The influence of particle breakage on the location of the critical state line of sands. *Soils Found.* **2011**, *51*, 591–600. [CrossRef]
37. Yu, F.W. Particle breakage and the critical state of sands. *Géotechnique* **2017**, *67*, 713–719. [CrossRef]
38. Zhao, B.; Wang, J.; Andò, E.; Viggiani, G.; Coop, M.R. Investigation of particle breakage under one-dimensional compression of sand using X-ray microtomography. *Can. Geotech. J.* **2020**, *57*, 754–762. [CrossRef]
39. De Bono, J.P.; McDowell, G.R. The fractal micro mechanics of normal compression. *Comput. Geotech.* **2016**, *78*, 11–24. [CrossRef]
40. Huang, J.Y.; Hu, S.S.; Xu, S.L.; Luo, S.N. Fractal crushing of granular materials under confined compression at different strain rates. *Int. J. Impact Eng.* **2017**, *106*, 259–265. [CrossRef]
41. Zhou, W.; Wang, D.; Ma, G.; Cao, X.; Hu, C.; Wu, W. Discrete element modeling of particle breakage considering different fragment replacement modes. *Powder Technol.* **2020**, *360*, 312–323. [CrossRef]
42. Hardin, B.O. Crushing of soil particles. *J. Geotech. Eng.* **1985**, *111*, 1177–1192. [CrossRef]

43. Afshar, T.; Disfani, M.M.; Narsilio, G.A.; Arulrajah, A. Post-breakage changes in particle properties using synchrotron tomography. *Powder Technol.* **2018**, *325*, 530–544. [CrossRef]
44. Zhao, B.; Wang, J.; Coop, M.R.; Viggiani, G.; Jiang, M. An investigation of single sand particle fracture using X-ray micro-tomography. *Géotechnique* **2015**, *65*, 625–641. [CrossRef]
45. Negussey, D.; Wijewickreme, W.K.D.; Vaid, Y.P. Constant-volume friction angle of granular materials. *Can. Geotech. J.* **1988**, *25*, 50–55. [CrossRef]
46. Wang, G.; Wang, Z.; Ye, Q.; Wei, X. Particle breakage and deformation behavior of carbonate sand under drained and undrained triaxial compression. *Int. J. Geomech.* **2020**, *20*, 04020012. [CrossRef]
47. Jia, Y.; Xu, B.; Chi, S.; Xiang, B.; Xiao, D.; Zhou, Y. Particle breakage of rockfill material during triaxial tests under complex stress paths. *Int. J. Geomech.* **2019**, *19*, 04019124. [CrossRef]
48. Ciantia, M.O.; Arroyo, M.; O'Sullivan, C.; Gens, A.; Liu, T. Grading evolution and critical state in a discrete numerical model of Fontainebleau sand. *Géotechnique* **2019**, *69*, 1–15. [CrossRef]
49. Xiao, Y.; Wang, C.; Zhang, Z.; Liu, H.; Yin, Z.-Y. Constitutive modeling for two sands under high pressure. *Int. J. Geomech.* **2021**, *21*, 04021042. [CrossRef]
50. McDowell, G.R.; Bolton, M.D. On the micromechanics of crushable aggregates. *Géotechnique* **1998**, *48*, 667–679. [CrossRef]
51. Xiao, Y.; Yuan, Z.; Lv, Y.; Wang, L.; Liu, H. Fractal crushing of carbonate and quartz sands along the specimen height under impact loading. *Constr. Build. Mater.* **2018**, *182*, 188–199. [CrossRef]

Journal of
Marine Science and Engineering

MDPI

Article

Permeability of High Clay Content Dredger Fill Treated by Step Vacuum Preloading: Pore Distribution Analysis

Wenchong Shan, Xiaoqing Yuan *, Hui-e Chen, Xiaolin Li and Jinfeng Li

College of Construction Engineering, Jilin University, Changchun 130026, China; shanwc19@mails.jlu.edu.cn (W.S.)
* Correspondence: yuanxiaoqing@jlu.edu.cn; Tel.: +86-137-5656-7311

Abstract: In this study, the step vacuum preloading method was used to reinforce high clay content dredger fill in the laboratory. The pore structures and permeability characteristics of dredger fill under different vacuum pressures were tested. The correlation between the pore structure parameters and permeability coefficient was analyzed using the grey T's correlation analysis method. The research results indicate that the pore ratio, large pore (the diameter with a range of 4–40 μm) content, and permeability coefficient of dredger fill decreased with the increase in vacuum pressures, but the decrease rates of the pore ratio were different at various sampling locations. The contents of micropores (the diameter with a range of <0.04 μm) and small pores (the diameter with a range of 0.04–0.4 μm) increased with the increase in vacuum pressure. The results of the correlation analysis showed that a large pore content had a strong correlation with the permeability coefficient and could be used to describe the permeability characteristics of soil. The research results can provide reference for the improvement of the reinforcement method and for the evaluation of the reinforcement effect of dredger fill in engineering practice.

Keywords: step vacuum preloading; high-clay content dredger fill; pore distribution; permeability; correlation analysis

check for updates

Citation: Shan, W.; Yuan, X.; Chen, H.-e.; Li, X.; Li, J. Permeability of High Clay Content Dredger Fill Treated by Step Vacuum Preloading: Pore Distribution Analysis. *J. Mar. Sci. Eng.* **2023**, *11*, 1714. https://doi.org/10.3390/jmse11091714

Academic Editor: José António Correia

Received: 19 July 2023
Revised: 20 August 2023
Accepted: 25 August 2023
Published: 30 August 2023

1. Introduction

When the traditional vacuum preloading method is used to consolidate high clay content dredger fill, the drainage capacity of the prefabricated vertical drain (PVD) will be weakened [1–3] due to the blockage caused by fine particles. Based on the vacuum preloading method, the step vacuum preloading (SVP) is applied to solve the problem [4–8]. Before the application of vacuum pressure, the dredger fill to be treated is characterized by a higher moisture content, higher porosity, and lower strength. After being reinforced, the soil will have a certain bearing capacity and its permeability will be changed greatly [9]. The permeability coefficient is an important parameter to evaluate the consolidation effects and the property of soil [10,11].

With the rapid development of reclamation engineering, the permeability features of soil treated by the vacuum preloading method have been studied extensively in recent years [12–18]. The influence of permeability on settlement has been discussed for the accurate prediction of settlement and proper evaluation of the reinforcement effect of the soil. Based on field data, Zhuang et al. [10] established a numerical model to analyze the relationship between the settlement and the permeability coefficient of a stabilized soft soil site. Wu et al. [9] discussed the change in permeability coefficient with depths and locations from the PVD after the dredger fill was treated by the laboratory vacuum preloading method. Li et al. [19] adopted a five-level (10, 20, 40, 60, and 80 kPa) vacuum preloading plan on a dredger fill in the laboratory. During the test, the permeability coefficients, pore ratio, settlement, and pore water pressure of the soil were measured. The results proved that the SVP was effective at reinforcing dredger fill with high clay content. The permeability of soil is controlled by its pore characteristics [20–23], and the flow of fluid in soil is determined by the geometric spatial characteristics of the pores. Porosity is the primary factor affecting the permeability of soil, and pore size distribution plays an important role in fluid flow in soil [24–27]. During the process of SVP, the characteristics of pores change significantly, causing the permeability of the dredger fill to change accordingly. So, it is necessary to analyze the permeability of soil from the aspect of pore characteristics. At present, many scholars have investigated the pore characteristics of soil treated by vacuum preloading, and have discussed the changes of porosity [28,29], pore size distribution [30], and pore directionality [31] during reinforcement. However, the study on the dynamic change of permeability coefficient during vacuum preloading process from the aspect of pore features is limited, the correlations between the pore parameters and permeability during the SVP process have seldom been studied. Therefore, it is necessary to study the evolution mechanism of permeability characteristics for dredger fill treated by the SVP method, so as to provide a reference for the improvement of the SVP method and for the evaluation of the consolidation efficiency of dredger fill.

In this study, SVP was used to treat high clay content dredger fill in the laboratory. During the consolidation process, the pore features changed all the time. Correspondingly, the permeability characteristics of the dredger fill changed. To clarify the variation of pore parameters and permeability with time and space during the SVP process, a mercury intrusion porosimetry test (MIP) and permeability test were conducted on samples obtained at different times and from the different locations of the test bucket. Then, the relationship between pore parameters and permeability coefficient of the soil was analyzed.

2. Materials and Methods

2.1. Sample Properties

The studied soil was obtained from Nangang Industrial Zone, Binhai New Area in Tianjin (Figure 1). The site was filled in 2016 and not subjected to any reinforcement process. The dredger fill had a thickness of 8–9 m in the sampling location, and its properties were relatively uniform. During sampling, the surface hard layer was removed first, and the soils with original moisture content of 30–80% were dug and put into a bag. The sampling depth was 1 m.

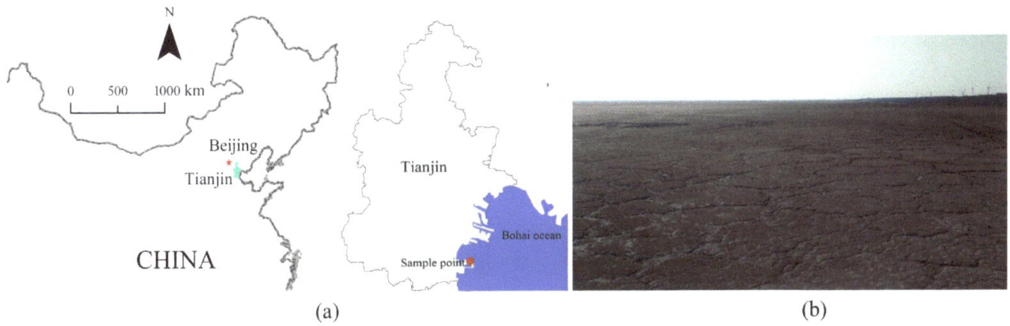

Figure 1. Sampling location: (**a**) Map of Tianjin, China; (**b**) realistic view of sampling location.

The soluble salt content of the soil is listed in Table 1. The total soluble salt content was 1.756%, and the main ions were Na^+ and Cl^-. The basic properties of the soil are showed in Table 2. The content of clay particles was more than 50%, so the soil belonged to high-clay content dredger fill. According to the ASTM D2487-17, the soil was classified as lean clay. The mineral compositions of the soil were measured by X-ray diffraction analysis, and the results are shown in Figure 2. Quartz had the highest content of 36.1%, followed by the illite-smectite mixed layer (I/S).

Table 1. Soluble salt content of the soil.

Component	Total	K^+	Na^+	Cl^-	Ca^{2+}	Mg^{2+}	SO_4^{2-}	CO_3^{2-}	HCO_3^-
Percentage (%)	1.756	0.029	0.525	0.951	0.016	0.022	0.156	0.000	0.025
Method	Water bath evaporation	Flame photometer		Silver nitrate titration	EDTA complex titration			Neutralisation titration	

Table 2. Basic properties of the soil.

Granulometric Composition			Physical Properties			
Size Fraction (mm) Percentage			Density	Liquid Limit	Plastic Limit	Plastic Index
>0.075	0.005–0.075	<0.005	ρ_S (kg/m^3)	w_L (%)	w_P (%)	I_P (%)
0.09%	47.91%	52.06%	2740	44.62	26.38	18.24

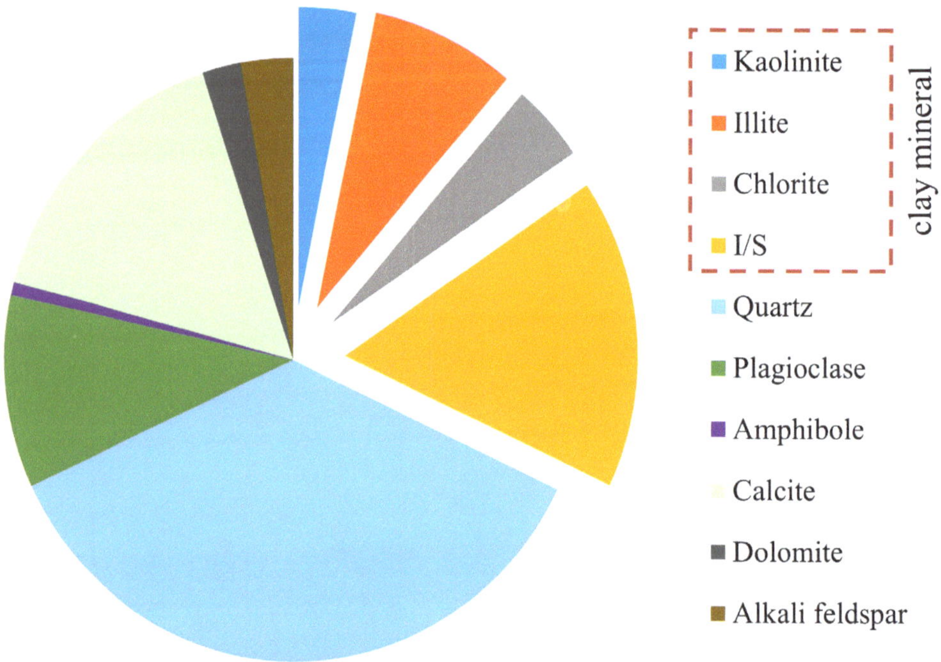

Figure 2. Mineral composition of the dredger fill.

2.2. SVP Model Test

The equipment used for the SVP test was a self-designed bucket. Figure 3 shows its schematic diagram. Two pore water pressure gauges were arranged at different positions of the bucket bottom. In the inner wall of the bucket, the settlement gauge was installed to monitor the settlement of the soil surface. A vacuum gauge was installed at the center of PVD to determine the vacuum pressure of the SVP test. Water collection devices were used to measure the displacement. The testing soil was sealed by a sealing film, and the vacuum pump, sealing film, and soil−water separator worked together to control the levels of vacuum pressure (through Valve 3 of the soil−water separator). The test has been described in detail by Shan et al. [8] and Li et al. [19].

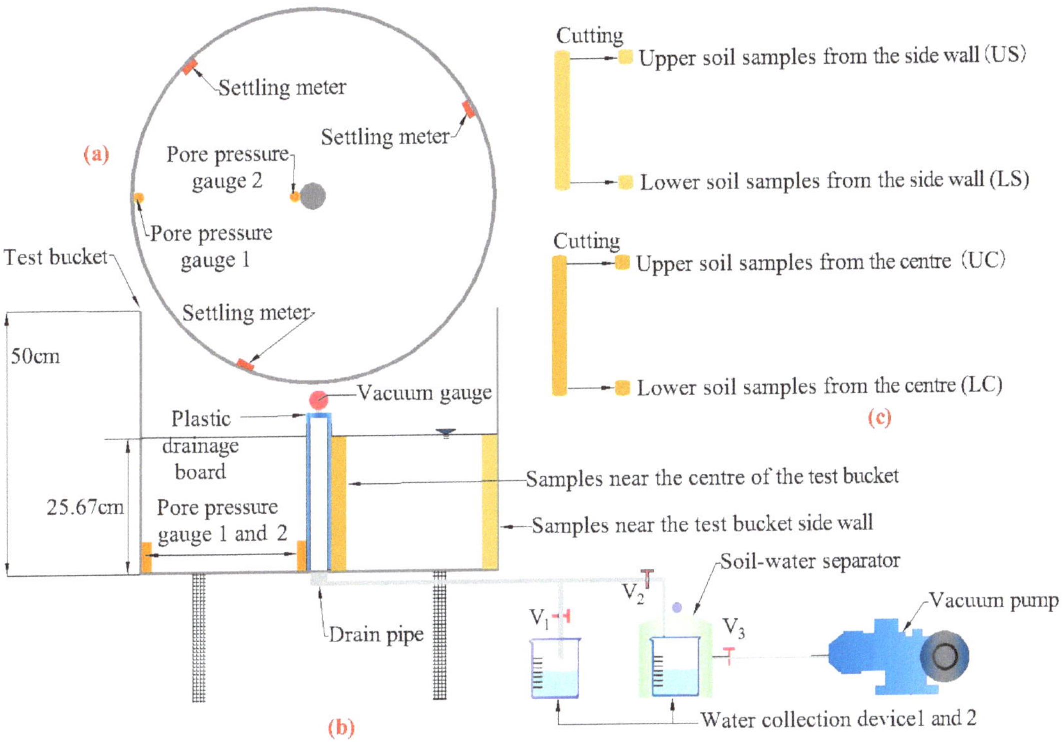

Figure 3. Test equipment and sampling positions: (**a**) plan view of the SVP test equipment; (**b**) front view of the SVP test equipment and sampling positions; (**c**) sampling positions.

The original soil underwent a natural sedimentation process in which some of the water was discharged. So, the moisture content of the soil was uneven. According to the study of Yuan et al. [4,28], the moisture content of the sample used in the SVP test was controlled as 120%. The SVP test lasted for 117 days. The corresponding time nodes of each experimental stage are shown in Figure 4a. There were two stages in the SVP test, namely, self-weight consolidation (SWC) stage and step vacuum pressure loading stage. The SWC stage included soil−water separation and self-weight sedimentation processes [32]. During the step vacuum pressure loading stage, five-level vacuum pressures (10, 20, 40, 60, and 80 kPa) were applied to consolidate the dredger fill.

SWC stage: As shown in Figure 4b, the soil sample was in a mud state at the beginning of the test. The initial height of the mud was 25.67 cm. At this stage, no water was drained, and the total weight of the mud remained unchanged. As time went on, the water and soil began to separate (Figure 4c), and this stage finished when the positions of water and soil surface and the readings of pore water pressure gauges were unchanged. After opening Valve 1, water was discharged through the PVD under gravity, and the stage of self-weight sedimentation started (Figure 4d). The ending criterion of this stage was the same as the soil−water separation stage. As the SWC stage finished, soil samples for the MIP test were taken from the UC, US, LC, and LS, as shown in Figure 3c.

Step vacuum pressure loading stage: five-level vacuum pressures (10 kPa to 20 kPa to 40 kPa to 60 kPa to 80 kPa) were applied to treat the dredger fill. Under a certain level of vacuum pressure, when the settlement of soil was stable and the pore water pressure changed from a stable state to a gradual increase, the vacuum pressure was then increased to the next level.

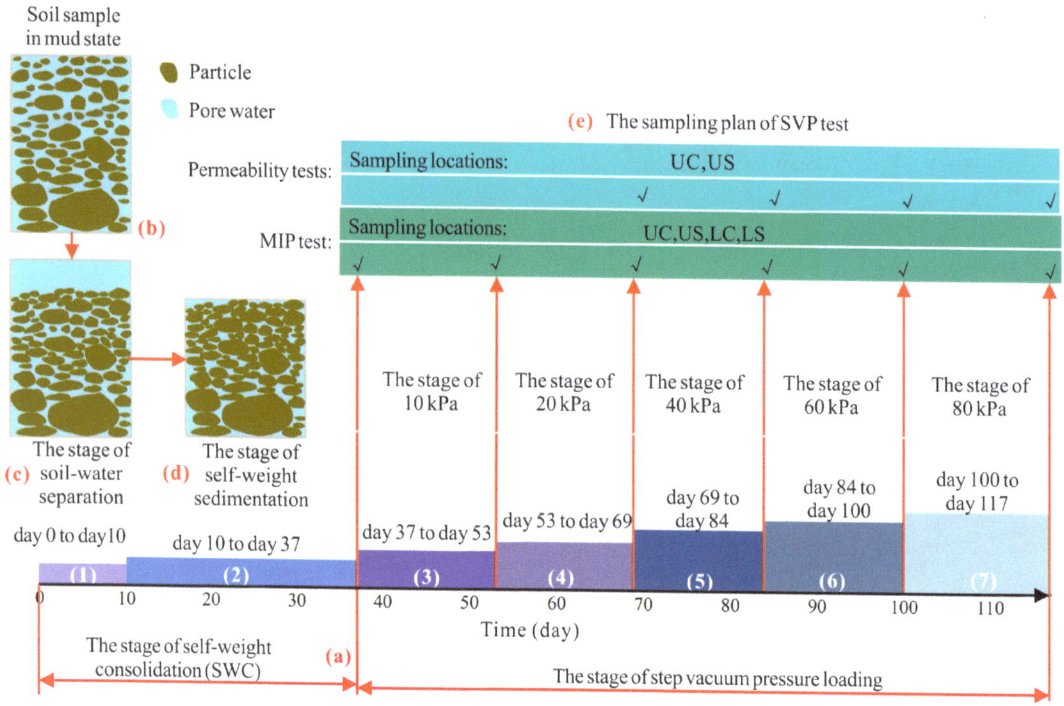

Figure 4. The experimental process and plan of SVP test: (**a**) the experimental process of SVP test; (**b**) SWC stage—the initial state of testing mud; (**c**) SWC stage—the stage of soil-water separation; (**d**) SWC stage—the stage of self-weight sedimentation; (**e**) vacuum pressure loading stage—the sampling plan for MIP test and permeability test.

The sampling plan for the MIP test and permeability test are shown in Figures 3c and 4e. It was difficult to collect samples for permeability test when the soil had a high moisture content. So, sampling was carried out from UC and US after the application of vacuum pressures of 20, 40, 60, and 80 kPa. The test stages and sampling plan are illustrated in Figure 4.

2.3. MIP Test

When the moisture content of the soil was high, it was difficult to prepare undisturbed specimens for the MIP test using traditional sampling methods. A self-designed and tried-and-tested sampler was used to prepare the specimens for the MIP test. Yuan et al. [28] demonstrated that the sampler could obtain the undisturbed soil samples with a high moisture content. We cut the samples into cylinders about 10 mm long with steel knives soaked in liquid nitrogen, and then frozen them in liquid nitrogen for 2 h. After being completely dried by a vacuum dryer, the samples were trimmed into cubes with a length of 10 mm, as shown in Figure 5b.

The MIP tests were carried out with an AutoPore IV 9500 Porosimeter (manufactured by Micromeritics Instrument Corporation, Norcross, GA, USA) (Figure 5a). Under different pressures, mercury entered the pores with different diameters. The volumes of injection mercury under each increment of pressure were recorded, and the pore diameters were calculated through Equation (1).

$$P = \frac{4\gamma \cos\theta}{d} \quad (1)$$

where d is the diameter of the pore, γ is the surface tension of mercury, θ is the contact angle between the particle and mercury, and P is the pressure.

Figure 5. Instrument and sample for MIP and SEM test: (**a**) AutoPore IV 9500 Porosimeter; (**b**) the sample for the MIP test.

2.4. Method of Correlation Analysis

The correlation analysis between the pore characteristics and permeability coefficient was carried out using Grey T's correlation analysis method. The basic idea of this method was to calculate the correlation level (r) according to the proximity of the relative change trend of the time series curves of factors, where the relative change trend refers to the ratio of the slope of curves of reference series and comparison series in the adjacent time period [33]. The positive or negative values of r did not indicate the magnitude, but only indicate whether the change trend of the comparison sequence and the reference sequence was consistent over time.

The method is described as follows:

$$\xi(X_i(k)) = \begin{Bmatrix} sgn(\Delta y_1(k) \cdot \Delta y_2(k)) \cdot \frac{\min(|\Delta y_1(k)|, |\Delta y_2(k)|)}{\max(|\Delta y_1(k)|, |\Delta y_2(k)|)} \\ 0(\Delta y_1(k) \cdot \Delta y_2(k) = 0) \end{Bmatrix} \tag{2}$$

$$r(X_1, X_2) = \frac{1}{n-1} \sum_{k=2}^{n} \xi(X_i(k)) \quad k = 2, 3, \cdots, n, \ i = 1, 2 \tag{3}$$

In Equation (2),

$$\Delta y_i = \{\Delta y_i(k) = y_i(k) - y_i(k-1), \ k = 2, 3, \cdots, n\}, \ i = 1, 2 \tag{4}$$

$$y_i = \{(X_i(k) - X_i(k-1))/D_i, k = 1, 2, \cdots, n\}, \ i = 1, 2 \tag{5}$$

$$D_i = \sum_{k=2}^{n} |X_i(k) - X_i(k-1)|/(n-1), \ i = 1, \ 2; k = 2, \ 3, \ 4, \ \cdots, n \tag{6}$$

where X is the study sequence. y is the sequence after standardization. Δy is the increment sequence. ξ is the correlation coefficient between sequences. r is the correlation level between the reference sequence and comparison sequence.

3. Results

3.1. Permeability Characteristics of Dredger Fill under Step Vacuum Pressures

The TST-55 permeameter (Nanjing Soil Instrument Factory Co., Ltd., Nanjing City, Jiangsu Province, China) was used for the permeability test. The permeability coefficient is calculated using Equation (7).

$$k = \frac{QL}{Aht} \tag{7}$$

where k is permeability coefficient, Q is the amount of seepage within time t, L is the length of seepage path, A is the cross-section area of the tested sample, and h is the height of the sample.

Figure 6 illustrates the permeability coefficients of the samples obtained from the soil under different SVP stages. It shows that the permeability coefficients present a decreasing tendency with the increase in vacuum pressure. At the end of 20 kPa vacuum pressure, the permeability coefficients of the UC and US samples were 5.46×10^{-8} and 8.12×10^{-8} m/s, respectively. When the vacuum pressure was 40 kPa, the permeability coefficients of UC and US both decreased significantly. With the vacuum pressure increased to 60 and 80 kPa, the permeability coefficients decreased slowly. During the SVP test, the permeability coefficient of the central soil sample was lower than that of the soil sample at the edge of test bucket. At the end of 80 kPa vacuum pressure, the permeability coefficients of the UC and US samples were 4.33×10^{-11} m/s and 1.88×10^{-10} m/s, respectively.

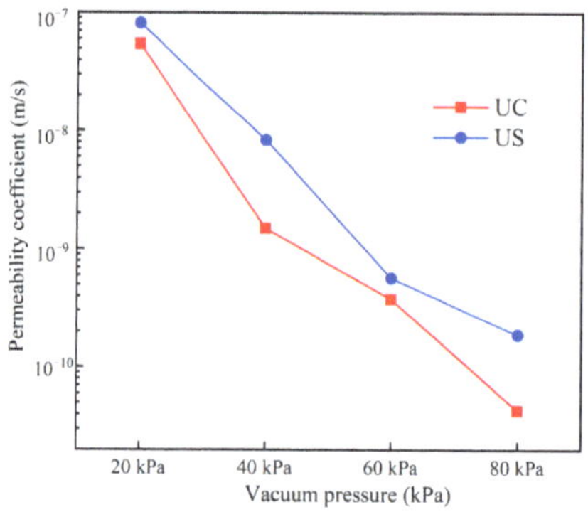

Figure 6. The permeability coefficients of the soil samples.

Under vacuum pressure, the soil near the PVD was consolidated first. Because of the long seepage path and hysteresis of vacuum pressure transmission, the soil near the side wall was consolidated slowly, resulting in a larger permeability coefficient for the US sample. During the SVP course, there was little difference in permeability coefficient between UC and US, indicating that the drainage capacity of the soil was almost the same.

3.2. Pore Ratio and Pore Size Distribution Analysis

Pore characteristics have an influence on the permeability of soil [25,26,34,35]. In the study, the pore features of the treated soil were analyzed from the aspects of pore ratio and pore size distribution.

3.2.1. Pore Ratio

According to the results of the MIP tests, the porosity (n) of the testing sample could be directly obtained, and its value was the percentage of the pore volume to the total volume of the testing sample. In engineering practice, pore ratio (e), the volume ratio of all the pores in the soil to the skeleton particles, is often used to describe the pore characteristics of soil. Figure 7 illustrates the pore ratio of the different samples. It decreased with time. At a certain time, the pore ratio of UC was the smallest, and the pore ratio of LS was the largest.

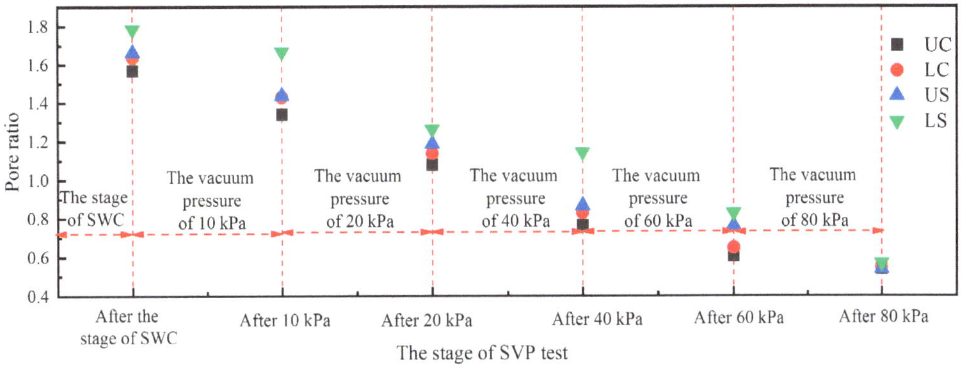

Figure 7. The pore ratio of the soil samples.

As the vacuum pressure increased from 0 to 40 kPa, the pore ratio of UC, LC, and US decreases rapidly. With the continuous increase in vacuum pressure, the pore ratio of the three sampling locations began to decrease at a lower rate. The pore ratio of LS decreased generally, and its decrease rate under the vacuum pressure of 60 and 80 kPa was the fasted, compared with the data of the other three sampling locations. When the stage of 80 kPa vacuum pressure ended, the pore ratio of the four sampling locations was almost the same, which was about 0.55, indicating that the SVP had a good consolidation effect on the high clay content dredger fill.

In the SVP test, the PVD was not only the seepage channel for water, but also the transmission channel for vacuum pressure. During the stage of SWC, the separated water on the upper soil discharged quickly through the PVD, leading to a small pore ratio of UC, which had been found and studied by Walker and Indraratna [36]. When the separated water was discharged completely, the vacuum pressure was loaded on the upper soil through the PVD. With the application of step vacuum pressure, the soil near the PVD consolidated first, causing the pore ratio of LC and UC to be small. Additionally, the upper soil was in contact with the sealing film, and the soil of US underwent vacuum pressure earlier than the soil of LS, leading to a smaller pore ratio of US than LS.

During a certain test stage, the consolidation degree of LS was the lowest among the four sampling locations. At the stage of SWC, the small hydraulic gradient in the soil at the bottom near the bucket wall led to a high pore ratio of LS. And under the vacuum pressure condition, the pore ratio of LS was large because it would take more time for the vacuum pressure to reach this part.

3.2.2. Pore Size Distribution

To describe the pores, the pore size needed to be defined. Here, the Expert Method (EM) was used to classify pores, and pore diameters of <0.04 μm, 0.04–0.4 μm, 0.4–4 μm, 4–40 μm, and >40 μm in the dredger fill were named the micropore, small pore, mesopore, large pore, and macropores, respectively [37]. Yuan et al. [30] used EM to describe the pore size change of the dredger fill during the SVP test, and proved its rationality in describing the pore size distribution of the dredger fill. The cumulative pore size distributions could be obtained based on the cumulative amount of mercury injection (Figure 8), and then the percentage of various types of pores could be determined (Figure 9).

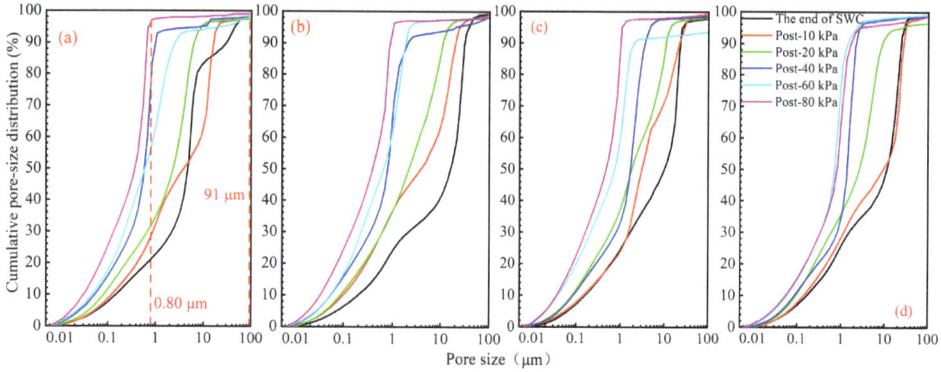

Figure 8. The curves of cumulative pore size distributions: (**a**) cumulative pore size distribution of samples from UC; (**b**) cumulative pore size distribution of samples from US; (**c**) cumulative pore size distribution of samples from LC; (**d**) cumulative pore size distribution of samples from LS.

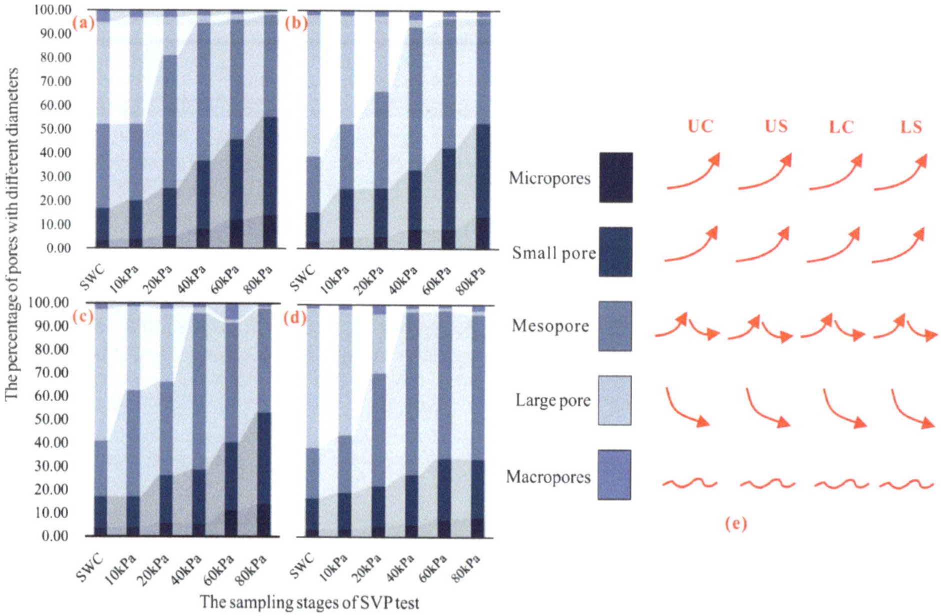

Figure 9. The percentage of pores with different diameters and its variation trend: (**a**) the percentage of pores with different diameters of UC; (**b**) the percentage of pores with different diameters of US; (**c**) the percentage of pores with different diameters of LC; (**d**) the percentage of pores with different diameters of LS; (**e**) the variation of different types of pores' percentage during the SVP test.

The curves in Figure 8 indicate that the pore size became small with the increase in vacuum pressure. In Figure 8a, with the vacuum pressure increasing from 10 kPa to 80 kPa, the largest diameter of pore decreased from 91 μm to 0.8 μm. Figure 9 shows that the content of large pores decreased with the increase in vacuum pressure, while the contents of the small pores and micropores increased. The percentage of mesopores first increased and then decreased. The change in various types of pores implied that the large pores were mainly compressed into mesopores, small pores, and micropores during the consolidation of soil. When the vacuum pressure was 40 kPa, the mesopores' content reached the peak value. When the vacuum pressure was 60 kPa and 80 kPa, the content of small pores and micropores increased, but the mesopores content decreased. So, under a high vacuum pressure, the mesopores were compressed into small pores and micropores.

In the early testing stages, the percentage of large pores in the UC sample was the lowest, and it was the highest for the LS sample. At the end of the experiment, the percentages of large pores at the four sampling locations were almost the same, with a value of 0.75. The change in large pore content of samples from different positions showed a similar trend; it dropped rapidly under the first three levels of vacuum pressures, then decreased slowly under a vacuum pressure of 60 kPa and 80 kPa.

3.3. Correlation Analysis

The pore ratio, percentage of large pore, mesopores, small pores, and micropores were selected as the comparison sequence, and the permeability coefficient of the soil was selected as the reference sequence. The selection of comparison sequence is consistent with permeability coefficient in terms of sampling location and testing stage. Table 3 lists the original data of the comparison sequence and reference sequence. Figure 10 lists the relationship between the pore parameters and the permeability coefficients.

Table 3. Original data: Reference sequence and comparison sequence.

Testing Stage: The Vacuum Pressure Value	Sampling Locations	Reference Sequence	Comparison Sequence				
		Permeability Coefficient (m/s)	Pore Ratio (-)	Micropore (%)	Small Pore (%)	Mesopore (%)	Large Pore (%)
20 kPa	UC	5.46×10^{-8}	1.08	5.67	19.59	55.78	15.93
	US	8.12×10^{-8}	1.19	5.39	20.14	40.49	31.58
40 kPa	UC	1.50×10^{-9}	0.76	8.60	28.20	57.78	3.07
	US	8.34×10^{-9}	0.87	8.51	24.65	59.75	3.26
60 kPa	UC	3.80×10^{-10}	0.61	12.28	33.54	50.23	2.37
	US	5.71×10^{-10}	0.76	8.64	33.79	54.39	0.96
80 kPa	UC	4.33×10^{-11}	0.54	14.27	41.10	42.81	0.7
	US	1.88×10^{-10}	0.54	13.56	39.18	44.19	0.744

In Figure 10, the negative value of the correlation degree with an underline indicates that the permeability coefficient is negatively related to the comparison sequences. The larger the absolute value of the correlation degree, the stronger the correlation between the permeability coefficient and the comparison sequence. As shown in Figure 10a, the large pore content, pore ratio of soil samples in US and UC, and the mesopore content of soils in US exhibited a positive correlation with the permeability coefficient. The large pore content of the soil samples from US demonstrated the highest r value among these relationships. However, the content of small pores and micropores in the soil samples from US and UC, as well as the mesopore content of the soil samples from UC were negatively related to the permeability coefficient.

At different sampling locations, the r values between the permeability coefficient and the pore ratio were lower than that between the permeability coefficient and the large pore content. In the soil, the pore ratio represented the total volume of pore water that could exist, so the value of r between the pore ratio and permeability coefficient was slightly high. The large pore was the main seepage channel for water, and its percentage had the most direct influence on the permeability of the soil. Therefore, the value of r between the large pore content and the permeability coefficient was the highest.

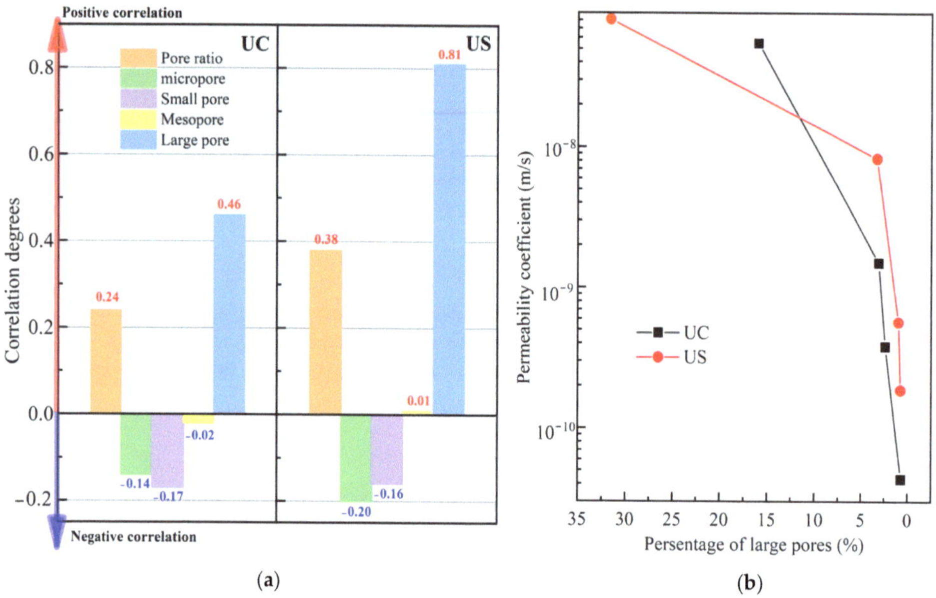

Figure 10. The relationship between pore parameters and permeability coefficients: (**a**) the correlation degrees between pore parameters and permeability coefficients; (**b**) the relationship curves between the permeability coefficient and large pore percentage of the soil samples.

As shown in Figure 10b, the permeability coefficients decreased with the decrease in large pore contents. According to the analysis of pore size distributions, the large pore content was high after the SWC stage. Under the vacuum pressure, the large pores facilitated the drainage of pore water. So, the soil was compressed and consolidated. As shown in Figure 11, taking scanning electron microscope (SEM) photos of the soil samples at UC under different vacuum pressures as an example, the large pores were transformed into mesopores, small pores, and micropores during the application of vacuum pressure. With the decrease in large pore content, the size of the seepage channels for water became small, the permeability of the soil decreased under the increasing vacuum pressure, and the drainage capacity of the soil decreased. The large pore content decreased significantly under pressures of 10, 20, and 40 kPa, as did the permeability coefficient. So, the strongest correlation was between the permeability coefficient and the large pore content.

The r between the large pore content of US and the permeability coefficient was greater than that of UC. The large pores of UC were compressed into mesopores, small pores, and micropores in the earlier stage of the SVP test. The large pore content of UC decreased from 15.93% to 0.7%. Meanwhile, the large pore content of US reduced quickly from 31.58% to 0.74%. The variation in large pore content of US was more significant than that of UC. So the large pore content of US had the largest correlation level.

The mesopore content of UC was negatively correlated with the permeability coefficient, but that of US presented a positive correlation. The correlation level was the sum of the proximity of the slope ratio between the comparison sequence and the reference sequence in the adjacent time. The mesopore content presented different trends before and after 40 kPa, while the permeability coefficient declined throughout the test, indicating that the mesopore content had a negative or positive correlation with the permeability coefficient at different testing stages. When the sum of the absolute value of the negative results was larger than the sum of the positive ones, r was negative; otherwise, r was positive.

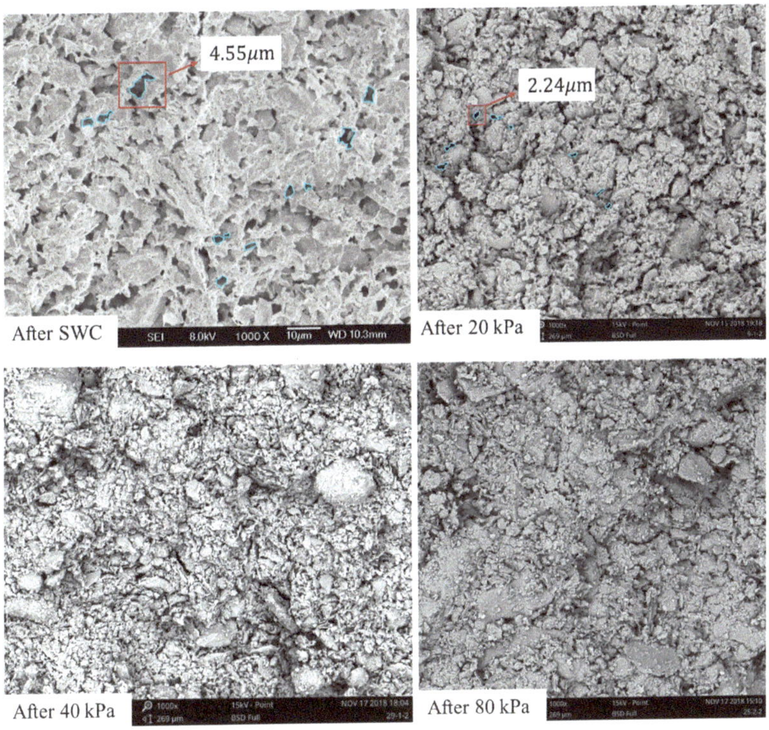

Figure 11. The SEM photos of soil samples from UC at the different sampling stages.

4. Discussion

According to the above analysis, the percentage of large pores has a strong correlation with the characteristics of soil permeability. In this section, we use it to discuss the change in permeability coefficient of soil during the SVP test.

Figure 12 shows the variation in large pore content of soil samples with time and location. The large pore content of the samples at different locations decreases with the increasing vacuum pressure.

At the beginning of SWC, due to the differences in size and mineral composition of particles, the flocculation and sedimentation of particles in the stage of soil−water separation caused the following sorting: sand and silt concentrated at the bottom, and clay fraction filled in the space among large particles or was distributed in the upper part of the soil. The separated water connected with the permeable part of the PVD. When Valve 1 was opened, the separated water in the upper soil discharged rapidly though PVD under gravity, and the fine particles migrated to the vicinity of the PVD by the flow of water. Therefore, the large pore content of UC was the lowest because of the pore filling effect of the fine particles. So, the permeability of soil at UC was poor. In general, the large pore content was high at the end of the SWC stage at all four sampling locations, and the presence of large pores was beneficial for the discharge of pore water under the vacuum pressure.

Under a vacuum pressure of 10 kPa, the change in large pore content of UC was smaller compared with the change at the testing stage of 20 kPa and 40 kPa. This is because the vacuum pressure of 10 kPa was too small to cause an obvious change in large pores. Therefore, the permeability of the UC sample decreased slowly at the initial stage of vacuum preloading, which was conducive to the drainage of pore water far from the center. The large pore contents in the LC and US samples decreased more rapidly compared with that in the LS samples. This discrepancy arose because it took time for vacuum pressure to transfer to the lower and marginal part of the soil, causing gradual consolidation of the deep soil far from the center. In contrast with the traditional vacuum preloading method, the SVP method utilizes a lower initial vacuum pressure when reinforcing dredger fill. These smaller pressures promote soil consolidation by restraining the migration of fine particles and preventing the blockage of PVD. This strategy is conducive to the discharge of pore water when higher vacuum pressures are subsequently applied, thereby improving the efficiency of soil consolidation.

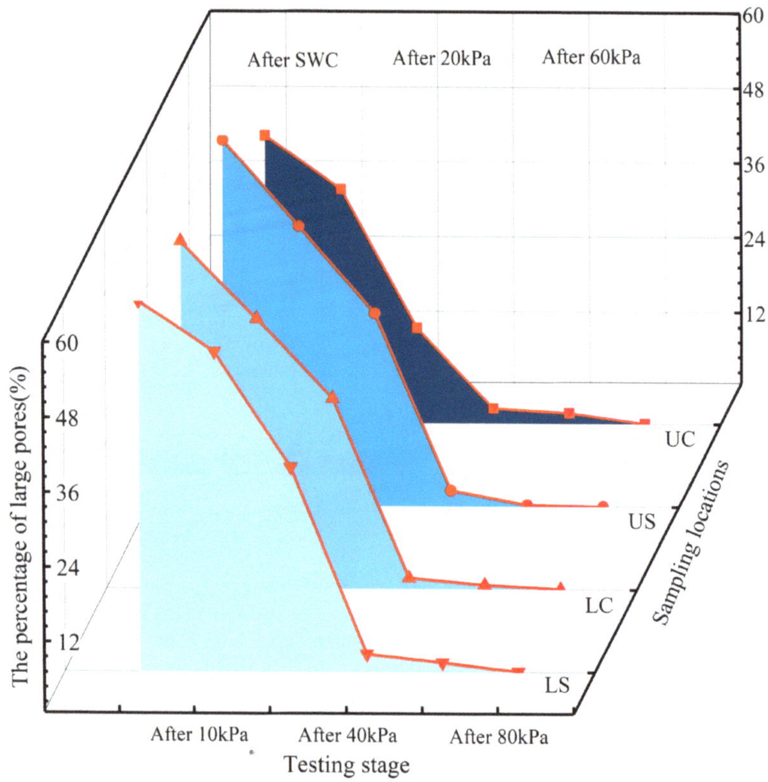

Figure 12. The large pore content of the soil samples.

When the vacuum pressure was 20 kPa, the change in large pore content of the four sampling locations increased. When the vacuum pressure was 40 kPa, the change decreased in the UC sample compared with that under the vacuum pressure of 20 kPa; but it increased in the US, LC, and LS samples. For the UC sample, at the end of SWC stage, it had the lowest large pore content compared with the samples from other three sampling locations, so after the reduction in large pore content under 10 kPa and 20 kPa vacuum pressures, it was difficult for the larger pore content of the UC sample to decrease continuously under 40 kPa vacuum pressure. During this stage, the mesopores content of the soil was the highest, and the large pore content was in the range of 2.40~3.60%. When the vacuum pressure was 60 kPa and 80 kPa, the change in large pore content was about 1%, but the contents of mesopores decreased quickly. The mesopores were transformed into small pores and micropores under high-level vacuum pressures of 60 kPa and 80 kPa. Due to the high vacuum pressure, the particle position constantly adjusted, resulting in the compact structure and low permeability coefficient of the soil.

The temporal and spatial variation of the pore characteristics and permeability coefficient showed the effectiveness of the SVP method in the reinforcement of high clay content dredger fill, and can provide a reference for the improvement of the reinforcement method.

5. Conclusions

The SVP test was conducted to consolidate dredger fill with a high clay content in the laboratory. Based on data obtained through MIP and permeability tests, the correlation between the pore parameters and permeability coefficient of soil was analyzed. The conclusions are as follows:

At the early stage of the SVP test, the permeability coefficient of soil samples decreased rapidly, followed by a slower decline as the pressure levels increased. The pore ratio of the samples from different locations decreased with the vacuum pressure. The pore ratio of the UC sample was the smallest, and it was the largest in the LS sample. Under a low vacuum pressure, the pore ratio decreased rapidly in the UC, US, and LC sampling locations. However, it became slow when the dredger fill was consolidated under vacuum pressure of 60 and 80 kPa. The pore ratio of the LS sample decreased with the vacuum pressure at a certain rate during the SVP test.

During the application of vacuum pressure, the large pores in the dredger fill gradually transformed into mesopores, small pores, and micropores. As the vacuum pressure increased to 60 kPa and 80 kPa, the mesopores were compressed into small pores and micropores.

The results of the correlation analysis indicate that among the various pore structure parameters, the large pore content demonstrated a strong correlation with the permeability coefficient of the soil. Therefore, the variation in large pore content with vacuum pressure could be used to describe the characteristics of soil permeability under vacuum pressure.

When acquiring soil samples for permeability tests becomes challenging, the permeability characteristics of the soil can be evaluated by analyzing the pore structure characteristics. The research results provide novel concepts for the establishment of permeability models, which could more accurately evaluate the settlement and stability of the soil during drainage consolidation.

Author Contributions: Conceptualization, H.-e.C.; methodology, H.-e.C. and W.S.; validation, H.-e.C., W.S., X.Y., X.L. and J.L.; formal analysis, W.S.; investigation, X.Y.; resources, X.Y.; data curation, X.L.; writing—original draft preparation, W.S.; writing—review and editing, W.S.; supervision, H.-e.C.; funding acquisition, H.-e.C., X.Y. and X.L. All authors have read and agreed to the published version of the manuscript.

Funding: This research was funded by the National Natural Science Foundation of China, grant number 42172301, 41602285, 51890914, and 42102346, respectively.

Institutional Review Board Statement: Not applicable.

Informed Consent Statement: Not applicable.

Data Availability Statement: The data will be shared upon reasonable request to the corresponding author.

Conflicts of Interest: The authors declare no conflict of interest.

References

1. Kianfar, K.; Indraratna, B.; Rujikiatkamjorn, C. Radial consolidation model incorporating the effects of vacuum preloading and non-Darcian flow. *Géotechnique* **2013**, *63*, 1060–1073. [CrossRef]
2. Geng, X.; Yu, H.-S. A large-strain radial consolidation theory for soft clays improved by vertical drains. *Géotechnique* **2017**, *67*, 1020–1028. [CrossRef]
3. Lei, H.; Lu, H.; Liu, J.; Zheng, G. Experimental Study of the Clogging of Dredger Fills under Vacuum Preloading. *Int. J. Géoméch.* **2017**, *17*, 04017117. [CrossRef]
4. Yuan, X.-Q.; Wang, Q.; Lu, W.-X.; Zhang, W.; Chen, H.-E.; Zhang, Y. Indoor simulation test of step vacuum preloading for high-clay content dredger fill. *Mar. Georesour. Geotechnol.* **2017**, *36*, 83–90. [CrossRef]
5. Lei, H.Y.; Qi, Z.Y.; Zhang, Z.P.; Zheng, G. New Vacuum-Preloading Technique for Ultrasoft-Soil Foundations Using Model Tests. *Int. J. Géoméch.* **2017**, *17*, 04017049. [CrossRef]

6. Fang, Y.; Guo, L.; Huang, J. Mechanism test on inhomogeneity of dredged fill during vacuum preloading consolidation. *Mar. Georesour. Geotechnol.* **2019**, *37*, 1007–1017. [CrossRef]
7. Zheng, G.; Liu, J.; Lei, H.; Rahman, M.S.; Tan, Z. Improvement of very soft ground by a high-efficiency vacuum preloading method: A case study. *Mar. Georesour. Geotechnol.* **2017**, *35*, 631–642. [CrossRef]
8. Shan, W.; Chen, H.-E.; Yuan, X.; Ma, W.; Li, H. Mechanism of pore water seepage in soil reinforced by step vacuum preloading. *Bull. Eng. Geol. Environ.* **2021**, *80*, 2777–2787. [CrossRef]
9. Wu, Y.J.; Yang, J.B. Laboratory Model Tests on Dredger Fills by Vacuum Preloading with No Sand Cushion. *Appl. Mech. Mater.* **2012**, *170–173*, 655–660. [CrossRef]
10. Zhuang, Y.; Cui, X.Y. Evaluation of Vacuum Preloading with Vertical Drains as a Soft Soil Improvement Measure. *Soil Mech. Found. Eng.* **2016**, *53*, 210–217. [CrossRef]
11. Agus, S.S.; Leong, E.-C.; Rahardjo, H. Estimating permeability functions of Singapore residual soils. *Eng. Geol.* **2005**, *78*, 119–133. [CrossRef]
12. Saowapakpiboon, J.; Bergado, D.; Voottipruex, P.; Lam, L.; Nakakuma, K. PVD improvement combined with surcharge and vacuum preloading including simulations. *Geotext. Geomembranes* **2011**, *29*, 74–82. [CrossRef]
13. Bin Xu, B.; Liang, A.H.; Li, M.Y. Effect of Variable Permeability Coefficient on Consolidation of Ultra-Soft Ground under Vacuum Preloading. *Appl. Mech. Mater.* **2014**, *580–583*, 56–60. [CrossRef]
14. Sun, L.; Jia, T.; Zhuo, R.; Yan, S.; Guo, B. Numerical Solutions for Consolidation of Under-Consolidated Dredger Fill under Vacuum Preloading. *J. Coast. Res.* **2015**, *73*, 277–282. [CrossRef]
15. Indraratna, B.; Zhong, R.; Fox, P.J.; Rujikiatkamjorn, C. Large-Strain Vacuum-Assisted Consolidation with Non-Darcian Radial Flow Incorporating Varying Permeability and Compressibility. *J. Geotech. Geoenviron. Eng.* **2017**, *143*, 04016088. [CrossRef]
16. Baral, P.; Rujikiatkamjorn, C.; Indraratna, B.; Kelly, R. Radial consolidation characteristics of soft undisturbed clay based on large specimens. *J. Rock Mech. Geotech. Eng.* **2018**, *10*, 1037–1045. [CrossRef]
17. Nguyen, B.-P.; Kim, Y.-T. Radial consolidation of PVD-Installed normally consolidated soil with discharge capacity reduction using large-strain theory. *Geotext. Geomembranes* **2019**, *47*, 243–254. [CrossRef]
18. Sun, H.-L.; Weng, Z.-Q.; Liu, S.-J.; Geng, X.-Y.; Pan, X.-D.; Cai, Y.-Q.; Shi, L. Compression and consolidation behaviors of lime-treated dredging slurry under vacuum pressure. *Eng. Geol.* **2020**, *270*, 105573. [CrossRef]
19. Li, J.; Chen, H.; Yuan, X.; Shan, W. Analysis of the Effectiveness of the Step Vacuum Preloading Method: A Case Study on High Clay Content Dredger Fill in Tianjin, China. *J. Mar. Sci. Eng.* **2020**, *8*, 38. [CrossRef]
20. Kozeny, J. Ueber kapillare Leitung des Wassers im Boden. *Sitzungsber Akad. Wiss. Wien* **1927**, *136*, 271–306.
21. Carman, P.C. Permeability of saturated sands, soils and clays. *J. Agric. Sci.* **1939**, *29*, 262–273. [CrossRef]
22. Zhai, Q.; Rahardjo, H.; Satyanaga, A. A pore-size distribution function based method for estimation of hydraulic properties of sandy soils. *Eng. Geol.* **2018**, *246*, 288–292. [CrossRef]
23. Romero, E.; Gens, A.; Lloret, A. Water permeability, water retention and microstructure of unsaturated compacted Boom clay. *Eng. Geol.* **1999**, *54*, 117–127. [CrossRef]
24. Tuli, A.; Hopmans, J.W.; Rolston, D.E.; Moldrup, P. Comparison of Air and Water Permeability between Disturbed and Undisturbed Soils. *Soil Sci. Soc. Am. J.* **2005**, *69*, 1361–1371. [CrossRef]
25. Liu, Y.; Jeng, D.-S. Pore scale study of the influence of particle geometry on soil permeability. *Adv. Water Resour.* **2019**, *129*, 232–249. [CrossRef]
26. Zhai, Q.; Rahardjo, H.; Satyanaga, A.; Priono; Dai, G.-L. -L. Role of the pore-size distribution function on water flow in unsaturated soil. *J. Zhejiang Univ. Sci. A* **2019**, *20*, 10–20. [CrossRef]
27. Deng, Y.; Yue, X.; Liu, S.; Chen, Y.; Zhang, D. Hydraulic conductivity of cement-stabilized marine clay with metakaolin and its correlation with pore size distribution. *Eng. Geol.* **2015**, *193*, 146–152. [CrossRef]
28. Yuan, X.-Q.; Wang, M.-L.; Yan, H.; Chen, J.-Q.; Tan, B.-M. 3D Visualization Study on Microstructure Variation of Dredger Fill by Step Vacuum Preloading. In Proceedings of the GeoShanghai 2018 International Conference: Fundamentals of Soil Behaviours, Shanghai, China, 27–30 May 2018; Springer: Singapore, 2018; pp. 226–236. [CrossRef]
29. Liu, J.; Lei, H.; Zheng, G.; Feng, S.; Rahman, M.S. Improved Synchronous and Alternate Vacuum Preloading Method for Newly Dredged Fills: Laboratory Model Study. *Int. J. Géoméch.* **2018**, *18*, 04018086. [CrossRef]
30. Yuan, X.Q.; Wang, Q.; Sun, T.; Xia, Y.B.; Song, J. Pore distribution characteristics of dredger fill during hierarchical vacuum preloading. *J. Jilin Univ. Earth Sci. Ed.* **2012**, *42*, 169–176. [CrossRef]
31. Wang, J.; Cai, Y.; Fu, H.; Hu, X.; Cai, Y.; Lin, H.; Zheng, W. Experimental study on a dredged fill ground improved by a two-stage vacuum preloading method. *Soils Found.* **2018**, *58*, 766–775. [CrossRef]
32. Liu, Y.; Xiao, S.F.; Wang, Q. Research on indoor scale-down test of dredger fill. *Rock Soil Mech.* **2004**, *25*, 518–521. [CrossRef]
33. Luo, N.; Zhong, W.; Mei, S. Relational Analysis between Technological Progress and Economic Growth: An Empirical Study in Counties from Jiangsu Province. In *Advances in Grey Systems Research. Understanding Complex Systems*; Springer: Berlin/Heidelberg, Germany, 2010; pp. 165–176. [CrossRef]
34. Burdine, N. Relative Permeability Calculations from Pore Size Distribution Data. *J. Pet. Technol.* **1953**, *5*, 71–78. [CrossRef]
35. Garciabengochea, I.; Altschaeffl, A.G.; Lovell, C.W. Pore distribution and permeability of silty clays. *J. Geotech. Eng. Div.* **1979**, *105*, 839–856. [CrossRef]

36. Walker, R.; Indraratna, B. Consolidation analysis of a stratified soil with vertical and horizontal drainage using the spectral method. *Géotechnique* **2009**, *59*, 439–449. [CrossRef]
37. Wang, Q.; Wang, J.P. A Study on fractal of porosity in the soils. *Chin. J. Geotech. Eng.* **2000**, *22*, 496–498.

Journal of *Marine Science and Engineering*

MDPI

Article

Bearing Characteristics of Tripod Bucket Foundation under Horizontal and Moment Load in Sand

Xin Wang [1], Xue-Liang Zhao [1,*], Shu-Huan Sui [1], Zi-Bei Hu [2], Wen-Ni Deng [1] and Qi-Ming Song [3]

[1] Key Laboratory of Concrete and Prestressed Concrete Structure of Ministry of Education, Southeast University, Nanjing 210096, China; wangxin_marine@163.com (X.W.); 220211261@seu.edu.cn (S.-H.S.); wenniedeng@163.com (W.-N.D.)
[2] Department of Industrial Design, Monash University, Melbourne 3800, Australia; zhuu0039@student.monash.edu
[3] Fujian Yongfu Power Engineering Co., Ltd., Fuzhou 350001, China; songqm@fjyongfu.com
* Correspondence: zhaoxl@seu.edu.cn; Tel.: +86-152-6145-1058

Abstract: Based on a series of physical model tests and numerical simulations, the bearing characteristics of a tripod-bucket foundation are investigated. It is found that with the decrease in aspect ratio (L/D), the rotation center of the foundation will decrease, and the displacement model change from rotation to uplift. Characteristics of earth pressure on the buckets from both finite element analysis and model tests are studied, which is used to explain the failure mechanism for tripod-bucket foundations with different L/D. A revised method is proposed to estimate the moment bearing capacity of the tripod bucket foundation under horizontal and moment load. This method is thought to be more convenient and applicable in the practice.

Keywords: tripod bucket foundation; bearing capacity; failure mechanism; model test; finite element analysis

check for updates

Citation: Wang, X.; Zhao, X.-L.; Sui, S.-H.; Hu, Z.-B.; Deng, W.-N.; Song, Q.-M. Bearing Characteristics of Tripod Bucket Foundation under Horizontal and Moment Load in Sand. *J. Mar. Sci. Eng.* **2023**, *11*, 1631. https://doi.org/10.3390/jmse11081631

Academic Editor: Joško Parunov

Received: 27 July 2023
Revised: 17 August 2023
Accepted: 19 August 2023
Published: 21 August 2023

1. Introduction

Offshore wind energy has been developing rapidly in recent years as a clean and sustainable energy resource [1,2]. However, the construction expense, especially the cost of the foundation, negatively affects the cost-effectiveness of offshore wind farms. The bucket foundation, which is installed by self-weight and suction pressure, is a good option for the offshore wind turbine foundation nowadays [3]. It is thought to be a cost-effective foundation type due to the speed of installation and reduction in material costs compared with other commonly used foundations. At present, tetrapod, tripod and single buckets are common types of bucket foundations. The tripod bucket foundation is made of three single buckets placed in a triangular shape and is thought to be more suitable for coastal areas with water depths greater than 20 m [4].

In the offshore environment, wind turbine foundations are subject to horizontal load (H) and the moment load (M) generated by wind, waves and currents [5–8]. Compared with offshore oil and gas platforms, the vertical load (V) from a wind turbine is relatively low [9]. Therefore, the combined H–M loads are critical for the foundation of offshore wind turbines [10,11]. To ensure safe operation and good serviceability of the wind turbines, the bearing characteristics of the tripod bucket foundation under combined horizontal load and moment load should be studied comprehensively.

A number of investigations into bearing capacity focusing on single bucket foundations [12–18] have been conducted. However, group effects on the bearing capacity of multiple foundations in sand are still not enough. Martin and Hazell [19] investigated the group effect of strip foundations in non-homogeneous clays and found that the interaction of the footings was beneficial for improving the vertical bearing capacity coefficient. Gourvenec and Jensen [20] conducted an in-depth analysis of the effect of barrel spacing (S) on the bearing capacity of the foundation, and the group effect of two pile group systems with an aspect ratio L/D (L is the embedment depth and D is the diameter) of 0.5 in uniform clay was explored. It is found that with the increase in foundation spacing and embedding depth, the combined bearing capacity of H–M loads increased significantly.

Bang et al. [21] conducted a series of centrifuge model tests. A three-dimensional analytical solution to evaluate the effect of the loading depth on the ultimate horizontal bearing capacity of suction piles was proposed. The results show that the location of the applied load influenced the modes of movement. When the loading depth increases, the ultimate horizontal bearing capacity of the suction piles would increase.

There are also some empirical formulae of bearing capacity proposed by theoretical analysis or numerical simulation. Based on the upper limit theory, Zhao et al. [22] found three critical positions S_1, S_2 and S_3 which have significant effects on the horizontal bearing capacity of the quadruple bucket foundation. Considering the influence of the load direction on the bearing capacity, the empirical formula of the relationship between the distance among piles, the load direction, and the horizontal bearing capacity was established. Kim and Oh [23] studied the group effect of a tripod bucket foundation in cohesive soil with a three-dimensional finite element method, and the efficiency coefficient was proposed. The results show that the group efficiency factor had little effect on the vertical bearing capacity with $S/D > 1.5$, whereas the factor of the horizontal and the moment capacity of the tripod bucket foundation tended to be larger. Compared with previous studies, Hung and Kim [24] proposed new equations to evaluate the vertical-, horizontal- and moment-bearing capacities of the bucket foundation considering the effects of the non-homogeneity of clay and embedment depth. The capacity envelopes under general combined loads were defined. Based on the finite element method, He et al. [25] simulated the bucket penetration in clay by reducing the friction factor between the bucket wall and the surrounding soil. The results indicated that the spacing between buckets was a key factor affecting the bearing capacity of the tripod bucket foundation, but its bearing capacity envelope was similar to the monopod bucket foundation.

Previous studies have shown that the bearing characteristics of foundations are significantly influenced by aspect ratios L/D [15], load location [21], load direction [22], spacing ratios S/D and group effects [19,23]. Additionally, most of the work focuses on the mechanical properties of the single bucket foundation or the mechanical properties of the bucket foundation in clay. However, the research on the mechanical properties of the tripod bucket foundation in sandy soil is not enough. There is little comparison and analysis between numerical simulation results and experimental results now. This study aims to analyze the effect of bucket spacing and aspect ratios on the bearing capacity of the tripod bucket foundation under the combined load of H–M based on a physical model test and three-dimensional finite element analysis.

2. Materials and Methods

2.1. Physical Model Test

A series of model tests were performed. Common sand was used in this study. Particle size distribution tests were conducted which showed it to be medium sand. Based on the relative porosity ratio measured on site, the relative density of the sand used in this experiment was 0.59, which was classified as medium-dense sand. The physical properties of the sand used in the model test are shown in Table 1. The tests were conducted in a model box with size of 1 m × 1 m × 1.2 m (length × width × height). Single-side drainage was adopted and pebbles, drainage pipe network and geotextile were laid at the bottom of the model box. The sandy soil was laid in the model box layer by layer and manual compaction was adopted to meet the test requirements. After the sand was laid, it was filled with water and allowed to stand for 3 days for consolidation. Q235 steel pipe was used as the bucket model because the deformation of the foundation was ignored in the test. The parameters of steel piles are shown in Table 1. To ensure the same model bucket weight and meet the requirements for eliminating the boundary effect of the model box [26], model sizes are calculated as shown in Table 2. The final distribution of the outer diameter of the bucket is 102 mm, 120 mm, 133 mm and the bucket length is 164.7 mm, 134 mm, and 112.6 mm, respectively, which is consistent with the range of aspect ratio of suction bucket foundation in practical engineering. Lifting lugs were set every 90 mm from the bottom of the tripod to apply horizontal load to the model as shown in Figure 1.

Table 1. Physical properties of sand and steel pipe.

Sand	Saturated Density γ_{sat} (kN/m^3)	Natural Density γ_{nat} (kN/m^3)	Friction Angel φ (°)	Poisson's Ratio v
	19.6	15.9	35	0.26
Steel pipe	Modulus of elasticity E_s (Mpa)		Unit weight γ_{steel} (kN/m^3)	Poisson's Ratio v
	2.06×10^5		78.5	0.31

Table 2. Bucket model size in test and numerical simulations.

Model	Diameter (mm)	Length (mm)	Aspect Ratio *L/D*	Wall Thickness (mm)
1	102	164.7	1.61	3
2	120	134	1.12	3
3	133	112.6	0.85	3

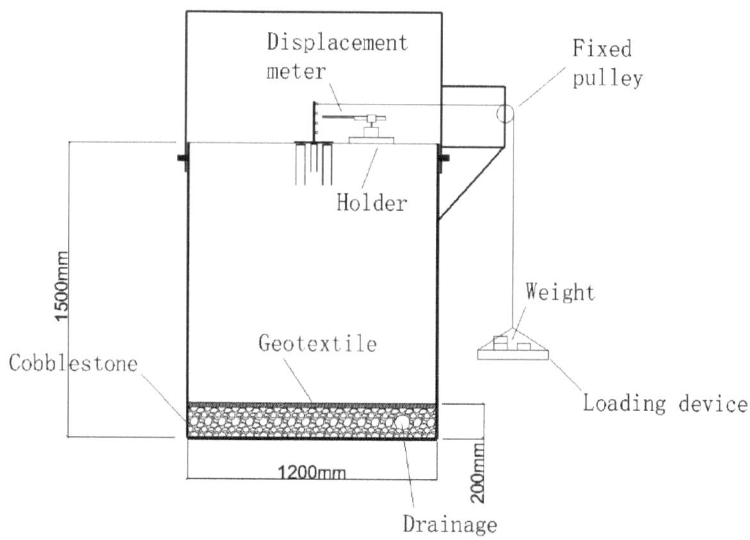

Figure 1. Schematic diagram of the model test.

Two bolt holes were set at the top of a single bucket model. One was used to connect with the tripod at the top of the model, and the other was used as an exhaust hole to facilitate the penetration and extraction of the bucket model. Earth pressure cells were placed symmetrically at 1/3 L and 2/3 L from the foundation top at the outside, as shown in Figure 2.

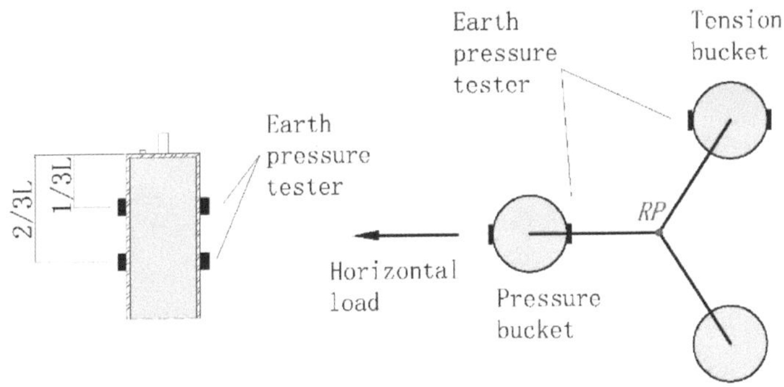

Figure 2. Arrangement of earth pressures in the model test.

When loading, ten steps were used with a load level difference of 1/10 of the estimated bearing capacity by adding weight blocks. Each level of loading was maintained for at least 5 min until the displacement remained almost stable. Both the earth pressure and displacement were measured and recorded to analyze the failure phenomenon and mechanism of the tripod bucket foundation in sand.

2.2. Numerical Simulation

The 3D Finite Element (FE) analyses were conducted with an elasto-plastic model and Mohr–Coulomb failure criterion. In practical engineering, the stiffness of the bucket is much higher than that of the soil, and it can be considered as a rigid foundation. This test does not consider the deformation of the bucket body under load but only considers the deformation between the bucket and the soil and the soil failure. The bucket foundation in the numerical simulation was modeled as a rigid body, and the parameters of the bucket foundation were shown in Table 2. The normally consolidated sand under drained conditions was modeled with the basic parameter same with the physical model test shown in Table 1. Brinkgreve et al. [27] confirmed that setting a small cohesion force and dilatancy angle was beneficial to improve the accuracy of the calculation. Young's modulus *Es* of the sand was set as 38.6 Mpa. The interface coefficient between the bucket and soil was set at 0.68, with the aim of increasing the flexibility of finite element mesh, reducing the sharp angle of strength reduction and avoiding non-physical stress results.

Figure 3 shows the arrangement of the tripod bucket foundation and boundary extensions adopted in this study. The size of the soil elements gradually increased from the bucket to the domain boundary. The soil was divided by a dense grid within a horizontal range of 3 *S* (three times the spacing between buckets) and a vertical range of 2 *L* (two times the length of the bucket) boundary, and the roughness coefficient is 0.35. The remaining parts of the soil were adaptively divided by software. The boundary with horizontal boundary extents of the bucket foundation model was 10 *S* and the vertical was 5 *L*, respectively, which was thought to be able to eliminate the boundary effects.

The connection between the individual bucket foundations as jacket structure in practice was numerically and rigidly simulated by using a Load Reference Point (LRP) at the top center of the triangle cross in this study [20,24,28]. The load was applied to LRP using the load-controlled method, which increased 1/10 of the estimated ultimate bearing capacity at every step. When the load-controlled method is used in the physical model experiment, the long interval between each stage of loading belongs to slow-loading. The sand has good permeability and sufficient time for drainage, so it is defined as drainage behavior. The ultimate bearing capacity was estimated according to the load-displacement curve. The bearing capacity was determined by the tangent intersection method or the phenomenon of pulling out [29].

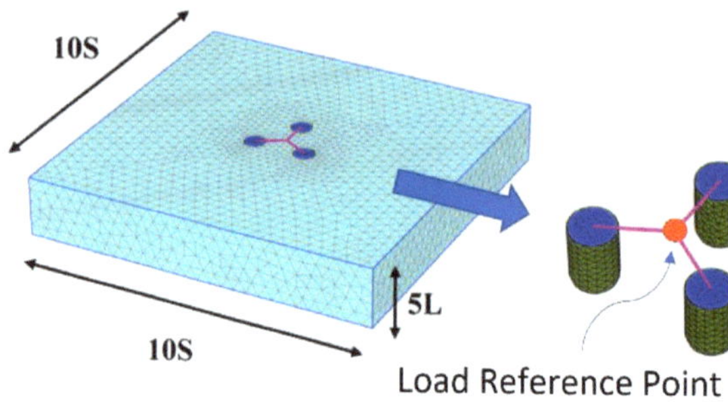

Figure 3. Geometry of the tripod bucket foundation and the load and displacement conventions.

2.3. Validation of Numerical Modeling

The numerical modeling adopted in this study was validated by the results from the physical model tests. Figure 4 shows comparisons of the horizontal bearing capacities from model tests and FE analysis with $S/D = 3$ and $L/D = 1.61$, 1.12 and 0.85, respectively. The maximum error may be caused by the longer load interval. Based on this validation, the numerical modeling adopted in this study was thought to be reliable to evaluate the bearing capacity of the tripod bucket foundation in sand.

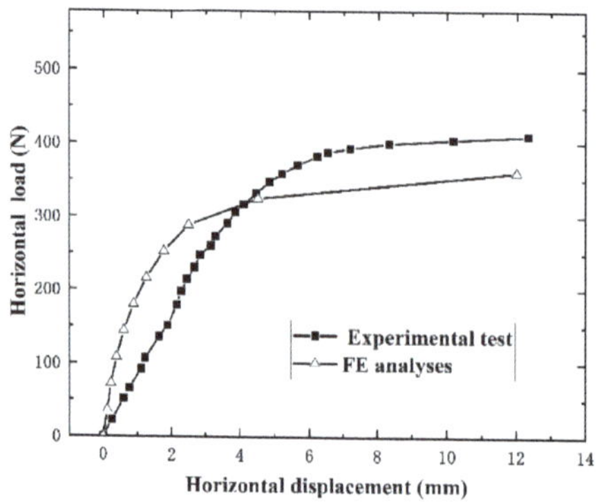

(a)

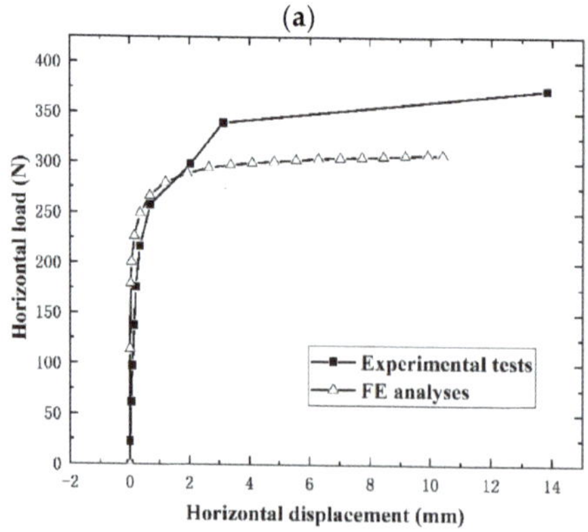

(b)

Figure 4. *Cont.*

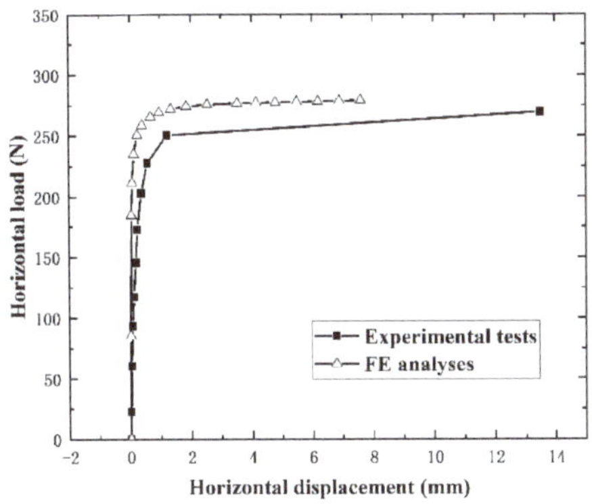

(c)

Figure 4. Comparison of load-displacement curves between FE analyses and model tests. (**a**) $S/D = 3$, $L/D = 1.61$; (**b**) $S/D = 3$, $L/D = 1.12$; (**c**) $S/D = 3$, $L/D = 0.85$.

3. Results

3.1. Failure Mechanism under H–M Loadings

According to the model tests, the failure process of the tripod bucket foundation can be divided into three stages, the initial stage, the intermediate stage and the failure stage.

In the initial stage, the horizontal displacement was very small and the soil around the bucket may be in the elastic state. Only micro-cracks behind the tension bucket could be observed. At this time, there was almost no obvious deformation around the compression bucket as shown in Figure 5a. As the load increased, the soil around the bucket gradually came to the plastic state. The cracks behind the tension bucket gradually expanded to a larger region at the rear side, and soil rise could be observed at the front side of the tension bucket. At this time, the tension bucket tended to pull up and rotate in the loading direction. Also, the cracks and soil deformation around the compression bucket were smaller than those around the tension bucket as seen in Figure 5b.

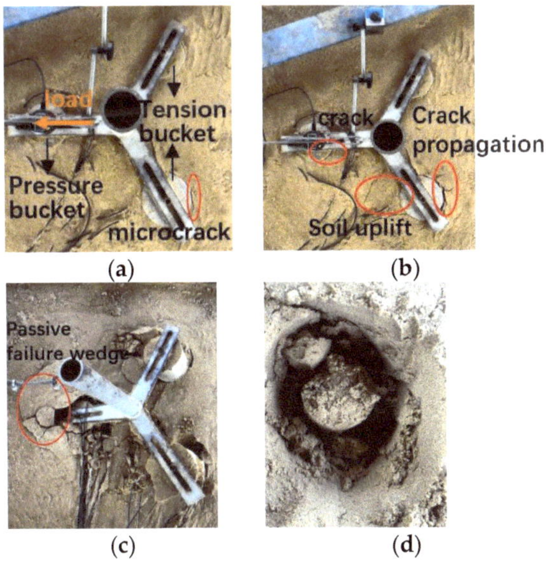

Figure 5. Failure process of the tripod bucket foundation under *H–M*. (**a**) Initial stage; (**b**) intermediate stage; (**c**) failure stage; (**d**) soil plug.

When the load increased to a certain value, the horizontal displacement of the bucket foundation started to increase dramatically. The soil in front of the compression bucket gradually rose to form a passive failure wedge. The foundation came to the state of complete instability, as shown in Figure 5c. The plastic failure area of the compression bucket had a wider scope, so the earth pressure in front of the compression bucket may provide greater horizontal resistance of the tripod bucket foundation. When the tension bucket was pulled up, it was observed that the soil plug fell off, as seen in Figure 5d, and the plastic failure area of the soil at the front and rear sides of the tension bucket was found to be limited to the layer close to the soil surface. It could be thought that the horizontal resistance provided by the tension bucket mainly came from the friction between the bucket wall and the sand.

According to the displacement vector diagram from FE analysis (Figure 6), with the decrease in L/D, the vertical uplift movement of the tension bucket was more obvious. This was consistent with the phenomenon observed in the model test that the soil layer of plastic failure was thinner; meanwhile, with the increase in the L/D (L/D = 1.61 compared to L/D = 1.12 and L/D = 0.85), the rotation trend of the compression bucket was more obvious and the tension bucket was also accompanied by the forward tilt rotation in the process of pulling up as shown in Figure 6.

In addition, according to the displacement analyses, the rotation center of the tripod bucket foundation was not located at the central axis of the compression bucket but deviated to the horizontal force direction. The specific results are shown in Table 3. It shows that with the decrease in L/D, the distance between the rotation center and axis changes only a little and the vertical position of the rotation center became deeper. This phenomenon may well explain that the main displacement of the tripod bucket foundation with a large aspect ratio was rotation and that with a small aspect ratio was horizontal displacement.

Table 3. Position of the rotating center of the foundation in the compression bucket.

L/*D*	Diameter (mm)	Deviation from the Axis (mm)	Ratio of Deviation Distance to Bucket Diameter (%)	Depth of the Rotation Center
1.61	102	14.4	14.1	0.86 *L*
1.12	120	16	13.3	0.97 *L*
0.85	133	14	10.5	>1 *L*

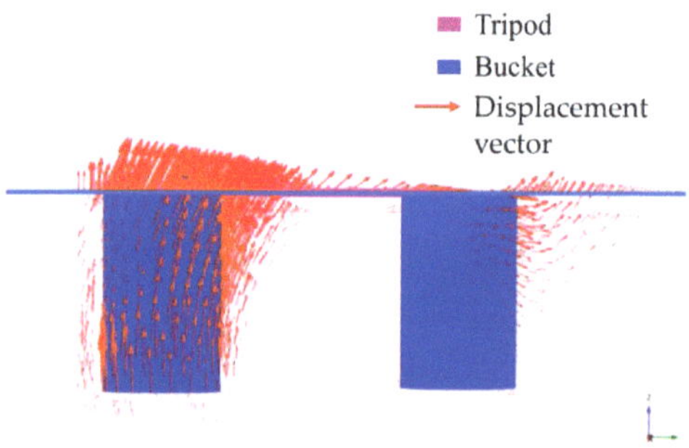

(a)

Figure 6. *Cont.*

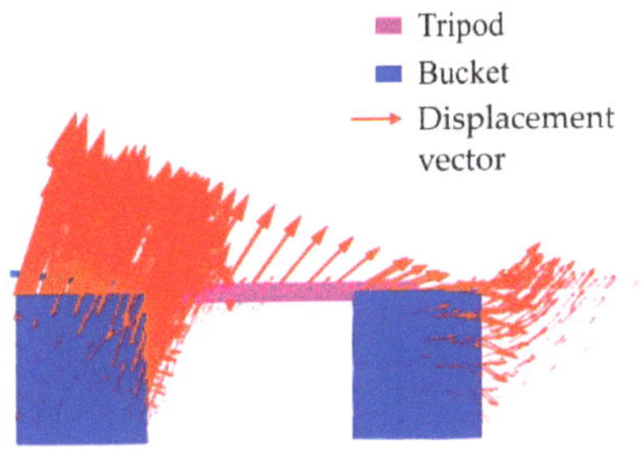

(**b**)

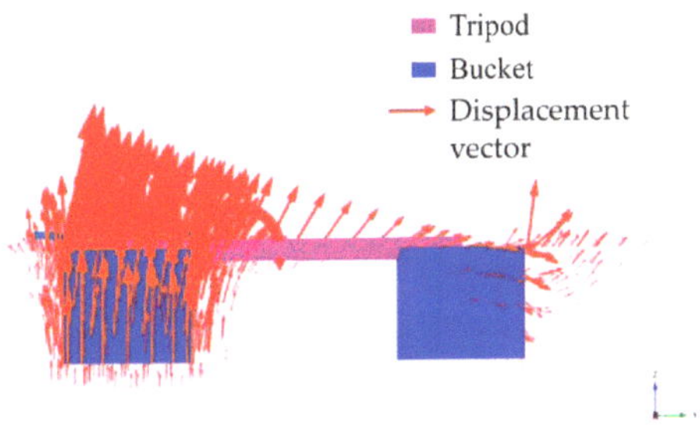

(**c**)

Figure 6. Displacement vector by FE analysis. (**a**) Displacement vector diagram of test 1 (L/D = 1.61); (**b**) displacement vector diagram of test 2 (L/D = 1.12); (**c**) displacement vector diagram of test 3 (L/D = 0.85).

3.2. Earth Pressure Analysis

In the model tests, the horizontal earth pressure was measured by the earth pressure cell placed in the contact surface between the outer side of the bucket model and the sand as shown in Figure 2. Figures 7–9 show the comparison results of soil pressure by model tests and FE analysis under the condition of L/D = 1.61, 0.85 and 1.12. According to the analyses of earth pressure, the failure mechanism of the tripod bucket foundation in the sand under the action of the horizontal and the bending moment can be further clarified.

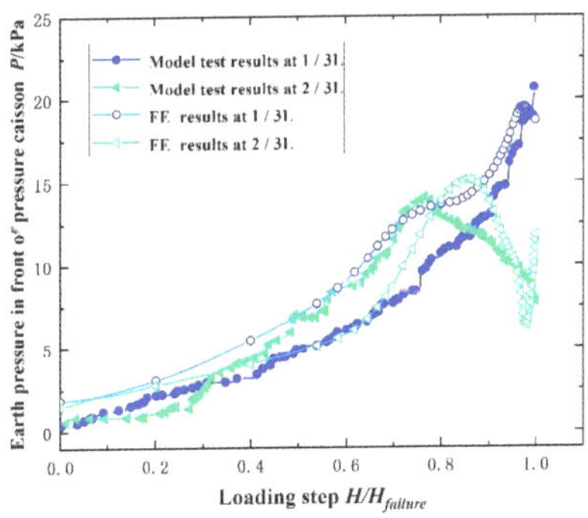

(**a**)

Figure 7. *Cont.*

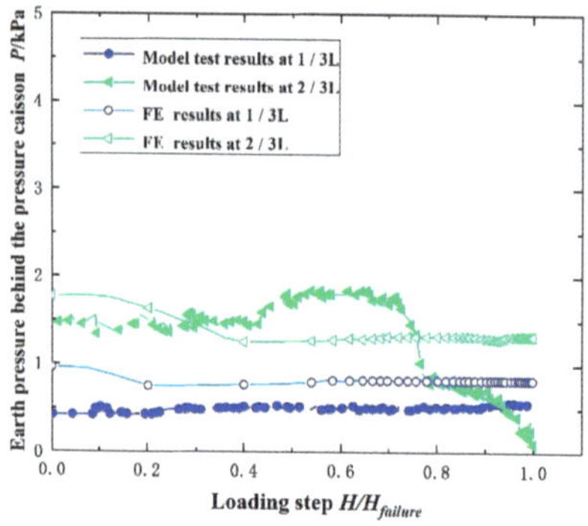

(**b**)

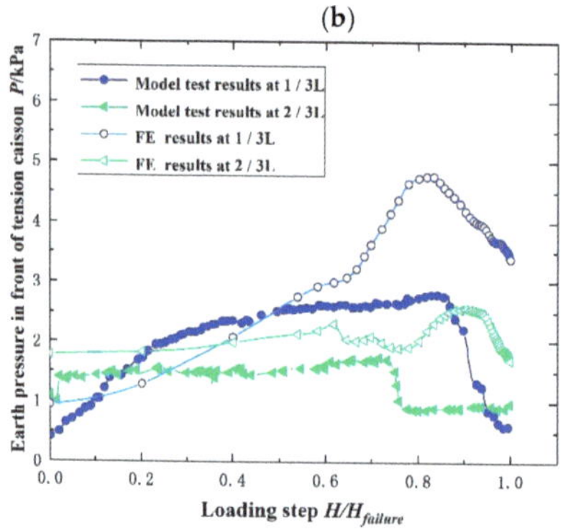

(**c**)

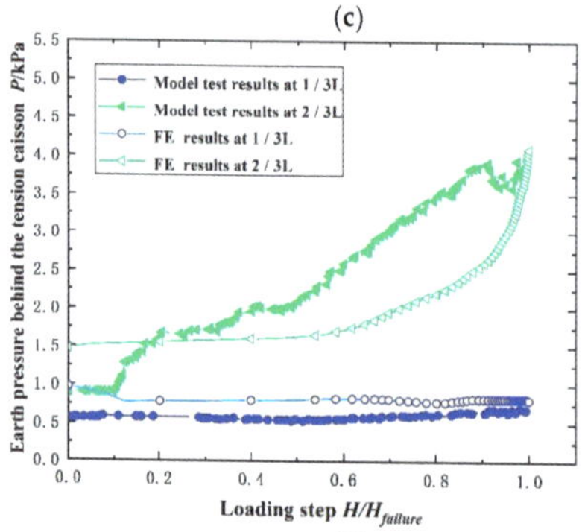

(**d**)

Figure 7. Distribution of earth pressure under the condition of $S/D = 3$ and $L/D = 1.61$. (**a**) Earth pressure in front of the compression bucket; (**b**) earth pressure behind the compression bucket; (**c**) earth pressure in front of the tension bucket; (**d**) earth pressure behind the tension bucket.

Figure 7a shows that the earth pressure in front of the compression bucket at $1/3\ L$ from the top of the bucket increased continuously with the increase in the horizontal load, while the earth pressure at $2/3\ L$ increased in the initial stage and decreased in the later stage. The reason was that the tripod bucket foundation would overturn and rotate under H–M and the rotation center was below $2/3\ L$ from the top of the bucket as shown in Table 3. Therefore, the soil at $1/3\ L$ and $2/3\ L$ in the front of the bucket was squeezed which resulted in the passive earth pressure. As the load increased, the rotation trend was more obvious, so the earth pressure caused by the squeezing effect increased. However, when the applied load increased to a certain value, the earth pressure at $2/3\ L$ began to decrease, which may be explained by the following two reasons. First, with the increase in load and deformation, the soil in front of the compression bucket changed from an elastic state to a plastic state due to the squeezing effect. Consequently, the earth pressure was cut down when the soil failed. Secondly, when the compression bucket is rotated, a void or loosen zone may be formed at the toe in front of the bucket. And the soil above (nearly 2/3 *L*) may fall down, which may cause the decrease in earth pressure.

Figure 7b shows the earth pressure distribution behind the compression bucket. Except for the model test at 2/3 *L*, the results show that the earth pressure behind the compression bucket was almost constant and small as the load increased. This indicated that the active earth pressure was sustained. When the load increased to about 0.7 of the ultimate capacity at 2/3 *L* in the model test, the earth pressure reduction may be caused by the separation between the earth pressure cell and soil due to the crack propagation behind the bucket.

Figure 7c shows the earth pressure results in front of the tension bucket. It can be seen that the earth pressure at 1/3 *L* increased first and then decreased, while the earth pressure at 2/3 *L* kept almost constant with slight fluctuations during the final steps. The tension bucket would rotate under *H*–*M*. The upper part of the bucket (about 1/3 *L*) tilted forward and the lower part (about 2/3 *L*) tilted back away from the soil. As a result, the soil at 2/3 *L* was in an active earth pressure zone and earth pressure kept a smaller value. It may slip slightly after being disturbed, resulting in the final decrease in earth pressure. The soil at 1/3 *L* was in a passive earth pressure area. With increasing loading, the earth pressure increased gradually until shear failure finally occurred.

It was noticeable that the results from model tests dropped more sharply, which may be caused by the existence of an earth pressure cell. The earth pressure cell was close to the model bucket occupying a certain volume, so the measured values were more vulnerable to cracks with increasing displacement compared with the results from the numerical simulation.

Figure 7d shows the earth pressure results behind the tension bucket. The earth pressure at 1/3 *L* tended to be constant and the values were smaller, which corresponded with the explanation in Figure 7c that soil at 1/3 *L* was in the active earth pressure zone and soil at 2/3 *L* was in the passive earth pressure zone. There was no shear failure of soil in the passive area, so earth pressure continued to increase as the load increased.

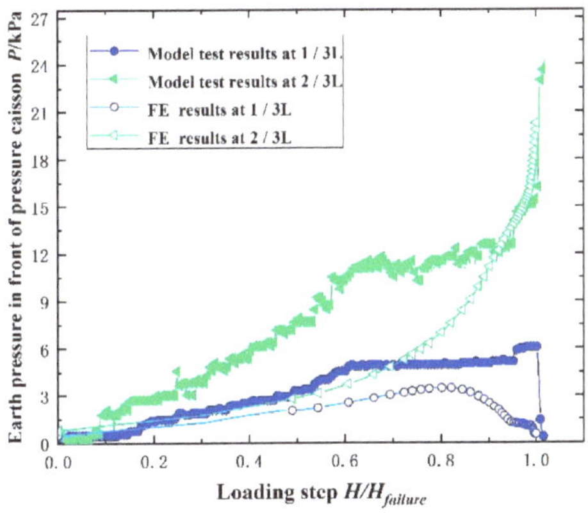

(a)

Figure 8. *Cont.*

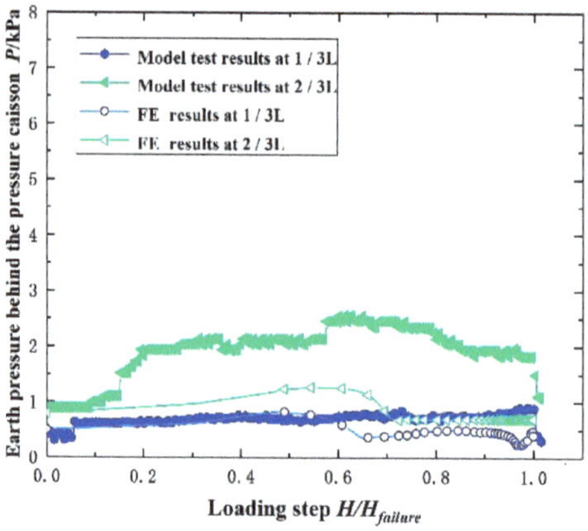

(**b**)

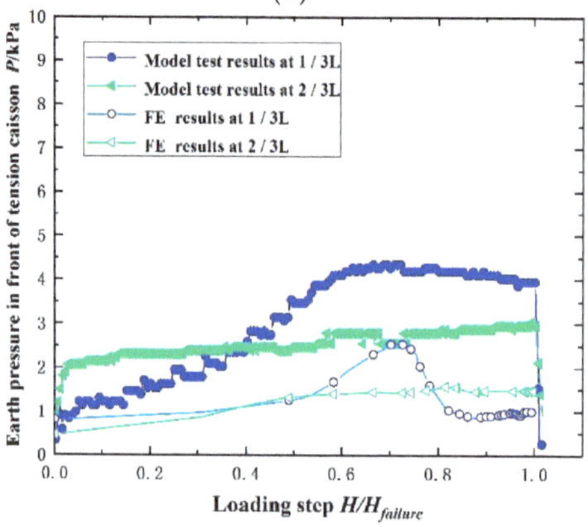

(**c**)

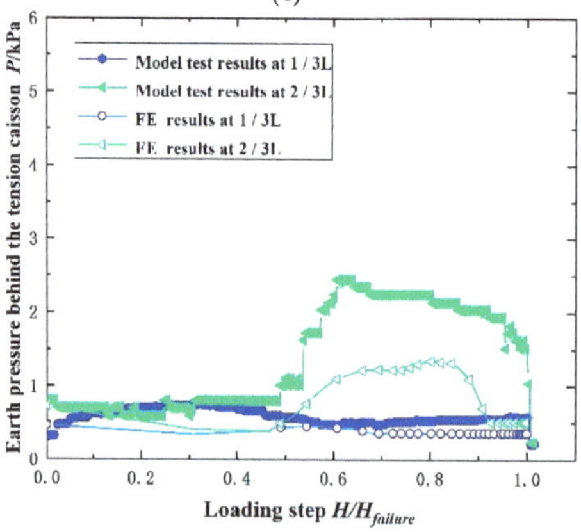

(**d**)

Figure 8. Distribution of earth pressure on bucket wall under the condition of $S/D = 3$ and $L/D = 0.85$. (**a**) Distribution of earth pressure in front of pressure bucket; (**b**) distribution of earth pressure behind pressure bucket; (**c**) distribution of earth pressure in front of tension bucket; (**d**) distribution of earth pressure behind tension bucket.

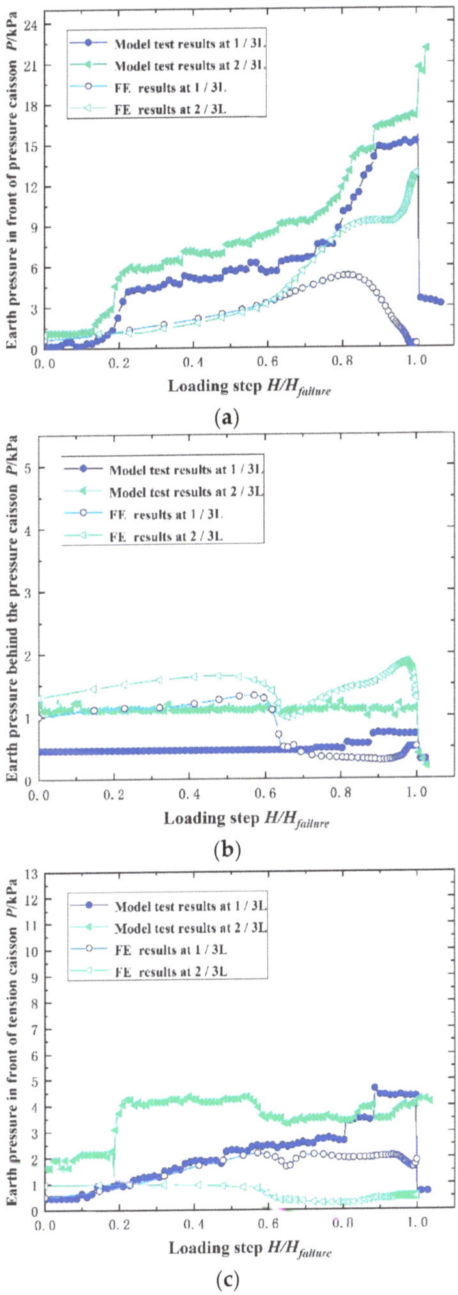

Figure 9. *Cont.*

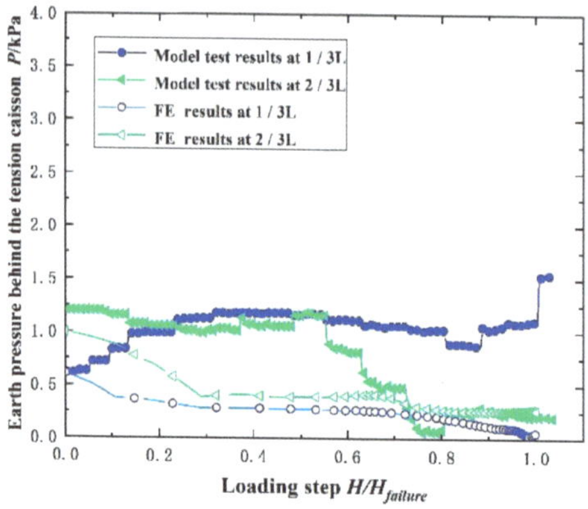

(d)

Figure 9. Distribution of earth pressure on bucket wall under the condition of S/D = 3 and L/D = 1.12. (**a**) Distribution of earth pressure in front of pressure bucket; (**b**) distribution of earth pressure behind pressure bucket; (**c**) distribution of earth pressure in front of tension bucket; (**d**) distribution of earth pressure behind tension bucket.

In comparison with Figure 7a, earth pressure in front of the compression bucket at 2/3 L kept increasing as shown in Figure 8a. This was mainly affected by the displacement behaviors. For the bucket foundation with a smaller L/D under H–M, its failure model can be compared with that of a wide shallow bucket foundation [30]. Horizontal displacement was dominant and the rotation center was located under the bottom of the bucket as shown in Table 3. The soil in front of the compression bucket was squeezed by the approximate translational motion of the bucket, which was difficult to cause sand leakage and local plastic failure.

The results in Figure 8b–d are similar to the earth pressure recorded in the test of S/D = 3 and L/D = 1.61. However, the results at 1/3 L and 2/3 L were closer, because the soil was more evenly squeezed. At the same time, the dominant horizontal displacement of the foundation was more likely to cause the separation between the earth pressure cell and soil, and so the earth pressure dropped dramatically.

Figure 9 shows the results of the earth pressure in the test of S/D = 3 and L/D = 1.12. Compared with the results in the test of S/D = 3 and L/D = 1.61, the horizontal displacement trend was more obvious. Compared with that of S/D = 3 and L/D = 0.85, the rotation trend was more obvious. The results in Figure 9a indicate that the bucket is subjected to passive earth pressure without obvious plastic failure. At the end of loading, there is a "steep change" in the earth pressure. In the model experiments, it is observed that the bucket maintains a small displacement until reaching the ultimate bearing capacity. At this moment, the tripod foundation loses stability suddenly and is pulled out. The difference between the earth pressure behind the compression bucket and the other two sets of tests is that the pressure decreases first and then increases when the loading comes to 0.6 times the ultimate capacity. The reason may be that there is a slight sliding of the soil behind the bucket. This results in a slight pressure decrease. In addition, the soil continues to compact in the later stage, thereby increasing the earth pressure. The distribution of earth pressure in front of the tension bucket is similar. However, the results behind the tension bucket are different. The earth pressure at 2/3 L from the top of the bucket begins to decrease in the later stage of loading, indicating that when the horizontal displacement is relatively large, there is a gap between the bucket wall and the soil behind it.

The size of the model used in the test was small and the volume occupied by the earth pressure cell cannot be ignored, so there were errors between the model tests and FE analysis. In fact, the bucket wall is circular arc-shaped, while the earth pressure measuring device is flat. The relatively bigger area of the measuring device may result in excessive and larger horizontal earth pressure measured. In numerical simulation, the measurement points did not take into account the size effect. So, the earth pressure measured by the big pressure measuring devices may be greater than that from the numerical simulation. If the size of the pressure measuring device can be reduced, it can be expected that the measurement results will be closer to the actual state. Basically, the trend of earth pressure distribution is the same. According to the results, the errors are acceptable.

Under *H–M* loadings, the instability mode of the tripod bucket foundation is manifested as follows. In the early stage, the foundation has a trend of translation and rotation. The rotation trend of the foundation with a large aspect ratio is more obvious. And for a small aspect ratio, the translation trend is more obvious. However, the final form is manifested as overturning failure. The specific failure mode is manifested as one bucket under compression and the other two under tension. The rotation center of the tripod bucket foundation is located near the axis of the compression bucket, and as the aspect ratio decreases, the position of the rotation center will be lower. Due to the center of rotation being located near the compression bucket, the deformation of the soil around the compression bucket is more obvious, and the earth pressure near the compression bucket is significantly higher than in other areas.

3.3. Moment Bearing Capacity of Tripod Bucket Foundation under M-H

Based on the experimental and numerical results, the failure mechanism of the tripod bucket foundation under *M-H* was actually in the form of one bucket under compression and the other two buckets under tension. According to the study of the failure mechanism, the tension bucket was pulled up with a certain forward inclination mode, and the compression bucket was compressed with a rotation mode. It was difficult to calculate the moment capacity directly. Hung and Kim [24] proposed a formula to calculate moment capacity based on the vertical capacity of the single bucket foundation, the length of the moment arm and a correction factor. The correction factor was determined as 1.1 by back-calculation.

In this paper, another formula was proposed to calculate moment capacity *M* as shown in Equation (1):

$$M_{0(T)} = 2f_M \times V_{0(S)} \times \frac{\sqrt{3}}{2}S \tag{1}$$

where f_M is the correction factor; S is the bucket spacing as shown in Figure 10; $V_{0(S)}$ is the uplift bearing capacity of a single tension bucket which can be calculated according to code API (American Petroleum Institute), as shown in Equations (2) and (3):

$$V_{0(S)} = f \times A_S \tag{2}$$

$$f = K \times p_0 \times \tan\delta \tag{3}$$

where f is the unit surface friction; A_S is the side surface of the bucket; K is the coefficient of lateral earth pressure and p_0 is the effective earth pressure.

Regarding the correction factor f_M, it was obtained by the back-calculation after applying horizontal force at 360 mm with L/D = 0.5, 0.85, 1.1, 1.5, 1.61, and S/D = 2, 2.5, 3. Finally, the parameter f_M was fitted according to Equation (4):

$$f_M = 0.47\ln\left(\frac{L}{D}\right) + 1.41f_M \tag{4}$$

The formula proposed in this paper was more convenient to calculate the moment capacity of the tripod bucket foundation in practice.

Table 4 shows the moment capacity of the tripod bucket foundation calculated by FE analysis and Equation (1), in which M_{0_FE} is the moment capacity from FE analyses, and M_{0_cal} is the calculated moment capacity using Equation (1) proposed in this study. According to the comparison results, the error was within 10%. The correction factor increased with the increase in L/D, instead of a constant value. The rotational trend of the foundation becomes apparent, accompanied by the complexity of the loading state, resulting in a more significant increase in bearing capacity. The bucket foundation with larger L/D tended to rotate under M-H, which was consistent with the phenomenon observed in the model test.

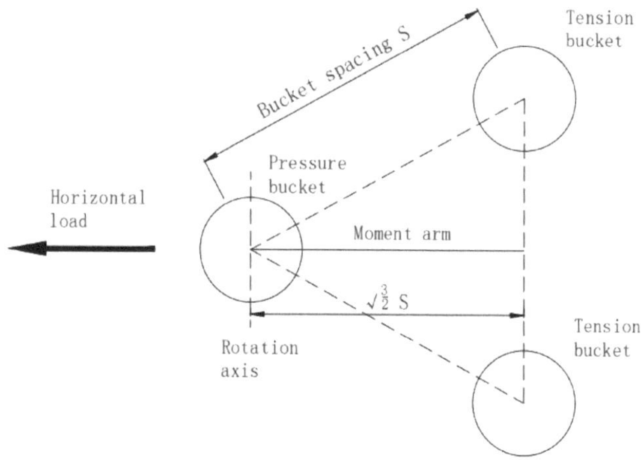

Figure 10. Definition of a moment arm.

Table 4. Comparison of FE results and calculated values for tripod bucket foundation with $S/D = 3$.

Test	D (mm)	$V_{0(S)}$ (N)	f_M	M_{0_FE} (N·cm)	M_{0_cal} (N·cm)	Error (%)
1	102	235.8	1.63	19,960	20,371	2.06
2	120	182.7	1.45	17,570	16,518	6.37
3	133	144	1.33	12,810	13,235	3.22

4. Discussion

In this study, physical model tests and numerical simulations were conducted to investigate the behaviors of the tripod bucket foundation in sand under horizontal and moment load. Failure mechanism and earth pressure characteristics were analyzed. Methods of estimating the moment bearing capacity were studied.

It should be noted that the results of this paper are based on a small scale model test, which may have a scaling effect. Further research on the bucket foundation of in situ tests under composite loading will be meaningful. In addition, the bearing characteristics of the tripod bucket foundation studied in this paper are in sandy soil conditions. In reality, the engineering geology where the wind farm is located may be clay soil, layered conditions or more comprehensive conditions. The bearing characteristics and behaviors of the tripod bucket foundation may be different in different conditions. It can be speculated that in clay, the ultimate failure mode of the tripod bucket foundation may still be one bucket under compression and the other two buckets under tension, but the overturning process may develop slowly rather than being directly pulled out in sand. In silty soil, considering the effect of accumulation of pore water pressure, the overall bearing capacity of the bucket foundation may decrease [18]. Bearing characteristics of tripod bucket foundations under layered soil conditions forms the ongoing work of the authors.

5. Conclusions

In this paper, bearing characteristics of tripod bucket foundation under horizontal and moment load in sand are investigated by physical model experiments and numerical simulations. Some main conclusions are as follows.

The failure process of the tripod bucket foundation under *H*–*M* can be divided into the initial stage, the intermediate stage and the failure stage. As L/D decreased, the rotation center lowered and the displacement mode varied from rotation to uplift.

Earth pressure measured by FE analysis and model tests were used to explore the failure mechanism of the tripod bucket foundation. It was found that the aspect ratio has a significant effect on the displacement mode and earth pressure characteristics. This was used to explain the failure mechanism for tripod bucket foundation.

An equation to calculate the moment capacity was proposed according to the uplift bearing capacity, the length of the moment arm and the correction factor. The correction factor was a function of L/D. It is thought to be a more convenient and practical method to estimate the moment capacity of the tripod bucket foundation.

Author Contributions: Conceptualization, X.-L.Z.; methodology, X.W.; software, X.W.; validation, X.W.; formal analysis, X.W.; investigation, X.W.; resources, X.-L.Z.; data curation, X.W.; writing—original draft preparation, X.W.; writing—review and editing, S.-H.S. and Z.-B.H.; visualization, S.-H.S. and Z.-B.H.; supervision, X.-L.Z., W.-N.D. and Q.-M.S.; project administration, X.-L.Z.; funding acquisition, X.-L.Z. All authors have read and agreed to the published version of the manuscript.

Funding: This research received no external funding.

Institutional Review Board Statement: Not applicable.

Informed Consent Statement: Not applicable.

Data Availability Statement: Not applicable.

Conflicts of Interest: The authors declare no conflict of interest.

References

1. Kim, J.-H.; Jeong, Y.-H.; Ha, J.-G.; Park, H.-J. Evaluation of Soil–Foundation–Structure Interaction for Large Diameter Monopile Foundation Focusing on Lateral Cyclic Loading. *J. Mar. Sci. Eng.* **2023**, *11*, 1303. [CrossRef]
2. Shi, Y.; Yao, W.; Yu, G. Dynamic Analysis on Pile Group Supported Offshore Wind Turbine under Wind and Wave Load. *J. Mar. Sci. Eng.* **2022**, *10*, 1024. [CrossRef]
3. Luke, A.M.; Rauch, A.F.; Olson, R.E.; Mecham, E.C. Components of suction caisson capacity measured in axial pullout tests. *Ocean Eng.* **2005**, *32*, 878–891. [CrossRef]
4. Standard, O. Design of offshore wind turbine structures. *Det Nor Ske Veritas* **2007**, *567–581*, 567–581.
5. Achmus, M.; Kuo, Y.-S.; Abdel-Rahman, K. Behavior of monopile foundations under cyclic lateral load. *Comput. Geotech.* **2009**, *36*, 725–735. [CrossRef]
6. Al-Ramthan, A.Q.O.; Aubeny, C.P. Numerical investigation of the performance of caissons in cohesive soils under cyclic loading. *Int. J. Geomech.* **2020**, *20*, 04020042. [CrossRef]
7. Ghorai, B.; Chatterjee, S. Estimation of installation resistance and subsequent short-term capacities of offshore skirted foundations in clay. *Int. J. Geomech.* **2020**, *20*, 04020133. [CrossRef]
8. Zhang, Y.; Bienen, B.; Cassidy, M.J.; Gourvenec, S. The undrained bearing capacity of a spudcan foundation under combined loading in soft clay. *Mar. Struct.* **2011**, *24*, 459–477. [CrossRef]
9. Kuo, Y.-S.; Achmus, M.; Abdel-Rahman, K. Minimum embedded length of cyclic horizontally loaded monopiles. *J. Geotech. Geoenviron. Eng.* **2012**, *138*, 357–363. [CrossRef]
10. Houlsby, G.T.; Ibsen, L.B.; Byrne, B.W. Suction caissons for wind turbines. In *Frontiers in Offshore Geotechnics, Proceedings of the ISFOG, Perth, WA, Australia, 19–21 September 2005*; CRC Press: Boca Raton, FL, USA, 2005; pp. 75–93.
11. Yun, G.; Bransby, M.F. The horizontal-moment capacity of embedded foundations in undrained soil. *Can. Geotech. J.* **2007**, *44*, 409–424. [CrossRef]
12. Achmus, M.; Akdag, C.; Thieken, K. Load-bearing behavior of suction bucket foundations in sand. *Appl. Ocean Res.* **2013**, *43*, 157–165. [CrossRef]
13. Bransby, F.; Randloph, M. The effect of embedment depth on the undrained response of skirted foundations to combined loading. *Soils Found.* **1999**, *39*, 19–33. [CrossRef]
14. Bransby, M.; Yun, G.-J. The undrained capacity of skirted strip foundations under combined loading. *Géotechnique* **2009**, *59*, 115–125. [CrossRef]
15. Gourvenec, S. Effect of embedment on the undrained capacity of shallow foundations under general loading. *Géotechnique* **2008**, *58*, 177–185. [CrossRef]

16. Ni, D.; Lu, M. Review on Bearing Capacity of Offshore Wind Turbine Single Pile and New Pile-barrel Combination Foundation. *Int. Core J. Eng.* **2022**, *8*, 433–442.
17. Sørensen, E.S.; Clausen, J.; Damkilde, L. Comparison of numerical formulations for the modeling of tensile loaded suction buckets. *Comput. Geotech.* **2017**, *83*, 198–208. [CrossRef]
18. Zhao, X.-L.; Wang, X.; Ding, P.-C.; Sui, S.-H.; Deng, W.-N. Development and Influence of Pore Pressure around a Bucket Foundation in Silty Soil. *J. Mar. Sci. Eng.* **2022**, *10*, 2020. [CrossRef]
19. Martin, C.; Hazell, E. Bearing capacity of parallel strip footings on non-homogeneous clay. In Proceedings of the International Symposium on Frontiers in Offshore Geotechnics, Perth, WA, Australia, 19–21 September 2005; pp. 427–433.
20. Gourvenec, S.; Jensen, K. Effect of embedment and spacing of cojoined skirted foundation systems on undrained limit states under general loading. *Int. J. Geomech.* **2009**, *9*, 267–279. [CrossRef]
21. Bang, S.; Leahy, J.C.; Cho, Y.; Kwon, O. Horizontal bearing capacity of suction piles in sand. *Transp. Res. Rec.* **2006**, *1975*, 21–27. [CrossRef]
22. Zhao, Z.; Li, D.; Zhang, F.; Qiu, Y. Ultimate lateral bearing capacity of tetrapod jacket foundation in clay. *Comput. Geotech.* **2017**, *84*, 164–173. [CrossRef]
23. Kim, S.-R.; Oh, M. Group effect on bearing capacities of tripod bucket foundations in undrained clay. *Ocean Eng.* **2014**, *79*, 1–9. [CrossRef]
24. Hung, L.C.; Kim, S.-R. Evaluation of undrained bearing capacities of bucket foundations under combined loads. *Mar. Georesour. Geotechnol.* **2014**, *32*, 76–92. [CrossRef]
25. He, B.; Jiang, J.; Cheng, J.; Zheng, J.; Wang, D. The capacities of tripod bucket foundation under uniaxial and combined loading. *Ocean Eng.* **2021**, *220*, 108400. [CrossRef]
26. Tran, N.X.; Kim, S.-R. Evaluation of horizontal and moment bearing capacities of tripod bucket foundations in sand. *Ocean Eng.* **2017**, *140*, 209–221. [CrossRef]
27. Brinkgreve, R.; Kumarswamy, S.; Swolfs, W.; Waterman, D.; Chesaru, A.; Bonnier, P. *PLAXIS 2016*; PLAXIS BV: Delft, The Netherlands, 2016.
28. Gourvenec, S.; Steinepreis, M. Undrained limit states of shallow foundations acting in consort. *Int. J. Geomech.* **2007**, *7*, 194–205. [CrossRef]
29. Kim, S.R. Evaluation of vertical and horizontal bearing capacities of bucket foundations in clay. *Ocean Eng.* **2012**, *52*, 75–82.
30. Palix, E.; Willems, T.; Kay, S. *Caisson capacity in clay: VHM resistance envelope—Part 1: 3D FEM numerical study. Frontiers in Offshore Geotechnics II*; CRC Press: Boca Raton, FL, USA, 2010; pp. 771–776.

Journal of
Marine Science and Engineering

MDPI

Article

Undrained Triaxial Shear Tests on Hydrate-Bearing Fine-Grained Sediments from the Shenhu Area of South China Sea

Ruchun Wei [1,2,3,4], Lele Liu [2,3,*], Chao Jia [1,4,*], Xiao Dong [1,4], Qingtao Bu [2,3], Yongchao Zhang [2,3], Changling Liu [2,3] and Nengyou Wu [2,3]

1 Institute of Marine Science and Technology, Shandong University, Qingdao 266232, China
2 Laboratory for Marine Mineral Resources, Qingdao National Laboratory for Marine Science and Technology, Qingdao 266237, China
3 Key Laboratory of Gas Hydrate, Ministry of Natural Resources, Qingdao Institute of Marine Geology, Qingdao 266237, China
4 The Key Laboratory of Geological Safety of Coastal Urban Underground Space, Ministry of Natural Resources, Shandong University, Qingdao 266232, China
* Correspondence: lele.liu@qnlm.ac (L.L.); jiachao@sdu.edu.cn (C.J.)

Abstract: Changes in undrained shear strength are important to the stability analysis of hydrate reservoirs during natural gas hydrate production. This study proposes a prediction model of undrained shear strength of hydrate-bearing fine-grained sediments based on the critical state theory. Several consolidated undrained triaxial shear tests are conducted on hydrate-bearing fine-grained samples from the Shenhu area of the South China Sea. The effects of effective consolidation stresses and hydrate saturations on the undrained shear strength are investigated. The results show that the undrained shear strength increases linearly with increasing effective consolidation stress. When the hydrate saturation is greater than the effective hydrate saturation, the undrained shear strength significantly increases with increasing hydrate saturation. The undrained shear strength of hydrate-bearing fine-grained sediments is a two-parameter function of effective hydrate saturation and a void ratio. The instability risk of the hydrate reservoir under undrained conditions is greater than that of under-drained or partially drained conditions. Furthermore, low-porosity reservoirs face more shear strength loss from hydrate decomposition yet lower risk than high-porosity ones. These results can improve the understanding of mechanical properties of hydrate-bearing fine-grained sediments under undrained conditions. This study also has implications for the design of marine structures in areas with hydrate-bearing sediment.

Keywords: natural gas hydrate; fine-grained sediment; critical state theory; effective hydrate saturation; stability analysis

Citation: Wei, R.; Liu, L.; Jia, C.; Dong, X.; Bu, Q.; Zhang, Y.; Liu, C.; Wu, N. Undrained Triaxial Shear Tests on Hydrate-Bearing Fine-Grained Sediments from the Shenhu Area of South China Sea. *J. Mar. Sci. Eng.* **2023**, *11*, 1604. https://doi.org/10.3390/jmse11081604

Academic Editor: Timothy S. Collett

Received: 18 July 2023
Revised: 7 August 2023
Accepted: 9 August 2023
Published: 16 August 2023

1. Introduction

Natural gas hydrates (NGHs) have been treated as a potential energy resource for decades because of their vast reserve and wide distribution in nature [1–3]. In recent years, several countries have carried out a series of production tests in the field to evaluate the possibility of different methods for the commercial production of NGHs [4–8]. The results clearly indicate that depressurization is a method with the highest production efficiency and the maximum probability of being commercialized [9–14]. The application effect of depressurization to marine NGHs is largely controlled by the hydraulic permeability of hydrate-bearing sediments which is highly stress dependent [15–19]. The stress of hydrate-bearing sediments generally increases during depressurization, and mechanical properties of hydrate-bearing sediments weaken due to NGHs dissociation [20–22]. In addition, NGHs dissociation can also be driven by climatic, oceanic, and geologic processes [23–27], and ultimately has a great potential to cause shear failures and trigger various geohazards [28–31]. In particular, for hydrate deposits in the Shenhu area of the

South China Sea [32,33], the fine-grained hydrate-bearing sediments are of low hydraulic permeabilities, and this implies that pore water and gas produced by NGHs dissociation due to underlying warm oil extraction and global warming can hardly flow out of pores in a short period, corresponding to an undrained condition [15,34–36]. Undrained shear properties of hydrate-bearing fine-grained sediments are of great significance to the stability of marine sediments and overburden structures.

Mechanical properties (e.g., shear strength, stiffness, and dilatancy) of hydrate-bearing sediments are largely dependent on hydrate saturation (i.e., a ratio of hydrate volume to the total pore volume) and hydrate pore habits [37–39]. With regard to mechanical properties, the hydrate pore habits mainly include pore-filling, load-bearing, and grain-cementing [40]. Debonding and/or crushing of NGHs within pores of coarse-grained sediments will occur during shearing [41]. However, the failure mechanism of hydrate-bearing fine-grained sediments has not been well understood since whether the cementation between clay and silt particles by NGHs exists or not remains elusive [42,43]. There is no doubt that NGHs behave as a solid filler of host sediments to reduce the effective void ratio (i.e., a ratio of fluid-occupied pore volume to the solid volume) of hydrate-bearing sediments [44–46], and the alteration of the effective void ratio affects the undrained shear strength of hydrate-bearing fine-grained sediments. For example, Yoneda et al. [47] found that there is a linear relationship between the undrained shear strength and the void ratio of fine-grained natural cores. Yun et al. [48] reported that the undrained shear strength of saturated natural cores acquired from the Gulf of Mexico is a function of water saturation. The enhancement effect to the effective void ratio due to NGHs dissociation is coupled with a reduction effect by compression deformation because of the increasing effective stress during depressurization. This makes it very challenging to predict the undrained shear strength of hydrate-bearing fine-grained sediments. Several prediction models have been proposed based on the Mohr-Coulomb criterion [49,50], and the models can give acceptable results in certain circumstances.

Critical state soil mechanics provides a theoretical framework to predict the undrained shear strength of saturated hydrate-free sediments with different void ratios [51]. The enhanced undrained shear strength of hydrate-bearing sediments can be captured by using an extended yield surface [52]. The extended yield surface in p'–q space is shown as a thick line in Figure 1, and the extension is an explicit product of multiple effects caused by the presence of NGHs. For details, Effect I represents that hydrate-free sediments can reach point A under an effective mean normal stress p'_0, indicating the yield surface of hydrate-free sediments shown as a thin solid line. The extension expressed as Effect II in the figure represents the observed strength enhancement due to the presence of NGHs within pores of marine sediments [53]. Since NGHs within pores have the potential to hinder volumetric deformation during consolidation [54], the void ratio of host sediments is generally larger than that of hydrate-free sediments when the effective mean normal stress is the same. This indicates that the yield surface of host sediments shown as a thin dashed line in Figure 1 is smaller than that of hydrate-free sediments, corresponding to the weakening effect (i.e., Effect III) of NGHs on the host sediments [55]. From this point of view, the intrinsic strength enhancement due to the presence of NGHs expressed as Effect IV in the figure is larger than the observed strength enhancement. For prediction model developments for mechanical properties (e.g., the undrained shear strength) of hydrate-bearing sediments, the intrinsic void ratio and shear strength enhancements due to the presence of NGHs within pores should be jointly considered.

To investigate the effects of effective consolidation stresses and hydrate saturations on the undrained shear strength on hydrate-bearing fine-grained sediments, this study performs undrained triaxial shear tests, and the experimental results are further used to validate a prediction model proposed based on the critical state theory in this study. The host sediments for hydrate formation are remodeled by using natural sediments acquired from the Shenhu area of the South China Sea, and the joint effect of hydrate is considered in the prediction model. This study has great potential to facilitate geotechnical designs of marine and submarine structures overlying hydrate-bearing sediments in nature.

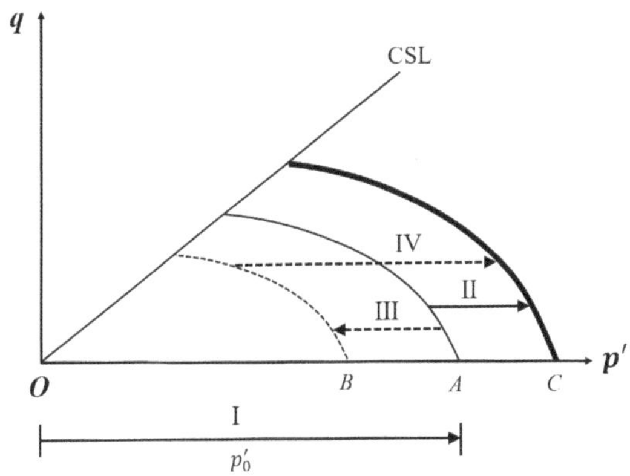

Figure 1. Yield surfaces in p'–q space for hydrate-free (the thin solid line) and hydrate-bearing (the thick solid line) sediments. The symbol p' stands for the effective mean normal stress, and q for the generalized shear stress. The abbreviation CSL represents the critical state line of soils.

2. Experimental Program

2.1. Experimental Apparatus and Materials

The experimental apparatus used in this study for undrained triaxial shear tests on hydrate-bearing sediments is upgraded from a commercialized triaxial shear test system manufactured by the Jiangsu Yongchang Science and Education Instrument Limited Company. A high-pressure test cell, resisting inner pressure up to 25 MPa, is applied instead of classic clear acrylic chambers, and an air bath is added to lower the temperature for hydrate formation. The load frame for triaxial shearing has a capacity of 250 kN, and the strain rate can be well controlled at a selected value ranging from 0.001 mm/min to 3 mm/min. A linear variable differential transformer (LVDT) produced by the Shanghai Tianmu Company is applied to measure the axial displacement, and the comprehensive accuracy is 0.1% of full scale (F.S). The axial force during triaxial shearing is measured by using a stress sensor with a comprehensive accuracy of 0.2% F.S. For more details about the apparatus, please refer to Dong et al. [56].

Clayey and silty sediments acquired from the Shenhu area in South China Sea are used to remodel host sediments for hydrate formation in this study. The grain size ranges from ~1 μm to 200 μm, and the median grain size is 13.9 μm. For more information about physical properties of marine sediments, please refer to Wei et al. [57]. Since methane hydrate formation within pores of fine-grained sediments is technically difficult and extremely time-consuming, tetrahydrofuran (THF) is selected as an analog of methane gas to form hydrate with pore water. Previous studies have shown that physical and mechanical properties of hydrate-bearing sediments are mostly dependent on hydrate occurrence characteristics [38,39,58]. The pore habit of THF hydrate is generally consistent with that of NGHs in nature [59]. Therefore, this analog is widely adopted in the gas hydrate community for experimental studies on mechanical properties of hydrate-bearing fine-grained sediments [60–62].

2.2. Procedure for Host Sediment Preparation

The procedure for host sediment preparation in this study is briefly summarized as follows: (i) The natural sediments are air-dried and then well mixed with some THF solution to acquire an initial solution content of 15% by weight; (ii) the moist sediments are sealed in a bag to distribute the solution for 12 h; (iii) the moist sediments are remodeled to form a cylindrical specimen with a diameter of 38 mm and a height of 76 mm as host sediments for hydrate formation; (iv) the vacuum method is applied to fully saturate the host sediments, and the saturation time is longer than 24 h. THF solutions with different mass ratios of THF and water are used to control hydrate saturation, and a void ratio of the host sediments is selected according to the porosity logging data.

2.3. Procedure for Hydrate Formation and Triaxial Shearing

In this study, consolidation is conducted before hydrate formation to avoid the hindering effect of pore hydrate on compression deformation. During the consolidation, the pore pressure is kept at the atmospheric pressure, and the confining pressure is enhanced by a rate of 100 kPa/min until a selected effective confining pressure is reached. The whole period of consolidation is generally longer than 24 h. After full consolidation, saturated host sediments are cooled, and the temperature is controlled at around 0 °C for 4 days to form THF hydrate. Then, the temperature is adjusted to 3 °C higher and kept for 2 days to eliminate potential ice within pores of hydrate-bearing sediments. For hydrate-free sediments, the consolidation and freezing procedures are also applied, but using pure water instead of a THF solution. Undrained triaxial shearing with a constant rate of 0.04%/min is performed on hydrate-bearing sediments, and the shearing is stopped when the axial strain reaches 15%. According to ASTM D2216 [63], the water content determined after triaxial shearing can be used to calculate the effective and intrinsic void ratios of hydrate-bearing sediments.

2.4. Test Design

To simulate the real case in the Shenhu area of the South China Sea [64,65], values of the initial porosity of host sediments are controlled as 0.525, 0.500, and 0.482, corresponding to the initial void ratios of 1.105, 1.000, and 0.932. To avoid over-consolidation, values of the effective confining pressure are set as 1.0 MPa, 2.0 MPa, and 3.0 MPa based on the void ratio vs. the consolidation stress curve [57]. Values of hydrate saturation are selected as 0%, 30%, and 50%, corresponding to the THF vs. water mass ratios of 0, 0.060, and 0.104 [15].

3. Results

3.1. Effects of Effective Consolidation Stress on the Undrained Shear Strength

Stress–strain curves under different conditions of the effective consolidation stress (i.e., confining pressure) are shown in Figure 2. It is obvious that all the curves are strain hardening, in which the generalized shear stress q (i.e., the deviatoric stress for triaxial shear tests) constantly increases with increasing axial strain ε_a. Values of the peak deviatoric stress q_f (i.e., two times the undrained shear strength S_u) under different conditions are summarized in Table 1, and the q_f-value increases with increasing effective consolidation stress. Values of the intrinsic void ratio e after triaxial shearing are also listed in Table 1. It is generally accepted that e-value decreases with increasing effective consolidation stress, leading to higher q_f-value of hydrate-bearing sediments. Figure 3 shows how the undrained shear strength of hydrate-free and hydrate-bearing sediments evolves with effective consolidation stress (i.e., confining pressure) and hydrate saturation. It is obvious that the undrained shear strength S_u linearly increases with increasing effective consolidation stress p'_0, and the fitted slope is 0.57, 0.78, and 1.67 for S_h = 0%, 30%, and 50%, respectively.

Table 1. Key information of undrained triaxial shear tests on hydrate-free and hydrate-bearing sediments in this study. (Note e_0 for the intrinsic void ratio of host sediments before consolidation, p'_0 for the effective consolidation stress, S_h for the hydrate saturation, q_f for the peak strength in stress–strain curves, and e for the intrinsic void ratio of sediments after triaxial shearing.).

e_0	p'_0 (MPa)	S_h (%)	q_f (MPa)	e
1.105	1	0	1.16	1.063
		30	1.86	1.120
		50	7.64	1.051

Table 1. *Cont.*

e_0	p'_0 (MPa)	S_h (%)	q_f (MPa)	e
		0	2.32	0.993
1	2	30	3.72	0.898
		50	11.30	0.924
		0	3.42	0.917
0.932	3	30	4.98	0.854
		50	14.32	0.890

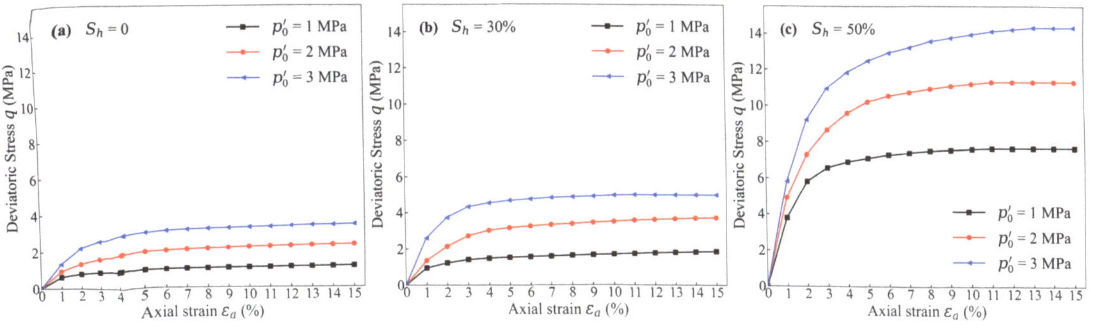

Figure 2. Stress–strain curves of hydrate-free (**a**) and hydrate-bearing sediments under different effective consolidation stresses (i.e., confining pressures) p'_0. The hydrate saturations S_h are 30% (**b**) and 50% (**c**).

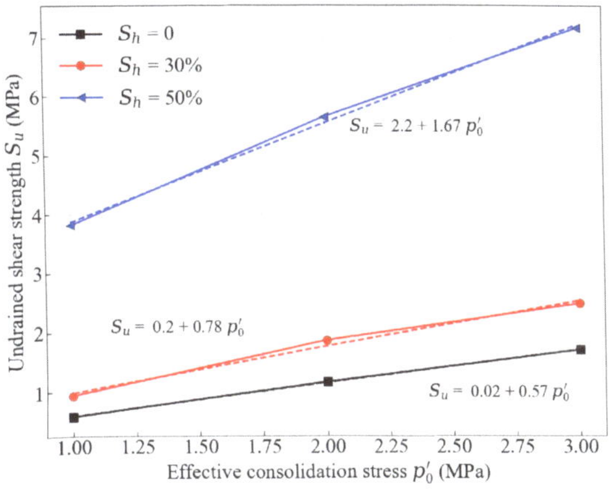

Figure 3. The undrained shear strength S_u of hydrate-free and hydrate-bearing sediments under different effective consolidation stresses p'_0.

3.2. Effects of Hydrate Saturation on the Undrained Shear Strength

Figure 4 shows the stress–strain curves of marine sediments with different hydrate saturations S_h. It is obvious that the q_f value increases with increasing S_h when the effective consolidation stress p'_0 is the same. For example, $q_f = 2.32$ MPa for $S_h = 0\%$ when $p'_0 = 2.0$ MPa, $q_f = 3.72$ MPa for $S_h = 30\%$, and $q_f = 11.3$ MPa for $S_h = 50\%$. The increasing hydrate saturation S_h reduces the effective void ratio of hydrate-bearing sediments, leading to increasing q_f. This is consistent with the results of Yang et al. [20]. In addition, the stress–strain curves for hydrate saturation equaling 50% are significantly higher than those for lower hydrate saturations. This indicates that the increasing hydrate saturation S_h

has a stronger effect on the elevating q_f of hydrate-bearing sediments than the increasing effective consolidation stress p'_0. Figure 5 shows the undrained shear strength S_u under different hydrate saturations S_h. It is clearly shown that the increase is nonlinear, and the increment of S_u for S_h increasing from 0% to 30% is significantly smaller than that for S_h increasing from 30% to 50%. This implies that there is a critical hydrate saturation S_{he} beyond which the undrained shear strength enhancement becomes much more obvious. The critical hydrate saturation is lower than 30%, and this is consistent with the results of De La Fuente et al. [66].

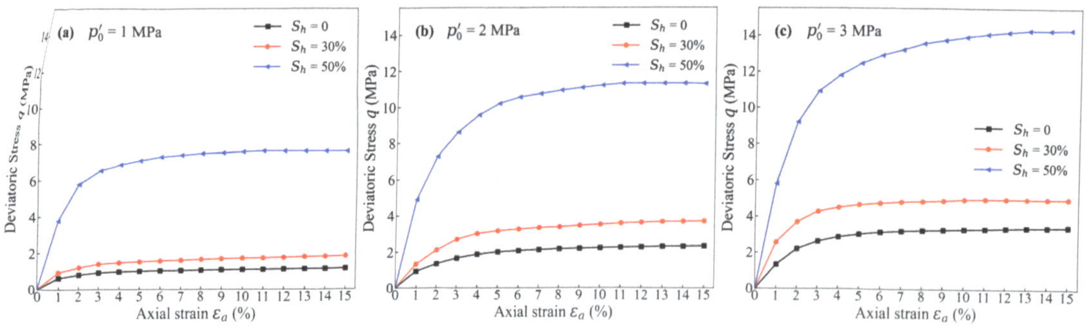

Figure 4. Stress–strain curves of marine sediments with different hydrate saturations S_h. p'_0 = 1.0 MPa (**a**), 2.0 MPa (**b**), and 3.0 MPa (**c**).

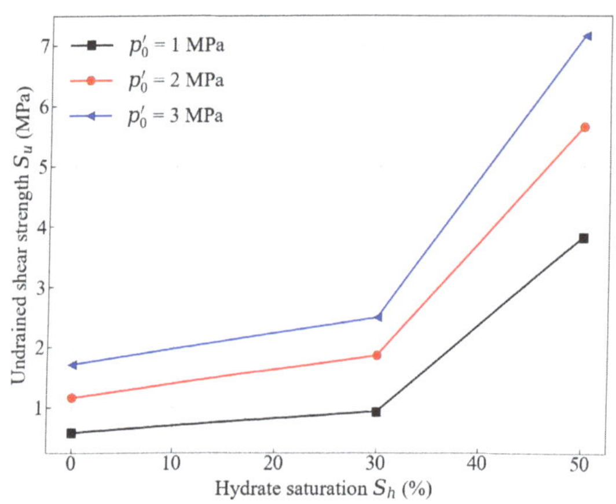

Figure 5. The undrained shear strength S_u of hydrate-free and hydrate-bearing sediments with different hydrate saturations S_h.

4. Development and Validation of a Theoretical Model for Predicting the Undrained Shear Strength

4.1. Development of the Prediction Model

According to critical state soil mechanics [51], the undrained shear strength S_u of saturated hydrate-free sediments can be predicted by:

$$S_u = \frac{M}{2} \exp\left(\frac{\Gamma - \nu}{\lambda}\right) \tag{1}$$

where M is stress ratio at the critical state; Γ is the specific volume intercept of the critical state line, and λ is the slope of the critical state line; ν represents specific volume and equals to $1 + e$. In addition, M, Γ, λ, and ν are dimensionless parameters.

Equation (1) represents that the undrained shear strength of saturated hydrate-free sediments depends on the void ratio regardless of the consolidation history. For saturated hydrate-bearing sediments, solid hydrate occupies the pore space and reduces the effective specific volume. However, not all the pore hydrate has obvious effects on mechanical properties of hydrate-bearing sediments, and an effective hydrate saturation S_{he} is adopted. Yan et al. [67] propose a simple equation to determine S_{he} as follows:

$$S_{he} = \langle S_h - \xi S_{hc} \rangle \tag{2}$$

where $\langle\ \rangle$ is the Macaulay bracket. When value of $S_h - \xi S_{hc}$ is larger than zero, the value of S_{he} is equal to that of $S_h - \xi S_{hc}$, while the value of S_{he} is equal to zero when the value of $S_h - \xi S_{hc}$ is lower than zero. Values of critical hydrate saturation S_{hc} are within a range from 25% to 40% [43], and S_{hc} = 25% in this study. Parameter ξ represents different effects of different hydrate pore habits, and ξ = 0 for cementing the hydrate while ξ = 1 for non-cementing (e.g., pore-filling and load-bearing) hydrate.

By treating pore hydrate as part of the solid skeleton, the effective void ratio of hydrate-bearing sediments e_h is calculated as $e \cdot (1 - S_{he})/(1 + e \cdot S_{he})$. Therefore, the effective specific volume v_h of hydrate-bearing sediments is expressed as:

$$v_h = \frac{1+e}{1+e \cdot S_{he}} \tag{3}$$

Combining Equations (1)–(3), a prediction model for the undrained shear strength of hydrate-bearing fine-grained sediments is proposed as:

$$\begin{cases} S_{hu} = \frac{M}{2}\exp\left(\frac{\Gamma - v_h}{\lambda}\right) = \frac{M}{2}\exp\left(\frac{\Gamma - \frac{1+e}{1+e \cdot S_{he}}}{\lambda}\right) \\ S_{he} = \langle S_h - \xi S_{hc} \rangle \end{cases} \tag{4}$$

Based on the critical state soil mechanics, when sediments reach the critical state, all the properties of structure within the sediments would be destroyed [51]. Therefore, parameters M, Γ, and λ are constant corresponding to the host sediments regardless of hydrate saturation [68,69]. Values of ξ and S_{hc} are related to hydrate pore habits [70]. Since it is difficult to form fully cementing hydrate within fine-grained sediments [43], ξ is equal to 1 in this study. Values of hydrate saturation S_h and effective void ratio e are measured during triaxial shear tests.

4.2. Validation of the Prediction Model

A series of consolidated undrained shear tests on hydrate-free sediments have been performed [57], and the results are used to fit for parameter value extraction (Figure 6). It is obvious that M = 1.28, Γ = 3.25, and λ = 0.175. These values are applied in this study, and the predicted results for hydrate-bearing sediments are shown in Figure 7. It is clearly shown that the predicted values are well consistent with measured values when the hydrate saturation is no larger than 30%. However, the predicted results are significantly higher than the test results for the specimen with a hydrate saturation of 50%. There are two possible reasons for this discrepancy: (i) the specimen with S_h = 50% has a relatively high density, and shear failure may occur before specimens reach the critical state; (ii) possible error in the critical hydrate saturation S_{hc} will lead to uncertainties in the calculated results of the effective hydrate saturation S_{he}.

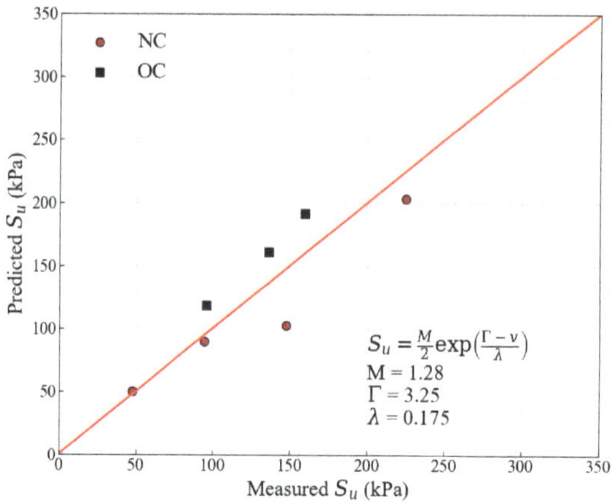

Figure 6. Predicted and measured values of the undrained shear strength S_u of hydrate-free sediments. The abbreviation NC stands for normal consolidated, and OC for over-consolidated. The measured values are from Wei et al. [57]. The model used in this figure is Equation (1).

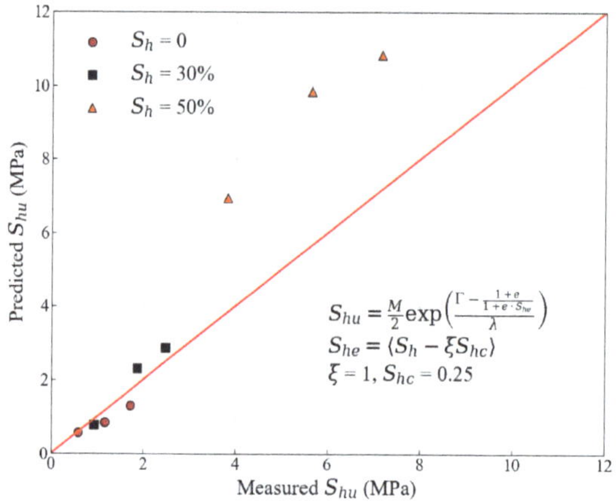

Figure 7. Predicted and measured values of the undrained shear strength S_{hu} of hydrate-bearing sediments in this study. The model used in this figure is Equation (4).

To further explain the effect of S_{hc} on the accuracy of the prediction model, values of the undrained shear strength with different critical hydrate saturations are calculated by using the prediction model, and the variation curves are shown in Figure 8. Mean void ratios of 1.078, 0.933, and 0.887 corresponding to p'_0 = 1.0 MPa, 2.0 MPa, and 3 MPa are used for calculation, and the arrows represent the overall increasing trend of S_{hc}. It is shown that the undrained shear strength decreases with increasing critical hydrate saturation. The test results with a hydrate saturation of 30% are on the curve with S_{hc} = 0.25, while the test results with a hydrate saturation of 50% are near the curve with S_{hc} = 0.30. Therefore, it can be inferred that the difference in prediction accuracy depends on the value of critical hydrate saturation S_{hc}. The results also indicate that mechanical properties of hydrate-bearing sediments are dependent on hydrate pore habits. For pore-filling hydrate, its effect only occurs when hydrate saturation S_h exceeds its effective hydrate saturation S_{he} [66]. The magnitude of effective hydrate saturation S_{he} is controlled by the critical hydrate saturation S_{hc} according to the proposed model in this study. Therefore, the value of critical hydrate saturation is crucial for the accuracy of model prediction results. Based on the comparison between the experimental data and theoretical calculation results shown in Figure 8, an empirical model is proposed to correct the effective hydrate saturation S_{he}:

$$S_{he} = \langle S_h - \xi(S_h + a) \cdot S_{hc} \rangle \tag{5}$$

where a is an empirical parameter, and a = 0.7 in this study.

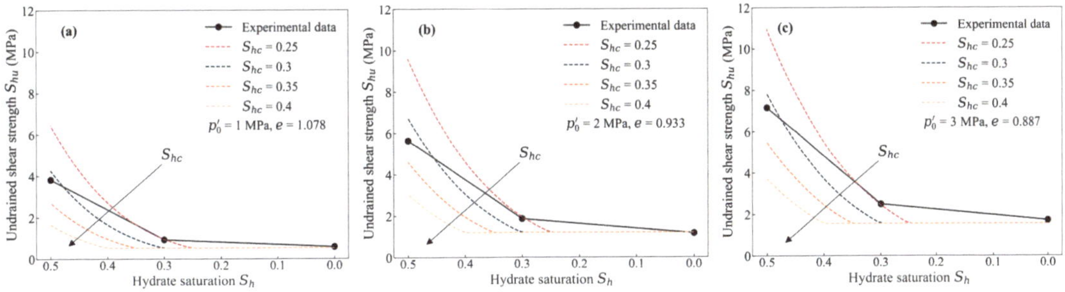

Figure 8. Effects of the critical hydrate saturation S_{hc} on the undrained shear strength of hydrate-bearing sediments S_{hu} under different consolidation stresses p'_0 = 1.0 MPa (**a**), 2.0 MPa (**b**), and 3.0 MPa (**c**). Colored dashed lines are drawn by using the prediction model of Equation (4).

Thus, the modified prediction model is expressed as:

$$\begin{cases} S_{hu} = \frac{M}{2} \exp\left(\frac{\Gamma - \frac{1+e}{1+e \cdot S_{he}}}{\lambda} \right) \\ S_{he} = \langle S_h - (S_h + a) S_{hc} \rangle \end{cases} \tag{6}$$

The prediction results calculated by using the modified prediction model are compared with the experimental data (Figure 9). It is shown that the corrected prediction results agree well with the experimental results. It is concluded that the undrained shear strength of hydrate-bearing fine-grained sediments is jointly affected by the effective hydrate saturation S_{he} and the effective void ratio e. The accuracy of the effective hydrate saturation S_{he} is dependent on value of the critical hydrate saturation S_{hc}.

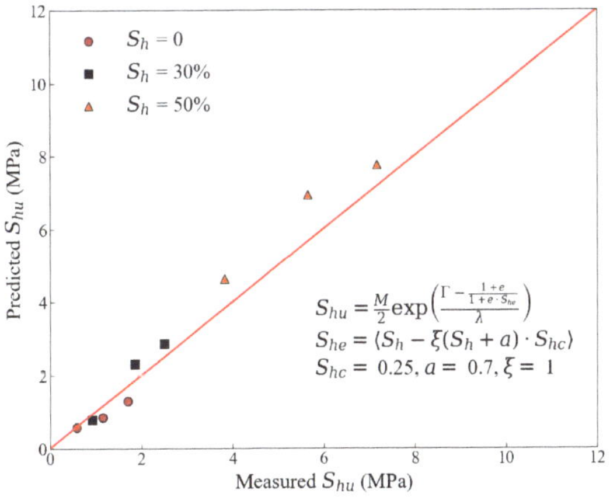

Figure 9. Predicted and measured values of the undrained shear strength of hydrate-bearing sediments S_{hu} in this study. The model used in this figure is Equation (6).

5. Implications to the Production of NGHs in Nature

Values of S_u with different hydrate saturations and porosities are calculated by using the modified prediction model, and the variation curves are shown in Figure 10. It is shown that hydrate dissociation obviously leads to a reduction in the undrained shear strength of hydrate-bearing sediments. However, if consolidation of the sediments is allowed, the increasing undrained shear strength due to the reduction in porosity would compensate for the loss in the undrained shear strength caused by hydrate decomposition. For sediments with an initial hydrate saturation of 0.5 and an initial porosity of 0.525, if the hydrate is completely decomposed and undergoes consolidated deformation with a final porosity of 0.425. The undrained shear strength before and after hydrate decomposition is almost the same. However, the equilibrium of undrained shear strength variation is not achieved for all conditions. For example, for sediments with an initial porosity of 0.475, and all other conditions being unchanged, the undrained shear strength would not exceed the initial state even if the consolidation is completed. Although the stability of the reservoirs benefits from the decrease in porosity, it will in turn greatly reduce the permeability limiting the extraction efficiency [71]. Therefore, further research is needed to explore the balance of stability and efficiency.

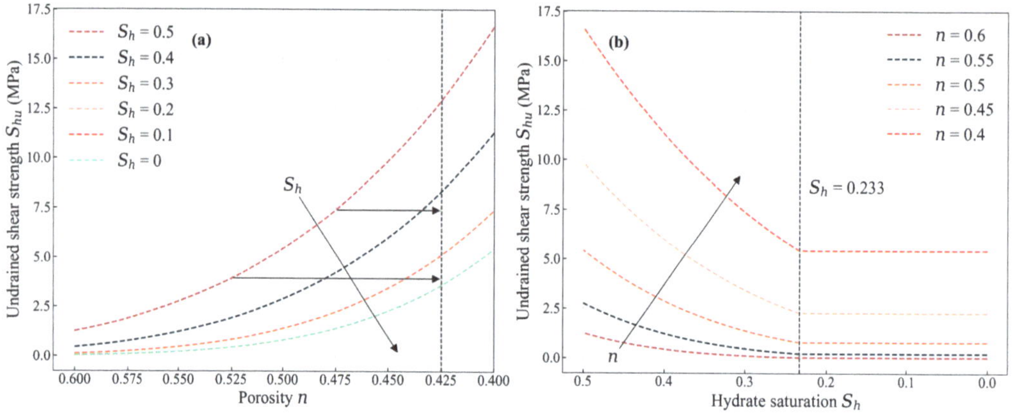

Figure 10. Calculated values of the undrained shear strength of hydrate-bearing sediments S_{hu} under different conditions of hydrate saturation (**a**) and porosity (**b**). Colored dashed lines are acquired by using Equation (6).

Under the condition that the porosity remains constant with hydrate saturation decreasing, the lower the porosity, the more obvious the effect of hydrate decomposition on its undrained shear strength (Figure 10b). Furthermore, the continued decrease in hydrate saturation does not cause a further decrease in the undrained shear strength when the hydrate saturation is lower than 0.233. The hydrate decomposition leads to a decrease in the density of sediments and thus its undrained shear strength. There is a lower limit of the undrained shear strength. This lower limit is related to the intrinsic porosity of host sediments. The magnitude of the lower limit of sediments with low porosity is greater than that of sediments with high porosity. Therefore, although the loss of undrained shear strength of reservoirs with low porosity is greater than that of reservoirs with high porosity during hydrate decomposition, the former has a lower instability risk than the latter.

To summarize, the results are of great potential to the stability analysis of hydrate reservoirs in marine and cold environments, and the design of structures for production [72,73]. The instability risk of the hydrate reservoir under undrained conditions is greater than that of under-drained or partially drained conditions. This is because there is a strength compensation caused by consolidated deformation although hydrate decomposition causes a decrease in the shear strength of reservoirs. For example, hydrate reservoirs in the South

China Sea are mainly composed of fine-grained sediments [33,74]. Since the fine-grained sediments are of low hydraulic permeabilities, the excess pore pressure caused by changes in stress states tends to be accumulated within the reservoirs during NGH production. This, in turn, leads to decreasing effective stress which promotes the possibility of submarine slope instability.

6. Conclusions

A prediction model of undrained shear strength of hydrate-bearing fine-grained sediments is developed based on the critical state theory. Several consolidated undrained triaxial shear tests are conducted on remodeled specimens. The effects of effective consolidation stresses and hydrate saturations on the undrained shear strength are investigated. The prediction model is validated using the test results, and reasonably corrected. The main conclusions are as follows:

(1) Values of q_f increase with increasing effective consolidation stress and hydrate saturation. There is a linear relationship between undrained shear strength and effective consolidation stress, while when the hydrate saturation is greater than the effective hydrate saturation, the undrained shear strength significantly increases with increasing hydrate saturation.
(2) The undrained shear strength of hydrate-bearing fine-grained sediments is a two-parameter function of effective hydrate saturation and void ratio.
(3) The instability risk of the hydrate reservoir under undrained conditions is greater than that of under-drained or partially drained conditions. This is because of that although hydrate decomposition causes a decrease in the shear strength of reservoirs, there is a strength compensation caused by consolidated deformation.
(4) The decrease in the undrained shear strength caused by the hydrate decomposition has a lower limit. This lower limit is related to the own porosity of host sediments. Low porosity reservoirs face more shear strength loss from hydrate decomposition yet lower risk than high porosity ones.

Admittedly, there are uncertainties in the effect of hydrate occurrences on parameters ξ and S_{hc}. Especially, since all the hydrate-bearing specimens undergo consolidation before hydrate generation in the test procedure, the effect of consolidation stress on hydrate occurrences is not considered. In addition, the validation of hydrate saturations ranged from 0 to 0.5. The higher hydrate saturation is needed to be further verified. Therefore, it is necessary to carry out more consolidated undrained triaxial shear tests on hydrate-bearing fine-grained sediments to improve the prediction model.

Author Contributions: Conceptualization, R.W., L.L. and C.J.; methodology, R.W.; validation, R.W. and X.D.; formal analysis, L.L.; investigation, Y.Z.; resources, C.L.; data curation, X.D.; writing—original draft preparation, R.W.; writing—review and editing, L.L. and C.J.; visualization, R.W. and X.D.; supervision, Q.B., Y.Z. and C.L.; funding acquisition, L.L., Q.B. and N.W. All authors have read and agreed to the published version of the manuscript.

Funding: This study is jointly supported by the National Natural Science Foundation of Shandong Province (grant number ZR2022YQ54), the Open Research Fund Program of State key Laboratory of Hydroscience and Engineering (grant number sklhse-2022-D-03), the Marine S&T Fund of Shandong Province for Pilot National Laboratory for Marine Science and Technology (Qingdao) (grant number 2021QNLM020002), the Taishan Scholars Program (grant number tsqn202306297), and the China Geological Survey (Project DD20230064).

Institutional Review Board Statement: Not applicable.

Informed Consent Statement: Not applicable.

Data Availability Statement: All data have been provided in the paper.

Conflicts of Interest: The authors declare no conflict of interest.

Abbreviations

CSL	critical state line
CU	consolidated undrained
e	void ratio
v	specific volume, equals to 1 + e
v_h	specific volume of hydrate-bearing sediments
M	stress ratio at the critical state
NC	normally consolidated
NGHs	natural gas hydrates
OC	over consolidated
p'_0	effective confining pressure
p'	mean effective stress, equal to $(\sigma'_1 + 2\sigma'_3)/3$
q	deviatoric stress, equal to $\sigma_1 - \sigma_3$
q_f	deviatoric stress at failure
SCS	South China Sea
S_h	hydrate saturation
S_{hc}	critical hydrate saturation
S_{he}	effective hydrate saturation
S_u	undrained shear strength, equal to $q_f/2$
S_{hu}	undrained shear strength of hydrate-bearing sediments
w	water content
ε_a	axial strain
Γ	specific volume intercept of the critical state line
λ	slope of the critical state line

References

1. Kvenvolden, K.A. Gas hydrates-geological perspective and global change. *Rev. Geophys.* **1993**, *31*, 173–187. [CrossRef]
2. Sloan, E.D. Fundamental principles and applications of natural gas hydrates. *Nature* **2003**, *426*, 353–359. [CrossRef] [PubMed]
3. Collett, T.; Bahk, J.-J.; Baker, R.; Boswell, R.; Divins, D.; Frye, M.; Goldberg, D.; Husebø, J.; Koh, C.; Malone, M.; et al. Methane Hydrates in Nature—Current Knowledge and Challenges. *J. Chem. Eng. Data* **2015**, *60*, 319–329. [CrossRef]
4. Wang, Z.; Zhang, Y.; Peng, Z.; Shan, Z.; Sun, B.; Sun, J. Recent Advances in Methods of Gas Recovery from Hydrate-Bearing Sediments: A Review. *Energy Fuels* **2022**, *36*, 5550–5593. [CrossRef]
5. Kurihara, M.; Sato, A.; Funatsu, K.; Ouchi, H.; Yamamoto, K.; Numasawa, M.; Ebinuma, T.; Narita, H.; Masuda, Y.; Dallimore, S.R.; et al. Analysis of Production Data for 2007/2008 Mallik Gas Hydrate Production Tests in Canada. In *Proceedings of the All Days*; SPE-132155-MS; SPE: Richardson, TX, USA, 2010.
6. Rossi, F.; Gambelli, A.M. Thermodynamic phase equilibrium of single-guest hydrate and formation data of hydrate in presence of chemical additives: A review. *Fluid Phase Equilibria* **2021**, *536*, 112958. [CrossRef]
7. Mahmood, M.N.; Guo, B. Gas production from marine gas hydrate reservoirs using geothermal-assisted depressurization method. *Adv. Geo-Energy Res.* **2023**, *7*, 90–98. [CrossRef]
8. Liu, L.; Lu, X.; Zhang, X.; Liu, C.; Du, B. Numerical simulations for analyzing deformation characteristics of hydrate-bearing sediments during depressurization. *Adv. Geo-Energy Res.* **2017**, *1*, 135–147. [CrossRef]
9. Ye, J.; Qin, X.; Xie, W.; Lu, H.; Ma, B.; Qiu, H.; Liang, J.; Lu, J.; Kuang, Z.; Lu, C.; et al. The second natural gas hydrate production test in the South China Sea. *China Geol.* **2020**, *3*, 197–209. [CrossRef]
10. Li, J.; Ye, J.; Qin, X.; Qiu, H.; Wu, N.; Lu, H.; Xie, W.; Lu, J.; Peng, F.; Xu, Z.; et al. The first offshore natural gas hydrate production test in South China Sea. *China Geol.* **2018**, *1*, 5–16. [CrossRef]
11. Yamamoto, K.; Terao, Y.; Fujii, T.; Ikawa, T.; Seki, M.; Matsuzawa, M.; Kanno, T. Operational overview of the first offshore production test of methane hydrates in the Eastern Nankai Trough. In Proceedings of the Offshore Technology Conference, Houston, TX, USA, 5–8 May 2014; pp. 1–15.
12. Anderson, B.; Hancock, S.; Wilson, S.; Enger, C.; Collett, T.; Boswell, R.; Hunter, R. Formation pressure testing at the Mount Elbert Gas Hydrate Stratigraphic Test Well, Alaska North Slope: Operational summary, history matching, and interpretations. *Mar. Pet. Geol.* **2011**, *28*, 478–492. [CrossRef]
13. Boswell, R.; Schoderbek, D.; Collett, T.S.; Ohtsuki, S.; White, M.; Anderson, B.J. The Iġnik Sikumi Field Experiment, Alaska North Slope: Design, Operations, and Implications for CO_2–CH_4 Exchange in Gas Hydrate Reservoirs. *Energy Fuels* **2017**, *31*, 140–153. [CrossRef]
14. Yamamoto, K.; Wang, X.-X.; Tamaki, M.; Suzuki, K. The second offshore production of methane hydrate in the Nankai Trough and gas production behavior from a heterogeneous methane hydrate reservoir. *RSC Adv.* **2019**, *9*, 25987–26013. [CrossRef] [PubMed]
15. Zhang, Z.; Liu, L.; Ning, F.; Liu, Z.; Sun, J.; Li, X.; Sun, J.; Hyodo, M.; Liu, C. Effect of stress on permeability of clay silty cores recovered from the Shenhu hydrate area of the South China Sea. *J. Nat. Gas Sci. Eng.* **2022**, *99*, 104421. [CrossRef]

16. Huang, L.; Su, Z.; Wu, N.; Cheng, J. Analysis on geologic conditions affecting the performance of gas production from hydrate deposits. *Mar. Pet. Geol.* **2016**, *77*, 19–29. [CrossRef]
17. Okwananke, A.; Hassanpouryouzband, A.; Vasheghani Farahani, M.; Yang, J.; Tohidi, B.; Chuvilin, E.; Istomin, V.; Bukhanov, B. Methane recovery from gas hydrate-bearing sediments: An experimental study on the gas permeation characteristics under varying pressure. *J. Pet. Sci. Eng.* **2019**, *180*, 435–444. [CrossRef]
18. Dai, S.; Kim, J.; Xu, Y.; Waite, W.F.; Jang, J.; Yoneda, J.; Collett, T.S.; Kumar, P. Permeability anisotropy and relative permeability in sediments from the National Gas Hydrate Program Expedition 02, offshore India. *Mar. Pet. Geol.* **2019**, *108*, 705–713. [CrossRef]
19. Santamarina, J.C.; Dai, S.; Terzariol, M.; Jang, J.; Waite, W.F.; Winters, W.J.; Nagao, J.; Yoneda, J.; Konno, Y.; Fujii, T.; et al. Hydro-bio-geomechanical properties of hydrate-bearing sediments from Nankai Trough. *Mar. Pet. Geol.* **2015**, *66*, 434–450. [CrossRef]
20. Yang, D.; Yan, R.; Yan, M.; Lu, D.; Wei, C. Geomechanical properties of artificial methane hydrate-bearing fine-grained sediments. *Gas Sci. Eng.* **2023**, *109*, 104852. [CrossRef]
21. Konno, Y.; Fujii, T.; Sato, A.; Akamine, K.; Naiki, M.; Masuda, Y.; Yamamoto, K.; Nagao, J. Key Findings of the World's First Offshore Methane Hydrate Production Test off the Coast of Japan: Toward Future Commercial Production. *Energy Fuels* **2017**, *31*, 2607–2616. [CrossRef]
22. Kato, A.; Konno, Y.; Yoneda, J.; Kida, M.; Oshima, M.; Jin, Y.; Nagao, J.; Tenma, N. Evaluation of failure modes and undrained shear strength by cone penetrometer for Natural Gas hydrate-bearing pressure-core sediment samples recovered from the Krishna–Godavari Basin, offshore India. *Mar. Pet. Geol.* **2019**, *108*, 502–511. [CrossRef]
23. Biastoch, A.; Treude, T.; Rüpke, L.H.; Riebesell, U.; Roth, C.; Burwicz, E.B.; Park, W.; Latif, M.; Böning, C.W.; Madec, G.; et al. Rising Arctic Ocean temperatures cause gas hydrate destabilization and ocean acidification. *Geophys. Res. Lett.* **2011**, *38*, L08602. [CrossRef]
24. Phrampus, B.J.; Hornbach, M.J. Recent changes to the Gulf Stream causing widespread gas hydrate destabilization. *Nature* **2012**, *490*, 527–530. [CrossRef] [PubMed]
25. Wallmann, K.; Riedel, M.; Hong, W.L.; Patton, H.; Hubbard, A.; Pape, T.; Hsu, C.W.; Schmidt, C.; Johnson, J.E.; Torres, M.E.; et al. Gas hydrate dissociation off Svalbard induced by isostatic rebound rather than global warming. *Nat. Commun.* **2018**, *9*, 83. [CrossRef]
26. Burton, Z.F.M.; Kroeger, K.F.; Hosford Scheirer, A.; Seol, Y.; Burgreen-Chan, B.; Graham, S.A. Tectonic Uplift Destabilizes Subsea Gas Hydrate: A Model Example From Hikurangi Margin, New Zealand. *Geophys. Res. Lett.* **2020**, *47*, e2020GL087150. [CrossRef]
27. Burton, Z.F.M.; Dafov, L.N. Salt Diapir-Driven Recycling of Gas Hydrate. *Geochem. Geophys. Geosyst.* **2023**, *24*, e2022GC010704. [CrossRef]
28. Sultan, N.; Cochonat, P.; Foucher, J.-P.; Mienert, J. Effect of gas hydrates melting on seafloor slope instability. *Mar. Geol.* **2004**, *213*, 379–401. [CrossRef]
29. Dhakal, S.; Gupta, I. Slope instability of submarine sediments due to hydrate dissociation: A case study of Northern Cascadia Margin. *Geoenergy Sci. Eng.* **2023**, *223*, 211558. [CrossRef]
30. Maslin, M.; Owen, M.; Betts, R.; Day, S.; Dunkley Jones, T.; Ridgwell, A. Gas hydrates: Past and future geohazard? *Philos. Trans. R. Soc. A Math. Phys. Eng. Sci.* **2010**, *368*, 2369–2393. [CrossRef]
31. Sun, Y.; Lu, H.; Lu, C.; Li, S.; Lv, X. Hydrate dissociation induced by gas diffusion from pore water to drilling fluid in a cold wellbore. *Adv. Geo-Energy Res.* **2018**, *2*, 410–417. [CrossRef]
32. Bai, C.; Su, P.; Su, X.; Cui, H.; Shang, W.; Han, S.; Zhang, G. Characterization of the sediments in a gas hydrate reservoir in the northern South China Sea: Implications for gas hydrate accumulation. *Mar. Geol.* **2022**, *453*, 106912. [CrossRef]
33. Ning, F.; Liang, J.; Wu, N.; Zhu, Y.; Wu, S.; Liu, C.; Wei, C.; Wang, D.; Zhang, Z.; Xu, M.; et al. Reservoir characteristics of natural gas hydrates in China. *Nat. Gas Ind.* **2020**, *40*, 1–24. [CrossRef]
34. Wei, R.; Jia, C.; Liu, L.; Wu, N. Analysis of the Characteristics of Pore Pressure Coefficient for Two Different Hydrate-Bearing Sediments under Triaxial Shear. *J. Mar. Sci. Eng.* **2022**, *10*, 509. [CrossRef]
35. Wei, J.; Yang, L.; Liang, Q.; Liang, J.; Lu, J.; Zhang, W.; Zhang, X.; Lu, X. Geomechanical properties of gas hydrate-bearing sediments in Shenhu Area of the South China Sea. *Energy Rep.* **2021**, *7*, 8013–8020. [CrossRef]
36. Priest, J.A.; Hayley, J.L.; Smith, W.E.; Schultheiss, P.; Roberts, J. PCATS triaxial testing: Geomechanical properties of sediments from pressure cores recovered from the Bay of Bengal during expedition NGHP-02. *Mar. Pet. Geol.* **2019**, *108*, 424–438. [CrossRef]
37. Wu, Y.; Liao, J.; Zhang, W.; Cui, J. Characterization of stress–dilatancy behavior for methane hydrate-bearing sediments. *J. Nat. Gas Sci. Eng.* **2021**, *92*, 104000. [CrossRef]
38. Hyodo, M.; Li, Y.; Yoneda, J.; Nakata, Y.; Yoshimoto, N.; Nishimura, A.; Song, Y. Mechanical behavior of gas-saturated methane hydrate-bearing sediments. *J. Geophys. Res. Solid Earth* **2013**, *118*, 5185–5194. [CrossRef]
39. Hassanpouryouzband, A.; Joonaki, E.; Vasheghani Farahani, M.; Takeya, S.; Ruppel, C.; Yang, J.; English, N.J.; Schicks, J.M.; Edlmann, K.; Mehrabian, H.; et al. Gas hydrates in sustainable chemistry. *Chem. Soc. Rev.* **2020**, *49*, 5225–5309. [CrossRef]
40. Wu, P.; Li, Y.; Sun, X.; Liu, W.; Song, Y. Mechanical Characteristics of Hydrate-Bearing Sediment: A Review. *Energy Fuels* **2021**, *35*, 1041–1057. [CrossRef]
41. Pinkert, S.; Grozic, J.L.H. Failure Mechanisms in Cemented Hydrate-Bearing Sands. *J. Chem. Eng. Data* **2015**, *60*, 376–382. [CrossRef]

42. Wu, P.; Li, Y.; Wang, L.; Sun, X.; Wu, D.; He, Y.; Li, Q.; Song, Y. Hydrate-bearing sediment of the South China Sea: Microstructure and mechanical characteristics. *Eng. Geol.* **2022**, *307*, 106782. [CrossRef]
43. Waite, W.F.; Santamarina, J.C.; Cortes, D.D.; Dugan, B.; Espinoza, D.N.; Germaine, J.; Jang, J.; Jung, J.W.; Kneafsey, T.J.; Shin, H.; et al. Physical properties of hydrate-bearing sediments. *Rev. Geophys.* **2009**, *47*, RG4003. [CrossRef]
44. Zhan, L.; Liu, B.; Zhang, Y.; Lu, H. Rock Physics Modeling of Acoustic Properties in Gas Hydrate-Bearing Sediment. *J. Mar. Sci. Eng.* **2022**, *10*, 1076. [CrossRef]
45. Guo, Z.; Lv, X.; Liu, C.; Chen, H.; Cai, Z. Characterizing Gas Hydrate–Bearing Marine Sediments Using Elastic Properties—Part 1: Rock Physical Modeling and Inversion from Well Logs. *J. Mar. Sci. Eng.* **2022**, *10*, 1379. [CrossRef]
46. Guo, Z.; Lv, X.; Liu, C.; Chen, H.; Mei, L. Characterizing Gas Hydrate-Bearing Marine Sediments Using Elastic Properties—Part 2: Seismic Inversion Based on a Pore-Filling–Solid Matrix Decoupling Scheme. *J. Mar. Sci. Eng.* **2022**, *10*, 1497. [CrossRef]
47. Yoneda, J.; Masui, A.; Konno, Y.; Jin, Y.; Egawa, K.; Kida, M.; Ito, T.; Nagao, J.; Tenma, N. Mechanical properties of hydrate-bearing turbidite reservoir in the first gas production test site of the Eastern Nankai Trough. *Mar. Pet. Geol.* **2015**, *66*, 471–486. [CrossRef]
48. Yun, T.S.; Narsilio, G.A.; Carlos Santamarina, J. Physical characterization of core samples recovered from Gulf of Mexico. *Mar. Pet. Geol.* **2006**, *23*, 893–900. [CrossRef]
49. Yan, R.; Yu, H.; Yang, D.; Tang, H.; Zhang, Q. Shear strength and pore pressure characteristics of methane hydrate-bearing soil under undrained condition. *Int. J. Hydrogen Energy* **2023**, *48*, 12240–12256. [CrossRef]
50. Liu, W.; Chen, P.; Wu, P.; You, Z.; Wang, L.; Huang, L.; Zhang, S.; Yu, T.; Li, Y. Mechanical Characteristics of Underconsolidated Methane Hydrate-Bearing Clayed-Silty Sediments. *Energy Fuels* **2023**, *37*, 6503–6514. [CrossRef]
51. Wood, D.M. *Soil Behaviour and Critical State Soil Mechanics*; Cambridge University Press: Cambridge, UK, 1991; ISBN 9780521332491.
52. Uchida, S.; Soga, K.; Yamamoto, K. Critical state soil constitutive model for methane hydrate soil. *J. Geophys. Res. Solid Earth* **2012**, *117*. [CrossRef]
53. Kong, L.; Wang, X.; Hua, L.; Lin, X. A bounding surface model of gas-hydrate-bearing sediments. *Environ. Geotech.* **2023**, *10*, 196–205. [CrossRef]
54. Yoneda, J.; Oshima, M.; Kida, M.; Kato, A.; Konno, Y.; Jin, Y.; Tenma, N. Consolidation and hardening behavior of hydrate-bearing pressure-core sediments recovered from the Krishna–Godavari Basin, offshore India. *Mar. Pet. Geol.* **2019**, *108*, 512–523. [CrossRef]
55. Zhou, J.; Yang, Z.; Wei, C.; Chen, P.; Yan, R. Mechanical behavior of hydrate-bearing sands with fine particles under isotropic and triaxial compression. *J. Nat. Gas Sci. Eng.* **2021**, *92*, 103991. [CrossRef]
56. Dong, L.; Li, Y.; Liao, H.; Liu, C.; Chen, Q.; Hu, G.; Liu, L.; Meng, Q. Strength estimation for hydrate-bearing sediments based on triaxial shearing tests. *J. Pet. Sci. Eng.* **2020**, *184*, 106478. [CrossRef]
57. Wei, R.; Liu, L.; Jia, C.; Zhao, H.; Dong, X.; Bu, Q.; Liu, C.; Wu, N. Undrained Shear Properties of Shallow Clayey-Silty Sediments in the Shenhu Area of South China Sea. *Sustainability* **2023**, *15*, 1175. [CrossRef]
58. Lee, J.Y.; Yun, T.S.; Santamarina, J.C.; Ruppel, C. Observations related to tetrahydrofuran and methane hydrates for laboratory studies of hydrate-bearing sediments. *Geochem. Geophys. Geosyst.* **2007**, *8*. [CrossRef]
59. Liu, Z.; Kim, J.; Lei, L.; Ning, F.; Dai, S. Tetrahydrofuran Hydrate in Clayey Sediments—Laboratory Formation, Morphology, and Wave Characterization. *J. Geophys. Res. Solid Earth* **2019**, *124*, 3307–3319. [CrossRef]
60. Yun, T.S.; Santamarina, J.C.; Ruppel, C. Mechanical properties of sand, silt, and clay containing tetrahydrofuran hydrate. *J. Geophys. Res. Solid Earth* **2007**, *112*. [CrossRef]
61. Kim, J.; Seol, Y.; Dai, S. The coefficient of earth pressure at rest in hydrate-bearing sediments. *Acta Geotech.* **2021**, *16*, 2729–2739. [CrossRef]
62. Liu, Z.; Dai, S.; Ning, F.; Peng, L.; Wei, H.; Wei, C. Strength Estimation for Hydrate-Bearing Sediments From Direct Shear Tests of Hydrate-Bearing Sand and Silt. *Geophys. Res. Lett.* **2018**, *45*, 715–723. [CrossRef]
63. *Stand. ASTM D2216-19*; ASTM Standard Test Method for Laboratory Determination of Water (Moisture) Content of Soil and Rock by Mass. ASTM: West Conshohocken, PA, USA, 2019.
64. Wang, X.; Hutchinson, D.R.; Wu, S.; Yang, S.; Guo, Y. Elevated gas hydrate saturation within silt and silty clay sediments in the Shenhu area, South China Sea. *J. Geophys. Res.* **2011**, *116*, B05102. [CrossRef]
65. Kuang, Y.; Yang, L.; Li, Q.; Lv, X.; Li, Y.; Yu, B.; Leng, S.; Song, Y.; Zhao, J. Physical characteristic analysis of unconsolidated sediments containing gas hydrate recovered from the Shenhu Area of the South China sea. *J. Pet. Sci. Eng.* **2019**, *181*, 106173. [CrossRef]
66. De La Fuente, M.; Vaunat, J.; Marín-Moreno, H. Consolidation of gas hydrate-bearing sediments with hydrate dissociation. *E3S Web Conf.* **2020**, *205*, 11007. [CrossRef]
67. Yan, R.T.; Liang, W.Y.; Wei, C.F.; Wu, E.L. A constitutive model for gas hydrate-bearing sediments considering hydrate occurring habits. *Yantu Lixue/Rock Soil Mech.* **2017**, *38*, 10–18. [CrossRef]
68. Sultan, N.; Garziglia, S. Geomechanical constitutive modelling of gas-hydrate-bearing sediments. In Proceedings of the 7th International Conference on Gas Hydrates, Edinburgh, UK, 17–21 July 2011; pp. 1–11.
69. Fang, H.; Shi, K.; Yu, Y. A state-dependent subloading constitutive model with unified hardening function for gas hydrate-bearing sediments. *Int. J. Hydrogen Energy* **2022**, *47*, 4441–4471. [CrossRef]
70. Yan, R.; Wei, C. Constitutive Model for Gas Hydrate–Bearing Soils Considering Hydrate Occurrence Habits. *Int. J. Géoméch.* **2017**, *17*, 4017032. [CrossRef]

71. Zhao, Y.; Liu, J.; Sang, S.; Hua, L.; Kong, L.; Zeng, Z.; Yuan, Q. Experimental investigation on the permeability characteristics of methane hydrate-bearing clayey-silty sediments considering various factors. *Energy* **2023**, *269*, 126811. [CrossRef]
72. Vasheghani Farahani, M.; Hassanpouryouzband, A.; Yang, J.; Tohidi, B. Insights into the climate-driven evolution of gas hydrate-bearing permafrost sediments: Implications for prediction of environmental impacts and security of energy in cold regions. *RSC Adv.* **2021**, *11*, 14334–14346. [CrossRef]
73. Fetisov, V.; Ilyushin, Y.V.; Vasiliev, G.G.; Leonovich, I.A.; Müller, J.; Riazi, M.; Mohammadi, A.H. Development of the automated temperature control system of the main gas pipeline. *Sci. Rep.* **2023**, *13*, 3092. [CrossRef]
74. Li, J.; Lu, J.; Kang, D.; Ning, F.; Lu, H.; Kuang, Z.; Wang, D.; Liu, C.; Hu, G.; Wang, J.; et al. Lithological characteristics and hydrocarbon gas sources of gas hydrate-bearing sediments in the Shenhu area, South China Sea: Implications from the W01B and W02B sites. *Mar. Geol.* **2019**, *408*, 36–47. [CrossRef]

Journal of
Marine Science and Engineering

MDPI

Article

Numerical Investigation of Uplift Failure Mode and Capacity Estimation for Deep Helical Anchors in Sand

Chi Yuan [1], Dongxue Hao [2,3,*], Rong Chen [2,3,*] and Ning Zhang [1]

1 College of Architecture and Civil Engineering, Beijing University of Technology, Beijing 110124, China; yuanc@emails.bjut.edu.cn (C.Y.)
2 Key Lab of Electric Power Infrastructure Safety Assessment and Disaster Prevention of Jilin Province, Northeast Electric Power University, Jilin 132012, China
3 School of Civil Engineering and Architecture, Northeast Electric Power University, Jilin 132012, China
* Correspondence: 20102291@neepu.edu.cn (D.H.); 20112384@neepu.edu.cn (R.C.); Tel.: +86-432-6480-6481 (D.H.)

Abstract: The uplift capacity of helical anchors is generally taken as the control condition for design in different applications, including transmission tower foundations and offshore structures. However, it is difficult to identify the failure surface for a deep helical anchor, which may result in an incorrect assessment of uplift capability. This research proposes a new unified method to estimate the uplift capacity of deep single-helix and multi-helix anchors based on the investigation of failure mechanisms. The deep failure mode was identified by FEM analysis using a modified Mohr–Coulomb model considering the strain softening of sand, along with the coupled Eulerian–Lagrangian technique. Thereby, a simplified rupture surface is proposed, and the equations estimating the uplift capacity are presented by the limit equilibrium method. Two important factors—the lateral earth pressure coefficient and the average internal friction angle included in the equations—are discussed and determined. The comparisons with centrifugal tests verify the reasonability of the proposed method.

Keywords: helical anchor; CEL analysis; deep failure mode; uplift capacity

check for updates

Citation: Yuan, C.; Hao, D.; Chen, R.; Zhang, N. Numerical Investigation of Uplift Failure Mode and Capacity Estimation for Deep Helical Anchors in Sand. *J. Mar. Sci. Eng.* **2023**, *11*, 1547. https://doi.org/10.3390/jmse11081547

Academic Editor: Dong-Sheng Jeng

Received: 17 June 2023
Revised: 28 July 2023
Accepted: 29 July 2023
Published: 4 August 2023

1. Introduction

A helical anchor, consisting of one or more helical plates welded to a steel shaft, is a deep foundation system used to support or resist any load or application. Due to their rapid installation and immediate service, relatively large bearing capacity, and lack of environmental damage, they have been extensively employed as a foundation system for structures such as transmission towers, offshore platforms, and wind turbines [1,2]. In recent years, this type of foundation has been suggested as a potential alternative to driven piles in offshore renewable energy structures.

There are three failure modes for single-helix anchors as the embedment depth ratio (H/D) increases, where D is the helix diameter. Figure 1 shows the shallow failure mode [3–12], deep failure mode [11,13–16], and transition failure mode [4,13] of circular plate anchors or single-helix anchors, which have been observed in most investigations. The rupture surface extends continuously to the ground at a shallow plate depth, defined as general shear failure or the shallow failure mode. With the increase in plate depth, the transition failure mode is observed, often referred to as the shallow failure mode [4,13]. The shallow rupture surface has been assumed to be a cylinder [3], an inverted cone [7,8,13], or a log-spiral surface [5,10,17]. Only a few small-scale model tests showed a closed bulb (or balloon-shaped) rupture surface [13–16,18]. The rupture surface is limited under the ground at a deep depth, defined as local shear failure or the deep failure mode. Recently, a centrifuge test with $H/D = 7$ observed the deep rupture surface in medium dense sand [19]. Although it provided intuitive observation, there could be a discrepancy between the results of the half-anchor model and the full-anchor model due to the difference in sand–strongbox interaction from the internal friction of sand.

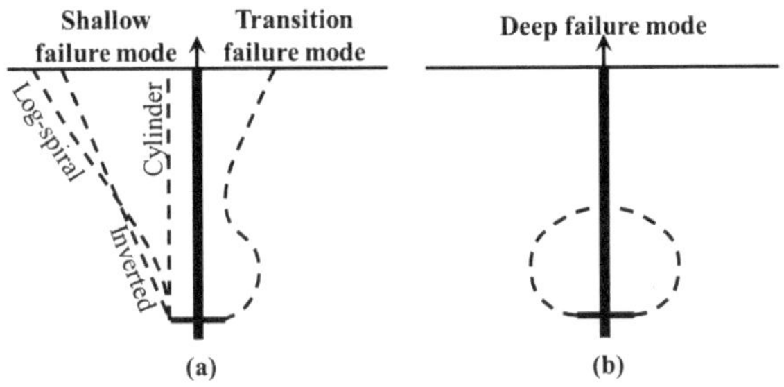

Figure 1. The failure mode of a single-helix or plate anchor. (**a**) Shallow and transition failure mode. (**b**) Deep failure mode.

It is generally believed that there are two failure modes for multi-helix anchors controlled by the helical plate space ratio (S/D)—that is, the cylindrical shear [4,20–22] and the individual bearing failure mode [23,24], as shown in Figure 2. The cylindrical shear failure mode means that the cylindrical rupture surface extends to the uppermost plate and then follows the single-helix anchor failure mode. The individual bearing failure mode means that each helical plate behaves independently of the others. The transition helix spacing between the two methods is usually regarded as $3D$ in some engineering manuals [23,25,26].

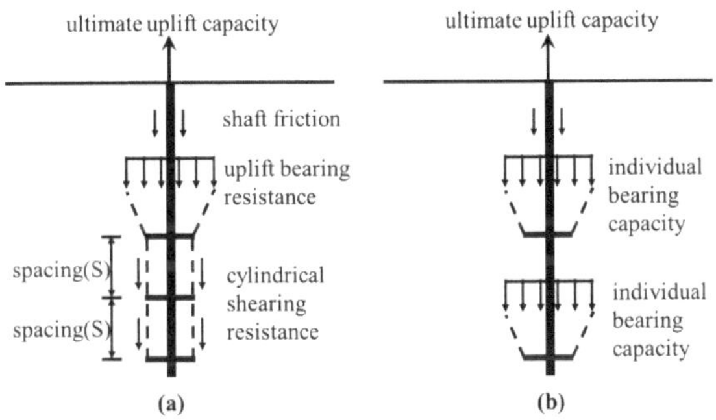

Figure 2. The failure mode of a multi-helix or plate anchor. (**a**) Cylindrical shear mode. (**b**) Individual bearing mode.

There are two issues that need further discussion for deep helical anchors in sand:

(1) The direct observations of rupture surfaces are limited, especially for the multi-helix anchors. Some observations on the rupture surface of single-helix or circular plate anchors in sand have been reported, but they mostly came from 1 g small-scale model tests. This may produce differences between the observed rupture surface and the actual state due to the low stress level, especially for deep anchors. Therefore, it is necessary to further study the failure mode of deep helical anchors, which is essential for the estimation of the uplift capacity.

(2) The estimation of uplift capacity for multi-helix anchors with transition helix spacing based on the two recognized failure modes (Figure 2) is inconsistent. That is, when the helix spacing is transition spacing, the uplift capacity calculated by the individual bearing method is higher than that calculated by the cylindrical shear method.

Numerical simulation is an effective option to analyze the failure mechanisms and bearing capacity of foundations. The finite element method has been used in previous studies [27–40]. Large deformation and mesh distortion are challenges for the FEM analysis of deep anchors, especially in sand. Hakeem et al. used the arbitrary Lagrangian–Eulerian (ALE) method to simulate a circular plate anchor subjected to centric loading with H/D = 1~20 [11]. For deep anchors, the load–displacement curve was approximately a straight line with a 0.4D maximum displacement, and the 25%D criterion was adopted to determine the ultimate uplift capacity; the authors proposed that this criterion is somewhat arbitrary.

Konkol et al. proposed that the ALE method is suitable for issues where soil displacement values are important, but the coupled Eulerian–Lagrangian (CEL) method is suitable for issues where mainly Lagrangian elements (piles, anchors, etc.) are analyzed [41]. Qiu et al. proposed that the CEL method can deal with large mesh distortions and contact problems [42]. It has proven successful in geotechnical problems such as a strip footing problem, installing a pile into the subsoil, and a ship running aground at an embankment [43,44]. In those studies, different constitutive models were used, such as the Mohr–Coulomb constitutive model, hardening soil model, and Drucker–Prager model. However, strain softening is important to incorporate into constitutive models of sand's behavior [45], so consideration of softening and shear swelling for large deformations may be more suitable for sand. Numerical results based on elastic–perfectly plastic or hardening soil models will overestimate the capacities of helical or circular anchors, especially in dense sand. Moreover, this method is widely used to study the load-bearing characteristics of plate or helical anchors in clay [42,44,46,47]. However, this method is very rarely used to study helical anchors in sand.

To solve the above issues, a numerical analysis was conducted, considering the impacts of embedment depths, helical plate spaces, and helical plate numbers. In this study, the coupled Eulerian–Lagrangian (CEL) approach in Abaqus/Explicit was employed for the large deformation analysis, combined with a modified Mohr–Coulomb model considering soil's strain softening and shear swelling, and the numerical model was validated by the centrifugal tests of the helical anchors (Section 2). Then, the deep failure model was presented according to the rupture surface observed via FEM (Section 3). Finally, an estimation of the uplift capacity was proposed by the limit equilibrium method, and the parameter sensitivity was studied (Section 4).

2. FEM Model and Validation

In the present study, the CEL approach in Abaqus was used to simulate the uplift behavior of helical anchors in sand, including the effects of density (D_r = 30%, 60%, and 100%), embedment ratio (H/D = 8~12), helix spacing ratio (S/D = 1.5~6), and helical plate number (n = 1~4).

2.1. FEM Model

The FEM models were constructed as illustrated in Figure 3. Helical plates were replaced with circular plates based on previous findings that the plates' geometrical shape has little influence on the uplift capacity [6,12], and this analysis does not focus on the stress of the anchor body. Only one-quarter of the anchor and soil domain was considered in this study in terms of axisymmetry. Boundary conditions were imposed on the two planes of symmetry by prescribing zero flow velocity as normal to these planes. The bottom of the computational domain was constrained against the flow in the vertical direction. The single-helix anchors had an embedment ratio of 8~12. The lowermost plate of the multi-helix anchors had the same embedment ratio of 12, varied spacing ratios of 1.5~6.0, and varied plate numbers of 2~4.

For comparison with the results of the centrifugal test in dense sand [12], the helix diameter D, the helix thickness t, and the shaft diameter of anchor d were 400 mm, 0.05D, and 0.235D, respectively. The anchor was modeled as a discrete rigid solid part meshed with the eight-node linear brick, reducing the integration element C3D8R. A reference point was set on the top of the anchor, with constrained horizontal displacement and axial rotation. The soil domain consisted of the eight-node reduced integration Eulerian element EC3D8R. A void layer with a $5D$ vertical distance above the ground was defined to allow the soil to heave and flow into the empty Eulerian elements during subsequent analysis. The computational domain size was $10D \times 2H$ (where H is the embedment depth of the anchor), which is sufficiently large to ignore the far-field boundary effects [43]. The mesh was densified in the zone around the anchor, from $5D$ above the uppermost plate to $5D$ below the lowermost plate vertically, and $4D$ from the shaft centerline horizontally. The minimum element size ΔB was in the vicinity of the plate. The contact between the anchor and the soil was automatically identified. A general contact was adopted, with "hard contact" for the normal contact and a penalty contact method for tangential contact.

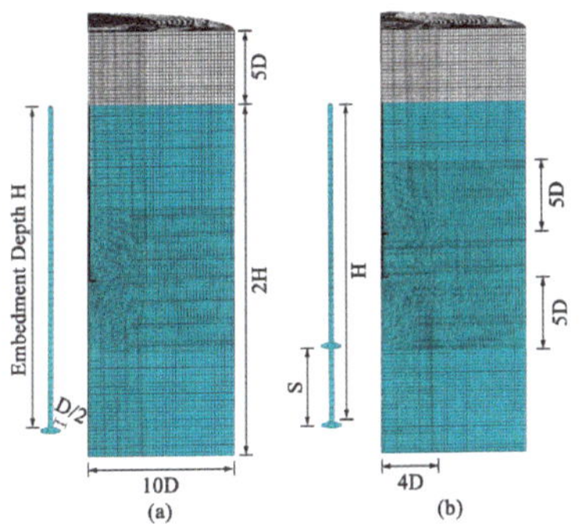

Figure 3. Numerical model. (**a**) Single-helix anchor. (**b**) Double-helix anchor.

Previously, the sand was represented using an elastoplastic Mohr–Coulomb constitutive model, which has limitations [48], and the sand softening was not reflected by the Mohr–Coulomb model in Abaqus. A modified Mohr–Coulomb model was used to consider the strain softening of sand, as shown in Figure 4.

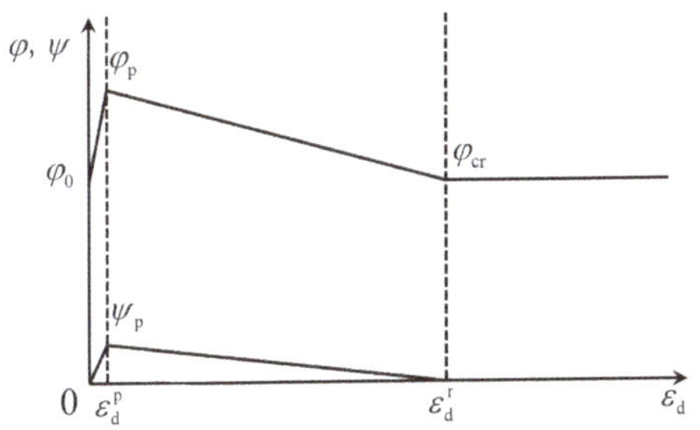

Figure 4. Diagrammatic sketch of the modified Mohr–Coulomb model.

The strength parameters, including the internal friction angle φ and dilatancy angle ψ, were simplified to increase linearly to the peak values and then decrease linearly to the critical state values with the increase in equivalent plastic strain ε_d, as shown in Equations (1) and (2), where the equivalent plastic strain can be calculated according to $\varepsilon = \frac{\sqrt{2}}{3}\left[(\varepsilon_1 - \varepsilon_2)^2 + (\varepsilon_2 - \varepsilon_3)^2 + (\varepsilon_3 - \varepsilon_1)^2\right]^{\frac{1}{2}}$.

$$\varphi = \begin{cases} \frac{\varphi_p - \varphi_0}{\varepsilon_d^p}\varepsilon_d + \varphi_0 & 0 \le \varepsilon_d < \varepsilon_d^p \\ \frac{\varphi_{cr} - \varphi_p}{\varepsilon_d^r - \varepsilon_d^p}\left(\varepsilon_d - \varepsilon_d^r\right) + \varphi_{cr} & \varepsilon_d^p \le \varepsilon_d < \varepsilon_d^r \\ \varphi_{cr} & \varepsilon_d > \varepsilon_d^r \end{cases} \tag{1}$$

$$\psi = \begin{cases} \frac{\psi_p}{\varepsilon_d^p}\varepsilon_d & 0 \le \varepsilon_d < \varepsilon_d^p \\ \frac{\psi_p}{\varepsilon_d^p - \varepsilon_d^r}(\varepsilon_d - \varepsilon_d^r) & \varepsilon_d^p \le \varepsilon_d < \varepsilon_d^r \\ 0 & \varepsilon_d > \varepsilon_d^r \end{cases} \tag{2}$$

The initial internal friction angle φ_0 is assumed to be the same as the critical internal friction angle φ_{cr}. The parameters contained in the model are Young's modulus E, Poisson's ratio v, the initial lateral pressure coefficient K_0, the peak internal friction angle φ_p, the peak dilatancy angle ψ_p, the critical internal friction angle φ_{cr}, the peak equivalent plastic strain ε_d^p, and the critical equivalent plastic strain ε_d^r.

The values of the modulus, peak friction angle, and dilatancy angle depend on confining pressure at the lowermost plate. For dense sand (D_r = 100%) used in the centrifugal test [12], Young's modulus E and the peak internal friction angle φ_p are taken from the literature [49] after being obtained from triaxial tests; that is, $E = 658\ p_a\ (\sigma_3/p_a)^{0.469}$ and $\varphi_p = 40.9–7.81\ g\ (\sigma_3/p_a)$, where p_a is the standard atmospheric pressure (101.3 kPa) and σ_3 is the confining pressure and takes the value of gravity stress $\gamma' H$, kPa. The values of φ_p herein are relatively close to those for dense uwa sand in the recent literature [50,51], and their difference is 1~2°. The values of E and φ_p for loose sand (D_r = 30%) and medium dense sand (D_r = 60%) were calculated proportionally based on the literature [52]. The densities of different compactness were 1.61 g/cm^3, 1.66 g/cm^3, and 1.75 g/cm^3, respectively. There was no groundwater influence, so the soil's effective unit weight γ' was 15.8 kN/m^3, 16.3 kN/m^3, and 17.2 kN/m^3, respectively. The peak dilatancy angle ψ_p was estimated by the equation $\varphi_p = 0.5\psi_p + \varphi_{cr}$ proposed by Bolton [53]. The validity of the loose sand and medium dense sand parameters was determined by comparison with the triaxial test results in reference [49]. The soil parameters for different embedment ratios are summarized in Table 1. Other parameters were determined based on reference [49]—that is, $\varphi_{cr} = 31°$, $v = 0.3$, $K_0 = 1 - \sin\varphi_{cr} = 0.485$, $\varepsilon_d^p = 2\%$, and $\varepsilon_d^r = 20\%$.

Table 1. Soil parameters.

D_r		30%			60%			100%	
H/D	E (MPa)	φ_p (°)	ψ_p (°)	E (MPa)	φ_p (°)	ψ_p (°)	E (MPa)	φ_p (°)	ψ_p (°)
8	24.27	34.77	7.54	35.47	38.31	12.26	50.4	42.9	23.8
9	25.64	34.65	7.3	37.48	38.07	11.86	53.26	42.5	23
10	26.94	34.54	7.08	39.38	37.86	11.5	55.96	42.14	22.28
10.5	27.57	34.49	6.98	40.29	37.76	11.34	57.25	41.98	21.95
12	29.35	34.36	6.71	42.89	37.49	10.89	60.95	41.52	21.05

2.2. Influence of Pullout Rate and Mesh Density

The pullout process of a plate anchor is essentially quasi-static in nature, while the Eulerian analysis is formulated in the framework of a dynamic explicit solution scheme instead of a static implicit framework [43]. To achieve a balance between matching the quasi-static state as closely as possible and reducing the computational time, a parametric study was carried out to investigate the effect of the pullout rate. Three pullout rates of 0.025 D/s, 0.05 D/s, and 0.1 D/s were considered, with a minimum element size $\Delta B = 0.1D$. Also, a mesh convergence study was performed to identify a suitable mesh density that gives sufficiently accurate results. Both studies were performed in the case of a single-helix anchor with $H/D = 9$ in dense sand.

The load–displacement curves of different pullout rates are illustrated in Figure 5a, with the minimum element size $\Delta B = 0.1D$. The pullout rate V has little influence on the uplift process. When the displacement is up to $0.5D$, the computational time of V_1 and

V_2 is about 12 times and 2 times that of V_3, respectively. Hence, considering the time consumption and stability, the pullout rate V_2 was adopted for all subsequent analyses.

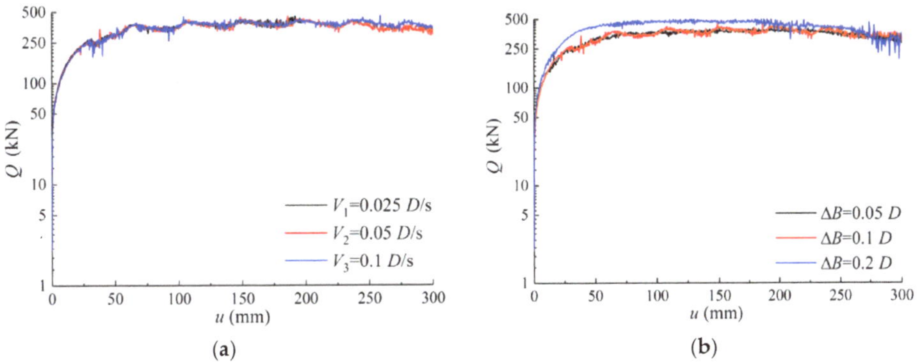

Figure 5. The curves of uplift resistance and displacement for the convergence study. (**a**) Pullout rate study. (**b**) Mesh convergence study.

Three finite element meshes with minimum element sizes ΔB near the anchor of $0.05D$, $0.1D$, and $0.2D$ were compared, and the element numbers were 613130, 280864, and 125856, respectively, where the pullout rate $V = 0.05\ D/\text{s}$. The load–displacement curves of different mesh densities are illustrated in Figure 5b. It can be seen that a finer mesh tends to give a smaller uplift capacity, and that the mesh with the minimum size $\Delta B = 0.1D$ is preferable because a further decrease in mesh size does not change the result. Therefore, meshing with a minimum element size of $0.1D$ was considered sufficient in terms of accuracy and was adopted for all subsequent analyses.

2.3. Determination and Verification of Uplift Capacity

The relationship between load and displacement of single-helix anchors (H/D = 8~12) in sand with different densities is presented in Figure 6. The relationship between load and displacement of multi-helix anchors in dense sand is shown in Figure 7. The characteristics of these curves can be divided into two types: those with an obvious peak point, and those without an obvious peak point. The peak values of the curves were undoubtedly taken as the ultimate uplift capacities, and the uplift capacities for the curves without obvious peak points were determined according to the curve development. The development of these curves presents three stages: the fast-rising straight-line stage, the curved stage, and a prolonged rising or stable stage (approximately a straight line). The starting point of the latter stationary section was regarded as the ultimate uplift capacity, Q_u, and the corresponding displacement is called failure displacement, uf. The points of Q_u are represented by circles in Figures 6 and 7.

The uplift capacity is normalized as the uplift capacity factor N_γ, as in Equation (3):

$$N_\gamma = Q_u/\gamma' AH \tag{3}$$

where γ' is the soil's effective unit weight, and A is the plate area, expressed as $A = \pi D^2/4$.

The uplift capacity factors N_γ for single-helix anchors from the FEM and centrifugal tests are plotted in Figure 8a, which shows that the uplift capacity factors of single-helix anchors are roughly constant as the embedment ratio H/D increases in loose sand, medium dense sand, and dense sand with $H/D > 9$. This feature is related to the failure mode, which will be explained later. The uplift capacity factors from the centrifugal test with D_r = 85.4~96.2% are encompassed between the numerical results of medium and dense sand. This result verifies the reliability of the FEM model. However, the FEM results overestimated the uplift capacities of helical anchors. This difference may be caused by the dilatancy angle determined by Bolton's equation [53], which is larger than the actual situation.

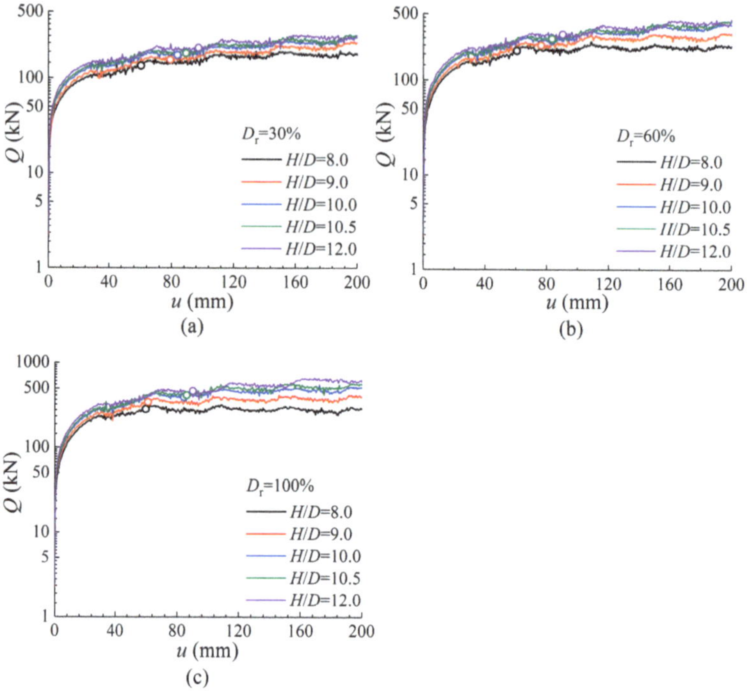

Figure 6. The relationship between load and displacement of single-helix anchors. (**a**) Loose and. (**b**) Medium dense sand. (**c**) Dense sand.

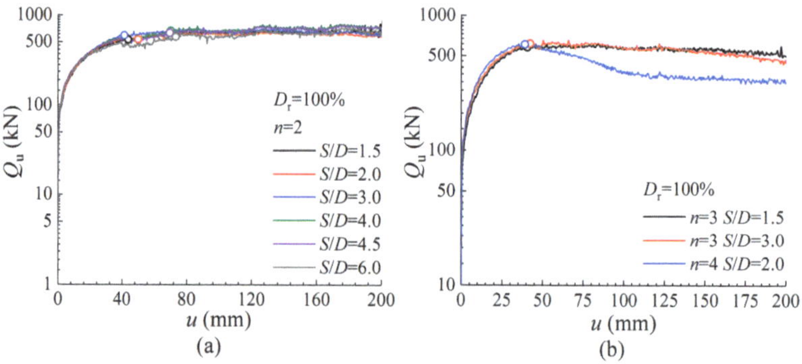

Figure 7. The relationship between load and displacement of multi-helix anchors. (**a**) Double-helix anchor. (**b**) Multi-helix anchor.

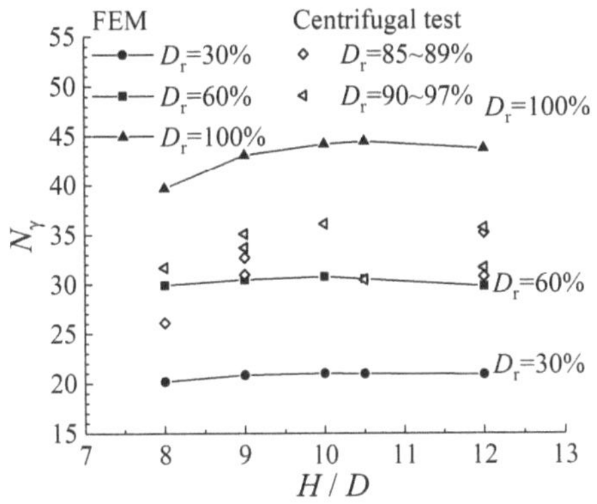

Figure 8. The comparison between the numerical and centrifugal test results.

3. Uplift Failure Mode

The effect of strain softening was incorporated into numerical analysis by utilizing the modified Mohr–Coulomb model described in Section 2, which allows for the variation of soil strength with strain and stress level and can show the subsequent soil strength mobilization and progressive failure. The failure surface was identified based on the contour of equivalent plastic strain at failure displacement, and the design method of the deep helical anchor is further proposed based on the visible deep failure mode in Section 4.

3.1. Process of Soil Strength Mobilization

Figure 9 shows the process of soil strength being mobilized while pulling the anchors, where u_f is the displacement corresponding to the ultimate uplift capacity defined above. The red zone corresponds to the peak internal friction angle, and the blue zone corresponds to the initial internal friction angle or the critical internal friction angle. The soil inside the red zone is undergoing softening, and the soil outside the red zone is experiencing hardening. The soil strength near the anchor plate is mobilized first and reaches its peak state. Then, the strength of this part of the soil reduces with the increase in the displacement or strain and reaches a critical state. The peak state boundary expands successively, and all of the soil within this boundary undergoes a softening process. This reflects the process of progressive failure of the foundation.

The local failure mode is present for different densities. As the displacement u increases, the influence range of the interaction between the anchor plate and soil expands gradually, and the soil strength of the partial zone around the plate is mobilized. The mobilized zone still expands outward gradually after the failure displacement u_f. Although a larger range of soil strength is mobilized, the soil strength near the plate decreases after peak strength. Hence, the overall uplift capacity shows a trend of slow increase, which corresponds to the load–displacement curve shown in Figure 6. The shapes of the mobilized zone in the deep underground area at failure displacement for different densities are similar, but the range becomes larger with the higher density. The very shallow soil near the ground surface is also mobilized due to shaft friction and low overburden pressure, which are not features of failure mode recognition.

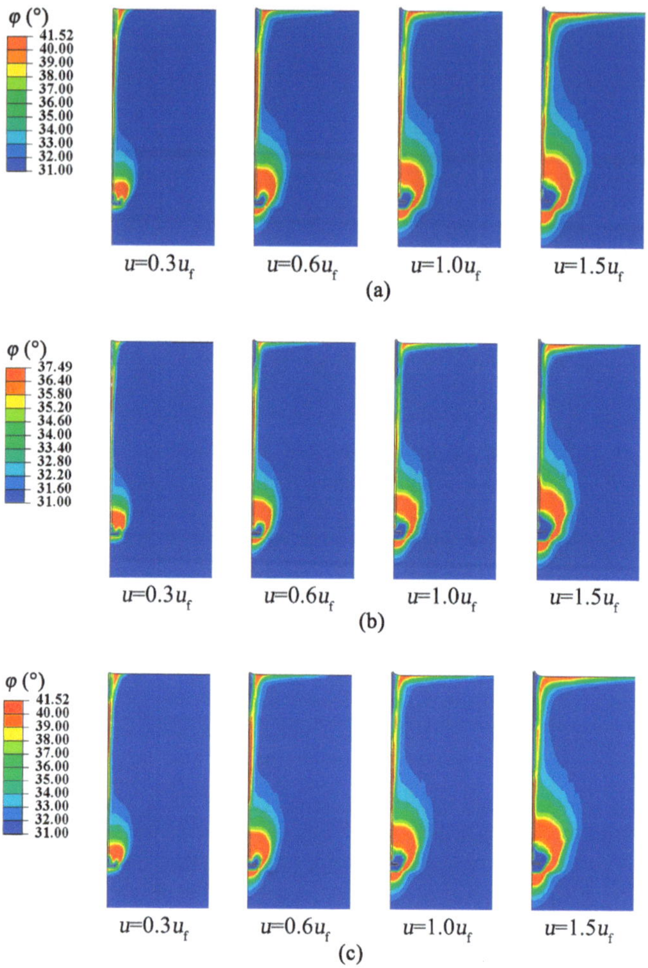

Figure 9. Contours of internal friction angle at different uplift displacements (H/D = 12). (**a**) Loose sand. (**b**) Medium dense sand. (**c**) Dense sand.

3.2. Deep Failure Behavior of Single-Helix Anchors

The equivalent plastic strain contours of the single-helix anchors at failure, with different embedment ratios H/D and relative densities, are shown in Figure 10. The white zone corresponds to the immobilized zone; the black zone corresponds to the softening zone, with an equivalent plastic strain greater than 2%; and the other zone corresponds to the hardening zone. The plastic zone surrounding the plate is limited below the ground surface and displays a closed bulb, which is the deep failure mode. However, the failure mode for the anchor with an embedment ratio of eight in dense sand displays the transitional mode, which will not be discussed in the following text.

For the deep failure mode, the influence of density on the mobilized zone range is greater than that of the embedment ratio. The boundary of the mobilized zone increases gradually with the increase in soil density. Its vertical height is 5D, 6D, and 7D, and the horizontal breadth is 2D, 2.5D, and 3D, in the loose, medium, and dense sand, respectively. However, the scope of the softening zone is hardly affected by density and embedment ratio. Its vertical height is approximately 2D, and the horizontal breadth is approximately

1.5D for different densities and embedment ratios. Due to the similarity of the failure modes, the dimensionless uplift capacities of deep single-helix anchors are approximately constant.

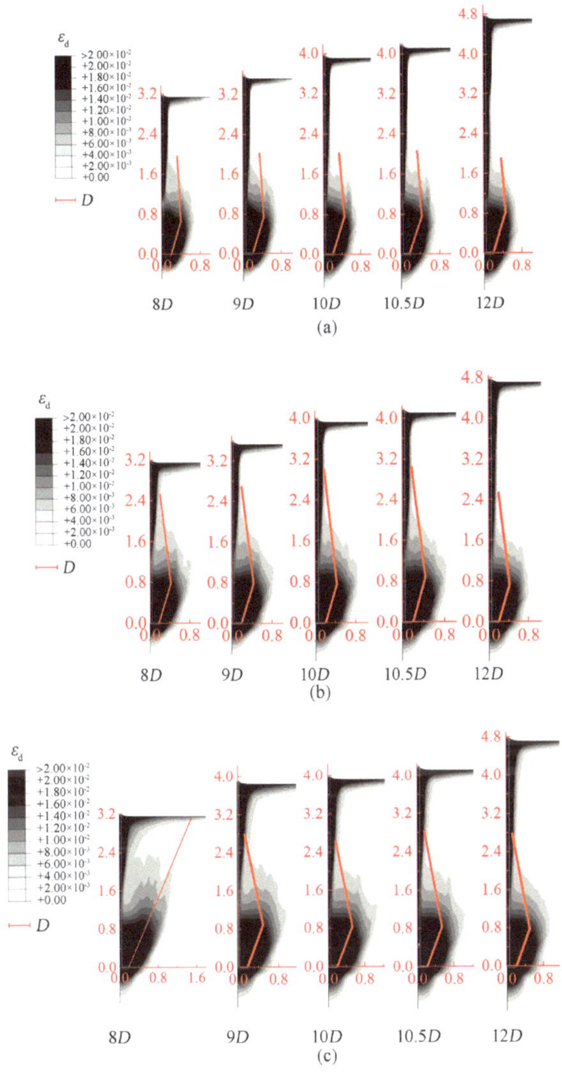

Figure 10. Single-helix anchor cases: contour of equivalent plastic strain and rupture surface (red line). (**a**) Loose sand. (**b**) Medium dense sand. (**c**) Dense sand.

For the convenience of calculation, Ghaly [13] assumed the deep failure mode as an inverted cone with a certain height. This assumption is similar to the experimental results observed by Motamedinia [18] and Salehzadeh [19]. The overburden pressure was simplified as a uniform load acting on the top of the inverted cone. Because the top of the inverted cone has a larger surface area than the potential region for load, this assumption can lead to excessive earth pressure. Liu [54] and Shi [55] proposed that the rupture surface can be determined by the points of maximum strain at varied depths based on the contour of plastic shear strain. This numerical analysis adopts the rule to identify the rupture surfaces of deep single-helix and multi-helix anchors. The deep rupture surface of the single-helix anchor can be simplified as two straight lines, as shown by the solid red line in Figure 10. From the plate edge to the boundary of the softening zone, there is an inverted truncated cone. The vertical distance between them is within (2~3)*D*. The inclination in the vertical direction rises from 15° to 20°. Then, an erected truncated cone extends from the boundary to the shaft. The vertical distance between them is within (4~6)*D*. The inclination in the vertical direction rises from 4° to 10°.

3.3. Deep Failure Behavior of Multi-Helix Anchors

Figure 11 shows the equivalent plastic strain at failure of multi-helix anchors buried in dense sand with an embedment ratio of 12*D* and various spacing ratios S/D and plate numbers n. It can be seen that the embedment depth of the uppermost plate controls the failure mode of the multi-helix anchor. Although the lowermost plate of all of these anchors is deeply embedded (12*D*), the mobilized zone of soil around the anchors extends to the ground, except for the double-helix anchors with S/D = 1.5~3 and the triple-helix anchor with S/D = 1.5. The softening zones are interconnected when the plate spacing is smaller than 6*D*. This failure mode is similar to the one assumed by the cylindrical shear method.

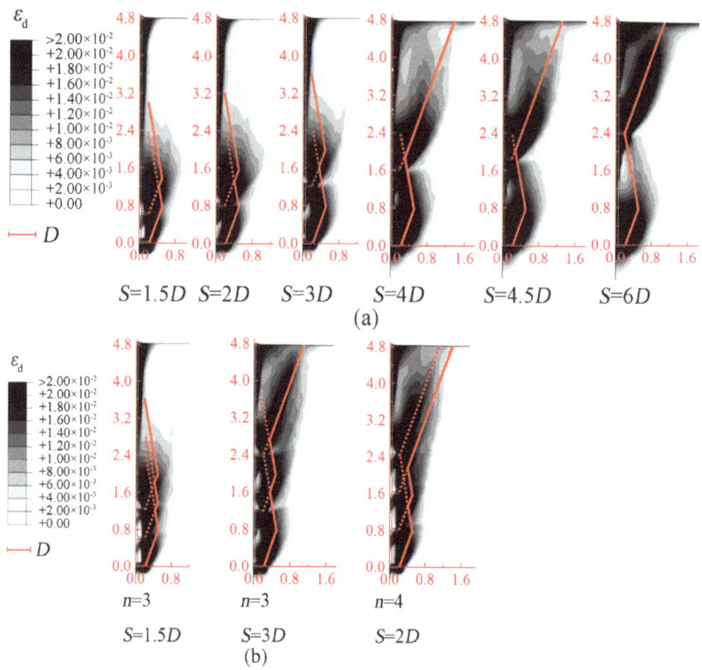

Figure 11. Multi-helix anchor cases: contour of equivalent plastic strain and rupture surface (red line) (dense sand). (**a**) Double-helix anchor. (**b**) Multi-helix anchor.

Compared with the failure mode of the single-helix anchor shown in Figure 11, it is worth noting that the helix spacing has little influence on the equivalent plastic strain range induced by each helical plate. For example, the soil mobilized zone of the double-helix anchor with S/D = 2 is just like a superposition of two single-helix anchors with H/D = 12 and H/D = 9. The upper plate experiences a transitional failure mode when the S/D = 4 for a double-helix anchor. When the S/D = 6 for the double-helix anchor, it is evident that the boundary of the soil mobilized zone caused by the lower plate bends towards the shaft. For multi-helix anchors, if more than one plate is at a shallow depth, the soil mobilized zone caused by the lowermost shallow plate will cover the mobilized zone of the other shallow plates, as with the quadruple-helix anchor shown in Figure 11b. The numerical observation confirms the conclusion from the centrifugal test that additional helices do not provide additional capacity if they are located within the region of soil mobilized by a lower helix [12]. As a result, the uplift resistance of the multi-helix anchor should be provided by the lowest shallow plate and each deep plate. The range of the soil plastic zone caused by each plate is only tied to the plate's embedment depth. The rupture surface of the multi-helix anchor can be regarded as the superposition of each helical plate's rupture surface when the above rule is adopted to identify the rupture surfaces [54,55], as shown by the solid red line in Figure 11.

4. Estimation of Uplift Capacity

It was possible to establish the link between the rupture surface and soil parameters, because varied soil characteristics were used. A simplified rupture surface was put forward according to the failure mode (Section 3). A unified calculation method using the limit equilibrium method (LEM) was proposed, including the lateral earth pressure coefficient, average internal friction angle, exponential decrease rate, and other parameters. The meanings of each symbol can be found in the Appendix A.

4.1. Simplified Rupture Surfaces

According to the failure mode determined from the above numerical results, the shallow and transition rupture surfaces of a single-helix anchor can be represented by one inverted truncated cone (inclined at ψ_p to the vertical), as illustrated in Figure 12a, which is the same as in previous experimental investigations [14,18,56,57]. The deep rupture surface of the single-helix anchor can be represented by one erected and one inverted truncated cone, as illustrated in Figure 12b. The inverted cone emerges from the plate edge with a vertical height of $2D$ and an inclination to the vertical of $\varphi_p/2$, and connected to the erected cone with a vertical height of $4D$ and an inclination to the vertical of $\psi_p/2$.

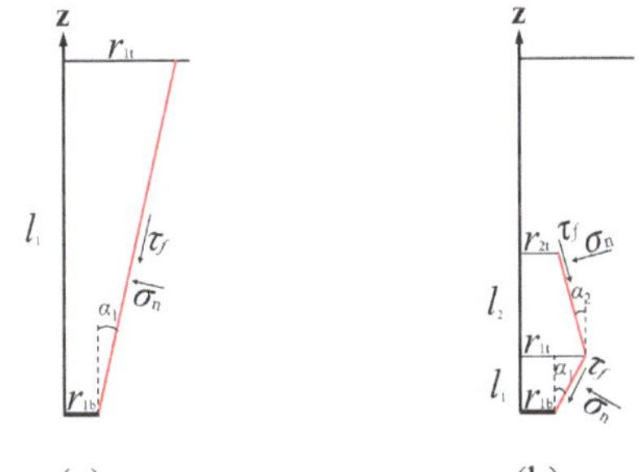

Figure 12. The simplified rupture surface of the single-helix anchor. (**a**) Shallow mode. (**b**) Deep mode.

The double-helix anchor is taken as an example. The deep rupture surface can be represented by two erected and two inverted truncated cones. If the rupture surface of the adjacent plate does not intersect, each plate can be calculated individually, as shown in Figure 13a. If the rupture surface of an adjacent plate intersects, its outermost contour is taken as the rupture surface, as shown in Figure 13b. Figure 13a,b cover the possible shallow and deep failure modes of the top helical plate simultaneously. The same simplification method can be applied to the multi-helix anchor. Thus, the rupture surface can be divided into several erected and inverted truncated cones. It is worth noting that the additional helical plates do not provide additional capacity if they are located within the region of soil mobilized by a lower plate.

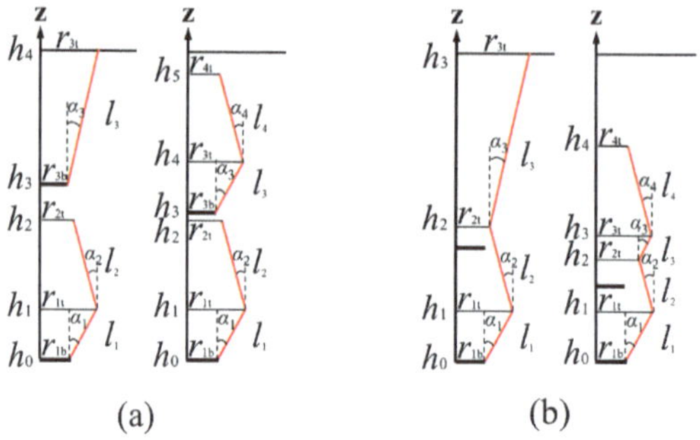

Figure 13. The simplified rupture surface of the double-helix anchor. (**a**) Individual mode. (**b**) Intersection mode.

4.2. A Unified Calculation Method

According to the limit equilibrium method, the ultimate uplift capacity of a single-helix anchor Q_u equals the sum of the vertical component of shearing resistance along each truncated cone and the soil weight, as shown in Equation (4):

$$Q_u = \sum (Q_{ui} + W_i) \tag{4}$$

where Q_{ui} is the uplift capacity of the ith truncated cone, as calculated using Equation (5):

$$Q_{ui} = \int_A \left(\tau_f \cos\alpha_i + (-1)^i \sigma_n \sin\alpha_i \right) dA \tag{5}$$

where σ_n is the normal stress on the rupture surface, $\sigma_n = K_u \gamma'(H - z)$, K_u is the lateral earth pressure coefficient, z is the vertical distance from the lowermost plate, H is the embedment depth of the plate, τ_f is the resisting shear stress on the rupture surface, $\tau_f = \sigma_n \tan\varphi^*_A$, and φ^*_A is the average internal friction angle. K_u and φ^*_A are determined as shown in the following section. The inclination angle to the vertical α_i for the deep plate is $\varphi_p/2$ when i is an odd number, and $\alpha_i = \psi_p/2$ when i is an even number. The shallow plate is assumed to be the inverted truncated cone, whose inclination angle is ψ_p.

W_i is the soil weight in the ith rupture surface of the truncated cone, as calculated using Equation (6):

$$W_i = \frac{\pi}{3} \gamma' l_i \left(r_{it}^2 + r_{ib}^2 + r_{it} r_{ib} \right) \tag{6}$$

where l_i is the vertical height of the ith rupture surface, and its bottom and top radii are r_{ib} and r_{it}, respectively. These parameters can be determined by geometry.

The above equations are also applied to the multi-helix anchor, since its failure mode is the superposition of the failure modes of each helical plate. Because the multi-helix anchor's failure mode is the superposition of each helical plate's failure mode, only the lateral earth pressure coefficient and the average internal friction angle of the single-helix anchor will be discussed.

4.3. Lateral Earth Pressure Coefficient

The effect of soil deformation will change the stress field around the plate, which is the significant difference between FEM and limit analysis. The stress distribution along the rupture surface of shallow circular anchors in sand was reflected by Cerfontaine [8] using FEM based on the hardening soil constitutive model, which can be described by a linearly increasing and then exponentially decreasing mathematical function. However, the stress analysis of deep anchors in sand is still unclear. This study investigated the stress distribution along the rupture surface of a deep anchor based on a modified Mohr–Coulomb model that incorporates the mobilized process of soil strength with plastic strain.

According to the Mohr–Coulomb limit equilibrium condition, the normal stress σ_n can be calculated by the maximum and minimum principal stresses and the internal friction angle of each element along the rupture surface. Then, the lateral earth pressure coefficient of deep anchor K_u along the rupture surface can be calculated by Equation (7):

$$K_u = \sigma_n / \gamma'(H - z) \quad (7)$$

where z is the vertical distance from the lowermost plate.

For shallow single-helix anchors, the lateral earth pressure coefficient K_n was proposed by Hao et al. [12], based on the assumption that the normal stress on the rupture surface remains in its initial state during pullout, as shown in Equation (8):

$$K_n = 1 - \frac{\sin \varphi_{cr}(1 + \cos 2\psi_p)}{2} \quad (8)$$

Then, K_u was standardized with K_n, and the distributions of K_u/K_n for different densities are shown in Figure 14.

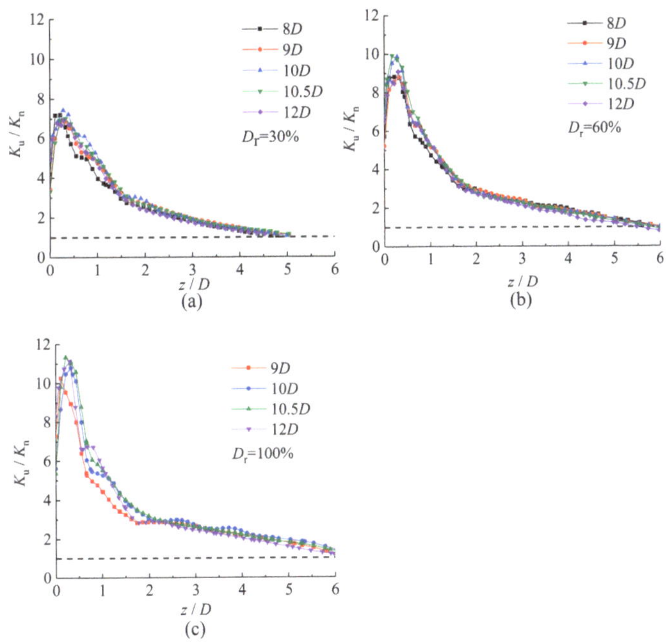

Figure 14. Distribution of normalized coefficients of lateral earth pressure along rupture surface. (**a**) Loose sand. (**b**) Medium dense sand. (**c**) Dense sand.

It can be seen from Figure 14 that the value of K_u increases linearly to the peak value $K_{u,peak}$ rapidly, and then it slowly decreases to close to the value of K_u. The trend of stress distribution is similar to the stress investigation around helical piles using the photoelasticity technique reported by Schiavon [16]. It is worth noting that the density has a more significant effect on the peak $K_{u,peak}$ instead of the embedment depth.

Considering the trend of K_u, two-stage functions are used to express the variation of K_u, as shown in Figure 15. Linear fitting is adopted before the peak value, which starts from the initial lateral pressure coefficient K_n to the peak lateral pressure coefficient $K_{u,peak}$. Then, exponential fitting is adopted after the peak value, which starts from the initial lateral pressure coefficient $K_{u,peak}$ to the lateral earth pressure coefficient proposed Hao [12] K_n, as shown in Equation (9):

$$K_u = \begin{cases} K_n + \left(K_{u,peak} - K_n\right)/z_p \cdot z & z \le z_p \\ K_n + \left(K_{u,peak} - K_n\right)e^{-\kappa(z-z_p)} & z > z_p \end{cases} \tag{9}$$

where z_p is the vertical distance of the peak value point from the plate, and κ is the exponential decrease rate. The position of z_p is always within 0.33*D* and has little relation to the relative density and embedment depth. Thus, z_p can be taken as 0.33*D* in Equation (9). The value of $K_{u,peak}$ in loose, medium, and dense sand is 3.5, 5, and 6.5, respectively. The value of κ is fitted to be 1.8.

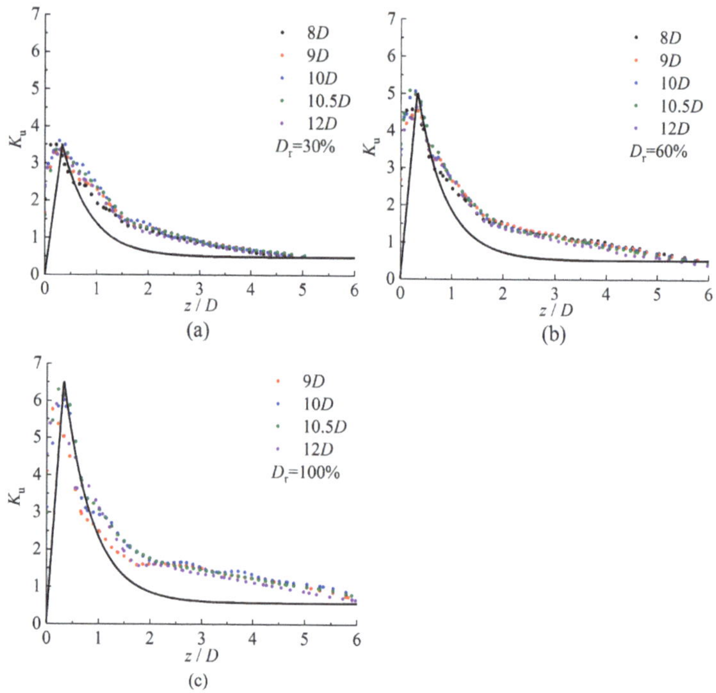

Figure 15. Simplified lateral earth pressure coefficient K_u. (**a**) Loose sand. (**b**) Medium dense sand. (**c**) Dense sand.

To simplify Equation (9), z_p can be taken as 0, because the value of z_p is small. The exponential fitting is adopted to express the variation of K_u, as shown in Equation (10). This part will be discussed in Section 4.5.

$$K_u = K_n + \left(K_{u,peak} - K_n\right)e^{-\kappa z} \tag{10}$$

4.4. Average Internal Friction Angle

According to the modified Mohr–Coulomb model, the internal friction angle is related to the equivalent plastic strain. Its distribution along the rupture surface is shown in Figure 16. The peak value point is vertically located 2*D* away from the plate, which corresponds to the boundary of the softening zone and is different from the position of the peak lateral earth pressure coefficient.

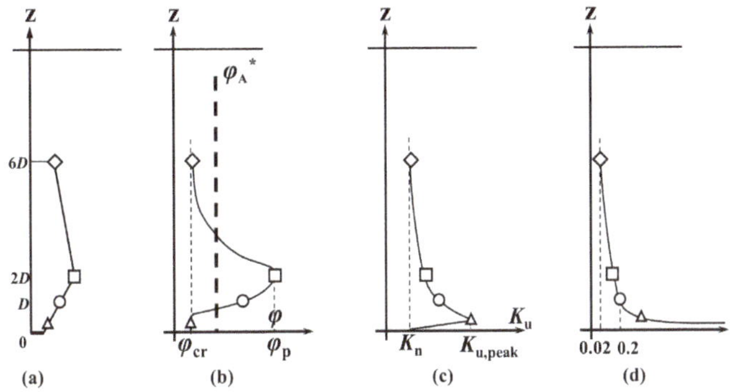

Figure 16. Distribution of internal friction angle, lateral earth pressure coefficient, and equivalent plastic strain along the rupture surface. (**a**) Rupture surface. (**b**) Internal friction angle. (**c**) Lateral earth pressure coefficient. (**d**) Equivalent plastic strain.

The internal friction angle near the plate equals the critical friction angle. As the distance from the plate increases, it begins to rise until it reaches the peak, and then it gradually decreases. This tendency makes the analysis more complex and realistic than FEM based on the HS small model [8]. The shear stress along the failure surface varies not only with the effective normal stress but also with the mobilization level of soil strength, which depends on plastic strain. Drescher [58] pointed out that numerical solutions to the bearing capacity problem at $\psi < \varphi$ predict a limit load higher than the estimates presented for Prandtl's failure mechanism. Davis [59–61] proposed Equation (11) to convert a non-associated plastic MC model into an equivalent associated plastic MC model [61]:

$$\tan\varphi^* = \frac{\sin\varphi_p \cos\psi_p}{1-\sin\varphi_p \sin\psi_p} \tag{11}$$

where φ^* is the internal friction angle of the equivalent associated plastic MC model.

For the convenience of calculation and application in LEM, an average internal friction angle φ^*_A was proposed. The internal friction angle along the rupture surface was extracted, and the average value was calculated. The comparison of φ^*_A from FEM with φ^* from Equation (11) [59] is shown in Figure 17. It can be seen that both values are similar for loose and medium dense sand, and the value of φ^* is 3° greater than that of φ^*_A for dense sand. For simplification and safety, the average friction angle can be estimated by Equation (11) directly for loose and medium dense sand, and by reducing the results of Equation (11) by 3° for dense sand.

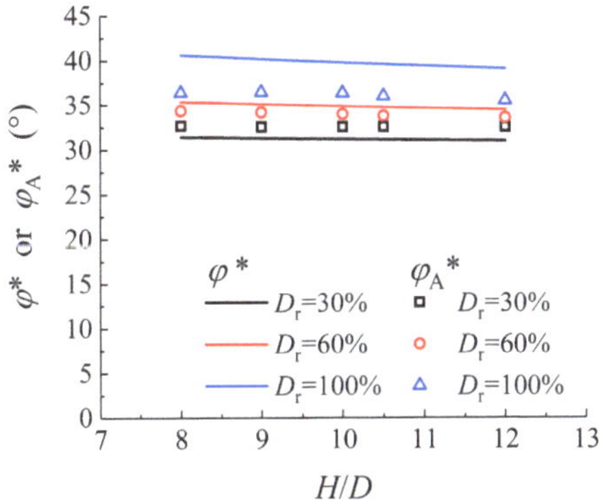

Figure 17. Comparison of the average internal friction angles from FEM and the calculation formula.

4.5. Comparison with Results

In Sections 4.1–4.4, the simplified rupture surfaces and the unified calculation method for helical anchors were discussed, and two important parameters were studied. So, the ultimate uplift capacity of a helical anchor can be estimated using Equations (4)–(11).

(1) Comparison with the FEM results

To verify the feasibility of the simplified rupture surface, the calculation results were compared with the finite element results, and the soil pressure coefficient was calculated using Equation (9). When the top anchor plate is a shallow embedment plate for multi-helix anchors, the lateral earth pressure coefficient is determined using Equation (8) without considering its changes. The comparison results are shown in Figure 18. The uplift capacity predicted by this method is about 5% lower on average than that of the finite element results. But for single helical anchors, the result with the largest deviation is found for $H/D = 9$ in loose sand, which is underestimated by up to 8.3%. For double-helix anchors, although it corresponds nicely when the spacing is bigger, it is 16.1% underestimated when the spacing is $3D$. As a result, this demonstrates that it is possible to calculate the uplift capacity of helical anchors using the simplified rupture mode presented in Section 4.1.

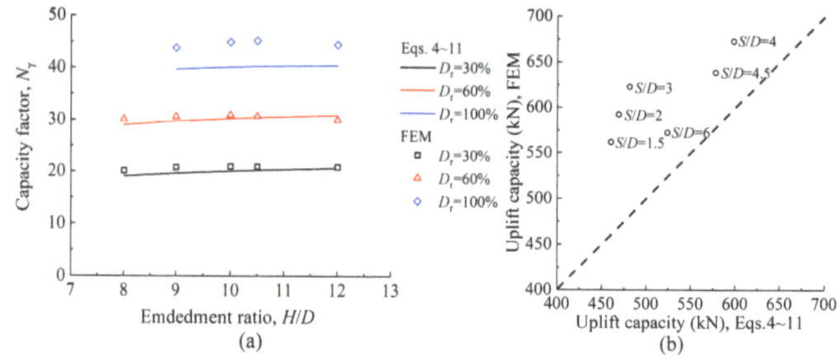

Figure 18. Comparison between the theoretical results and the FEM results. (**a**) Single-helix anchor. (**b**) Double-helix anchor.

(2) Comparison between Equations (9) and (10)

To verify the feasibility of the simplified calculation method, the uplift capacity of single-helix anchors was calculated using Equations (9) and (10) as the lateral earth pressure coefficient. The results are shown in Figure 19. It can be seen that the simplified calculation results using Equation (10) will reduce the estimated uplift capacity. The values of N_γ for loose sand, medium dense sand, and dense sand are reduced by an average of 7.5%, 9.3%, and 10.5%, respectively. However, for safety reasons, Equation (10) can be used instead of Equation (9) for calculation.

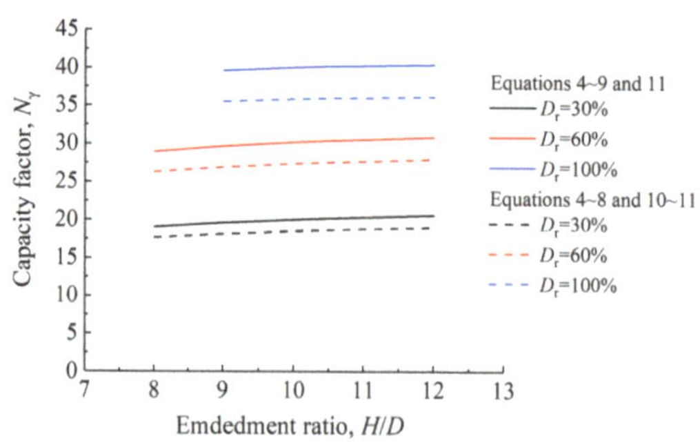

Figure 19. Comparison between Equations (9) and (10).

(3) Comparison with the test results

To verify the unified calculation method, the calculation results were compared with the centrifuge test results [12]. The soil characteristics were based on the values in Section 2.1, and the lateral earth pressure coefficient was calculated using Equation (10). When the top anchor plate is a shallow embedment plate for multi-helix anchors, the lateral earth pressure coefficient is determined using Equation (8) without considering its changes. The results of this theoretical method were compared with the test results, as shown in Figure 20. It can be seen that for single-helix anchors in dense sand, this theoretical method underestimates by up to 10.9% and overestimates by up to 6.7%. For multi-helix anchors in dense sand, this method underestimates by up to 17.0% and overestimates by up to 21.7%. Therefore, this method may be used to estimate the uplift capacity of helical anchors.

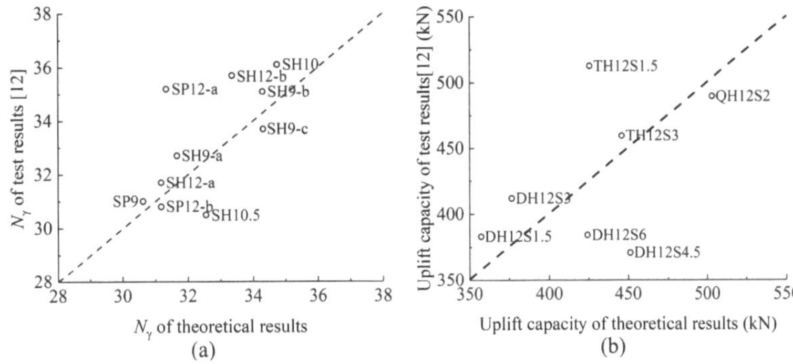

Figure 20. Comparison between the theoretical results and the centrifuge test results. (**a**) Single-helix anchor. (**b**) Multi-helix anchor.

5. Conclusions

The failure mechanism of deep anchors in sand is not well understood. The existing design methods of deep helical anchors in sand seldom consider the deep rupture surface. This study shows the possibility of simulating the pullout process of deep helical anchors in sand. Considering the strain softening of soil can help to more accurately simulate the uplift characteristics of deep helical anchors in sand, which can reflect the progressive failure process of deep anchors. In this process, the boundary of the plastic zone is a closed bulb and expands rapidly before the failure displacement and slowly after the failure displacement. A two-truncated-cone failure mode was identified based on the locus of maximum value of plastic strain at various depths for deep single-helix anchors, and on this basis, the analytical method of estimation for the ultimate uplift capacity of deep helical anchors is proposed.

The helix spacing has little influence on the equivalent plastic strain range induced by each helical plate. The boundary of the plastic zone for multi-helix anchors is a superposition of each helical plate. The distributions of the lateral earth pressure coefficient K_u can be simplified by a linearly increasing and then exponentially decreasing mathematical function. The peak value of K_u and the exponential decrease rate κ are almost independent of the embedment ratio of deep anchors, but they are related to the sand's compactness. The value of the average internal friction angle can be determined by the Davis formula.

This study proposes a unified calculation method for helical anchors in sand, and the parameters in this method were determined according to the results of numerical analysis. This innovative method can be suitable for single-helix or multi-helix anchors, as well as shallow- or deep-embedment helix anchors. It may be helpful in the design of helical anchors. These findings could improve the current design method of helical anchors, especially deep-embedment helical anchors.

Author Contributions: Conceptualization, formal analysis, and writing—review and editing, D.H. and C.Y.; investigation and writing—original draft preparation, C.Y. and N.Z.; investigation and data curation, C.Y. and R.C. All authors have read and agreed to the published version of the manuscript.

Funding: This research was funded by the National Natural Science Foundation of China (grant number 52078108).

Institutional Review Board Statement: Not applicable.

Informed Consent Statement: Not applicable.

Data Availability Statement: Due to the nature of this research, participants of this study did not agree for their data to be shared publicly, so supporting data is not available.

Conflicts of Interest: The authors declare no conflict of interest.

Appendix A

Table A1. Parameters in the paper.

Parameters	Definition
D	Helix diameter
t	Helix thickness
H	Embedment
S	Helix spacing
ΔB	Minimum element size
D_r	Density
φ_0	Initial internal friction angle
φ_{cr}	Critical internal friction angle
φ_p	Peak internal friction angle
ψ_p	Peak dilatancy angle
σ_3	Confining pressure
ε_d^r	Critical equivalent plastic strain
ε_d^p	Peak equivalent plastic strain
E	Young's modulus
v	Poisson's ratio
N_γ	Uplift capacity factors
γ'	Soil effective unit weight
A	Plate area
Q_u	Ultimate uplift capacity
u_f	Failure displacement
z	Vertical distance from the lowermost plate
z_p	Vertical distance of the peak value point from the plate
α	Inclination between the simplified rupture plane and the vertical direction
σ_n	Normal stress
K_0	Initial lateral pressure coefficient
K_u	Lateral earth pressure coefficient
K_n	Lateral earth pressure coefficient proposed by Hao [12]
$K_{u,peak}$	Peak lateral earth pressure coefficient
φ^*	Internal friction angle of the equivalent associated plastic MC model
φ^*_A	Average internal friction angle
W	Soil weight
l_i	Vertical height of the ith rupture surface
r_{ib}, r_{it}	Bottom and top radii, respectively

References

1. Hao, D.; Che, J.; Chen, R.; Zhang, X.; Yuan, C.; Chen, X. Experimental Investigation on Behavior of Single-Helix Anchor in Sand Subjected to Uplift Cyclic Loading. *J. Mar. Sci. Eng.* **2022**, *10*, 1338. [CrossRef]
2. Lin, Y.; Xiao, J.; Le, C.; Zhang, P.; Chen, Q.; Ding, H. Bearing Characteristics of Helical Pile Foundations for Offshore Wind Turbines in Sandy Soil. *J. Mar. Sci. Eng.* **2022**, *10*, 889. [CrossRef]
3. Meyerhof, G.; Adams, J. The ultimate uplift capacity of foundations. *Can. Geotech. J.* **1968**, *5*, 225–244. [CrossRef]

4. Mitsch, M.P.; Clemence, S.P. The uplift capacity of helix anchors in sand. In Proceedings of the Uplift Behavior of Anchor Foundations in Soil, New York, NY, USA, 24 October 1985; pp. 26–47.
5. Murray, E.; Geddes, J.D. Uplift of anchor plates in sand. *J. Geotech. Eng.* **1987**, *113*, 202–215. [CrossRef]
6. Ghaly, A.; Hanna, A.; Ranjan, G.; Hanna, M. Helical anchors in dry and submerged sand subjected to surcharge. *J. Geotech. Eng.* **1991**, *117*, 1463–1470. [CrossRef]
7. Giampa, J.R.; Bradshaw, A.S.; Schneider, J.A. Influence of dilation angle on drained shallow circular anchor uplift capacity. *Int. J. Geomech.* **2017**, *17*, 04016056. [CrossRef]
8. Cerfontaine, B.; Knappett, J.A.; Brown, M.J.; Bradshaw, A.S. Effect of soil deformability on the failure mechanism of shallow plate or screw anchors in sand. *Comput. Geotech.* **2019**, *109*, 34–45. [CrossRef]
9. Ghaly, A.; Hanna, A. Ultimate pullout resistance of single vertical anchors. *Can. Geotech. J.* **1994**, *31*, 661–672. [CrossRef]
10. Saeedy, H.S. Stability of circular vertical earth anchors. *Can. Geotech. J.* **1987**, *24*, 452–456. [CrossRef]
11. Al Hakeem, N.; Aubeny, C. Numerical investigation of uplift behavior of circular plate anchors in uniform sand. *J. Geotech. Geoenvironmental Eng.* **2019**, *145*, 04019039. [CrossRef]
12. Hao, D.; Wang, D.; O'Loughlin, C.D.; Gaudin, C. Tensile monotonic capacity of helical anchors in sand: Interaction between helices. *Can. Geotech. J.* **2019**, *56*, 1534–1543. [CrossRef]
13. Ghaly, A.; Hanna, A.; Hanna, M. Uplift behavior of screw anchors in sand. I: Dry sand. *J. Geotech. Eng.* **1991**, *117*, 773–793. [CrossRef]
14. Ilamparuthi, K.; Dickin, E.; Muthukrisnaiah, K. Experimental investigation of the uplift behaviour of circular plate anchors embedded in sand. *Can. Geotech. J.* **2002**, *39*, 648–664. [CrossRef]
15. Tappenden, K.M.; Sego, D.C. Predicting the axial capacity of screw piles installed in Canadian soils. In Proceedings of the Canadian Geotechnical Society (CGS), OttawaGeo2007 Conference, Ottawa, ON, Canada, 21–24 October 2007; pp. 1608–1615.
16. Schiavon, J.; Tsuha, C.; Esquivel, E. Experimental stress analysis in helical pile foundations by the photoelastic method. In *ICPMG2014–Physical Modelling in Geotechnics Proceedings of the 8th International Conference on Physical Modelling in Geotechnics, Perth, Australia, 14–17 January 2014*; CRC Press: Boca Raton, FL, USA, 2014; pp. 757–762.
17. Girout, R.; Blanc, M.; Dias, D.; Thorel, L. Numerical analysis of a geosynthetic-reinforced piled load transfer platform–validation on centrifuge test. *Geotext. Geomembr.* **2014**, *42*, 525–539. [CrossRef]
18. Motamedinia, H.; Hataf, N.; Habibagahi, G. A study on failure surface of helical anchors in sand by PIV/DIC technique. *Int. J. Civ. Eng.* **2019**, *17*, 1813–1827. [CrossRef]
19. Salehzadeh, H.; Nuri, H.; Rafsanjani, A.A.H. Failure Mechanism of Helical Anchors in Sand by Centrifuge Modeling and PIV. *Int. J. Geomech.* **2022**, *22*, 04022111. [CrossRef]
20. Das, B.M.; Shukla, S.K. *Earth Anchors*; J. Ross Publishing: Plantation, FL, USA, 2013.
21. Zhang, D.J.Y. Predicting Capacity of Helical Screw Piles in Alberta Soils. Ph.D. Thesis, University of Alberta, Edmonton, AB, Canada, 1999.
22. Clemence, S.P.; Pepe, F. Measurement of lateral stress around multihelix anchors in sand. *Geotech. Test. J.* **1984**, *7*, 145–152.
23. Chance. *Technical Design Manual Edition 4*; Intech Anchoring Systems: Caseyville, IL, USA, 2019.
24. Lutenegger, A.J. Behavior of multi-helix screw anchors in sand. In Proceedings of the 14th Pan-American Conference on Soil Mechanics and Geotechnical Engineering, Toronto, ON, Canada, 2–6 October 2011.
25. Council, I.C. *International Building Code (IBC)*; ICC: Washington, DC, USA, 2015.
26. Society, C.G. *Canadian Foundation Engineering Manual*, 4th ed.; Canadian Geotechnical Society: Saskatoon, SK, Canada, 2006; p. 488.
27. Kurian, N.P.; Shah, S.J. Studies on the behaviour of screw piles by the finite element method. *Can. Geotech. J.* **2009**, *46*, 627–638. [CrossRef]
28. Nasr, M. Performance-based design for helical piles. In *Contemporary Topics in Deep Foundations*; ASCE: Lawrence, KS, USA, 2009; pp. 496–503.
29. Salhi, L.; Nait-Rabah, O.; Deyrat, C.; Roos, C. Numerical modeling of single helical pile behavior under compressive loading in sand. *EJGE* **2013**, *18*, 4319–4338.
30. Gavin, K.; Doherty, P.; Tolooiyan, A. Field investigation of the axial resistance of helical piles in dense sand. *Can. Geotech. J.* **2014**, *51*, 1343–1354. [CrossRef]
31. Mittal, S.; Mukherjee, S. Behaviour of group of helical screw anchors under compressive loads. *Geotech. Geol. Eng.* **2015**, *33*, 575–592. [CrossRef]
32. Spagnoli, G.; Gavin, K. Helical piles as a novel foundation system for offshore piled facilities. In Proceedings of the Abu Dhabi International Petroleum Exhibition and Conference, Abu Dhabi, United Arab Emirates, 9–12 November 2015.
33. Akl, S.A.; Elhami, O.M.; Abu-keifa, M.A. Lateral performance of helical piles as foundations for offshore wind farms. In *Geo-Chicago 2016*; ASCE: Lawrence, KS, USA, 2016; pp. 449–458.
34. Fahmy, A.; El Naggar, M. Cyclic lateral performance of helical tapered piles in silty sand. *DFI J.-J. Deep. Found. Inst.* **2016**, *10*, 111–124. [CrossRef]
35. George, B.E.; Banerjee, S.; Gandhi, S.R. Numerical analysis of helical piles in cohesionless soil. *Int. J. Geotech. Eng.* **2017**, *14*, 361–375. [CrossRef]

36. Afzalian, M.; Medhi, B.; Eun, J.; Medhi, T. Retrofitting Uplift Capacity of Telecommunication Tower Foundation with Helical Piles in Dense Granular Soils. In Proceedings of the Geo-Congress 2020: Foundations, Soil Improvement, and Erosion, Minneapolis, MN, USA, 25–28 February 2020; pp. 239–245.
37. Alwalan, M.; El Naggar, M. Finite element analysis of helical piles subjected to axial impact loading. *Comput. Geotech.* **2020**, *123*, 103597. [CrossRef]
38. Cerfontaine, B.; Knappett, J.A.; Brown, M.J.; Davidson, C.S.; Al-Baghdadi, T.; Sharif, Y.U.; Brennan, A.; Augarde, C.; Coombs, W.M.; Wang, L. A finite element approach for determining the full load–displacement relationship of axially loaded shallow screw anchors, incorporating installation effects. *Can. Geotech. J.* **2021**, *58*, 565–582. [CrossRef]
39. Ho, H.M.; Malik, A.A.; Kuwano, J.; Rashid, H.M.A. Influence of helix bending deflection on the load transfer mechanism of screw piles in sand: Experimental and numerical investigations. *Soils Found.* **2021**, *61*, 874–885. [CrossRef]
40. Nowkandeh, M.J.; Choobbasti, A.J. Numerical study of single helical piles and helical pile groups under compressive loading in cohesive and cohesionless soils. *Bull. Eng. Geol. Environ.* **2021**, *80*, 4001–4023. [CrossRef]
41. Konkol, J. Numerical solutions for large deformation problems in geotechnical engineering. *PhD Interdiscip. J.* **2014**, 49–55.
42. Qiu, G.; Henke, S.; Grabe, J. Application of a Coupled Eulerian–Lagrangian approach on geomechanical problems involving large deformations. *Comput. Geotech.* **2011**, *38*, 30–39. [CrossRef]
43. Chen, Z.; Tho, K.K.; Leung, C.F.; Chow, Y.K. Influence of overburden pressure and soil rigidity on uplift behavior of square plate anchor in uniform clay. *Comput. Geotech.* **2013**, *52*, 71–81. [CrossRef]
44. Qiu, G.; Henke, S.; Grabe, J. Applications of Coupled Eulerian-Lagrangian method to geotechnical problems with large deformations. In Proceedings of the SIMULIA Customer Conference, London, UK, 18–21 May 2009; pp. 420–435.
45. Wood, D.M.; Belkheir, K. Strain softening and state parameter for sand modelling. *Géotechnique* **1994**, *44*, 335–339. [CrossRef]
46. Cheng, P.; Liu, Y.; Li, Y.P.; Yi, J.T. A large deformation finite element analysis of uplift behaviour for helical anchor in spatially variable clay. *Comput. Geotech.* **2022**, *141*, 104542. [CrossRef]
47. Cheng, P.; Guo, J.; Yao, K.; Chen, X. Numerical investigation on pullout capacity of helical piles under combined loading in spatially random clay. *Mar. Georesources Geotechnol.* **2022**, 1–14. [CrossRef]
48. Popa, H.; Batali, L. Using Finite Element Method in geotechnical design. Soil constitutive laws and calibration of the parameters. Retaining wall case study. *WSEAS Trans. Appl. Theor. Mech.* **2010**, *5*, 177–186.
49. Chen, R.; Fu, S.; Hao, D.; Shi, D. Scale effects of uplift capacity of circular anchors in dense sand. *Chin. J. Geotech. Eng.* **2019**, *41*, 78–85.
50. Chow, S.H.; Roy, A.; Herduin, M.; Heins, E.; King, L.; Bienen, B.; O'Loughlin, C.; Gaudin, C.; Cassidy, M. *Characterisation of UWA Superfine Silica Sand*; The University of Western Australia: Crawley, Australia, 2019.
51. Pei, H.; Wang, D.; Liu, Q. Numerical study of relationships between the cone resistances and footing bearing capacities in silica and calcareous sands. *Comput. Geotech.* **2023**, *155*, 105220. [CrossRef]
52. Hsu, S.; Liao, H. Uplift behaviour of cylindrical anchors in sand. *Can. Geotech. J.* **1998**, *35*, 70–80. [CrossRef]
53. Bolton, M. The strength and dilatancy of sands. *Geotechnique* **1986**, *36*, 65–78. [CrossRef]
54. Liu, J.; Liu, M.; Zhu, Z. Sand deformation around an uplift plate anchor. *J. Geotech. Geoenvironmental Eng.* **2012**, *138*, 728–737. [CrossRef]
55. Shi, D.; Yang, Y.; Deng, Y.; Xue, J. DEM modelling of screw pile penetration in loose granular assemblies considering the effect of drilling velocity ratio. *Granul. Matter* **2019**, *21*, 74. [CrossRef]
56. Nazir, R.; Chuan, H.S.; Niroumand, H.; Kassim, K.A. Performance of single vertical helical anchor embedded in dry sand. *Measurement* **2014**, *49*, 42–51. [CrossRef]
57. Srinivasan, V.; Ghosh, P.; Santhoshkumar, G. Experimental and Numerical Analysis of Interacting Circular Plate Anchors Embedded in Homogeneous and Layered Cohesionless Soil. *Int. J. Civ. Eng.* **2020**, *18*, 231–244. [CrossRef]
58. Drescher, A.; Detournay, E. Limit load in translational failure mechanisms for associative and non-associative materials. *Géotechnique* **1993**, *43*, 443–456. [CrossRef]
59. Davis, E. Theories of plasticity and the failure of soil masses. In *Soil Mechanics: Selected Topics*; Springer: Berlin/Heidelberg, Germany, 1969.
60. Vermeer, P. The orientation of shear bands in biaxial tests. *Geotechnique* **1990**, *40*, 223–236. [CrossRef]
61. Chen, X.; Wang, D.; Yu, Y.; Lyu, Y. A modified Davis approach for geotechnical stability analysis involving non-associated soil plasticity. *Géotechnique* **2020**, *70*, 1109–1119. [CrossRef]

Journal of
Marine Science and Engineering

MDPI

Article

Pullout Bearing Capacity of End-Bearing Torpedo Anchors in Cohesive Soil Seabed

Gang Li [1], Jinli Zhang [2,*], Jia Liu [3], Yu Xi [1] and Honggang Kou [1]

1 Shaanxi Key Laboratory of Safety and Durability of Concrete Structures, Xijing University, Xi'an 710123, China; t_bag945@126.com (G.L.); xiyu@xijing.edu.cn (Y.X.); 20190020@xijing.edu.cn (H.K.)
2 State Key Laboratory of Coastal and Offshore Engineering, Dalian University of Technology, Dalian 116024, China
3 School of Geological Engineering and Geomatics, Chang'an University, Xi'an 710054, China; 15929935077@163.com
* Correspondence: jlzhang@dlut.edu.cn

Abstract: As a new type of deep-sea anchoring foundation, thetorpedo anchor has the characteristics of simple installation, low cost, and high bearing capacity. Compared with the torpedo anchor without an anchor wing, the end-bearing torpedo anchor forms pile end resistance using a bearing plate, thus further improving its uplift bearing capacity. By conducting the pullout model test of torpedo anchors, we have compared and analyzed the effects of the pullout angle and bearing-plate radius on the bearing characteristics of T, EN3, EN4, and EC types of torpedo anchors. Based on the model test results, we established the V-H envelope of torpedo-anchor bearing capacity using the regression analysis method. The results show that when the displacement is small, the load-displacement curve of the torpedo anchor increases in an approximately linear mode, and the curve fluctuates and gradually entersa stable state with the gradual increase of the displacement. With the increasing pullout angle, the bearing capacity of the torpedo anchor increases first and then decreases. When the pullout angle is 45°, the torpedo anchor has the maximum bearing capacity. When the pullout angle is constant, the order of bearing capacity generated by different types of torpedo anchors is as follows: EC > EN4 > EN3 > T, and the bearing capacity rises with the increasing bearing-plate area. Through regression analysis, it is concluded that for the torpedo anchors of T, EN3, EN4, and EC types, the V-H envelope of bearing capacity shows an outward convex trend, and the determination coefficient reaches 0.930, indicating that the established V-H envelope is suitable for predicting the bearing capacity of torpedo anchors.

Keywords: end-bearing; torpedo anchor; bearing capacity; pullout angle; cohesive soil

Citation: Li, G.; Zhang, J.; Liu, J.; Xi, Y.; Kou, H. Pullout Bearing Capacity of End-Bearing Torpedo Anchors in Cohesive Soil Seabed. *J. Mar. Sci. Eng.* **2023**, *11*, 1548. https://doi.org/10.3390/jmse11081548

Academic Editor: Erkan Oterkus

Received: 16 July 2023
Revised: 28 July 2023
Accepted: 2 August 2023
Published: 4 August 2023

1. Introduction

With the exhaustion of offshore oil and gas resources, the exploitation of oil and gas has gradually advanced from offshore areas to the deep sea, and the traditional mooring system cannot meet the needs therefrom [1–9]. As a new type of deep-sea anchoring foundation, the torpedo anchor takes the shape of the cylinder as a whole; the conical anchor tip is 30°, the interior is filled with concrete or scrap metal, and its weight is about 100 t. During installation, the torpedo anchor falls freely to obtain kinetic energy and then penetrates the seabed at the water depth of 30 to 150 m, thereby ensuring a sufficient penetration depth and obtaining a higher bearing capacity. The torpedo anchor has the advantages of simple installation, low cost, and high bearing capacity. Therefore, this type of anchor has attracted great attention fromacademia and the engineering industry [10]. Yu et al. [11,12] concluded that the uplift bearing capacity of the torpedo anchor with four anchor wings is 1.9 times greater than that of the torpedo anchor without an anchor wing, and the uplift bearing capacity is larger when the pullout-load angle is between 30° and 45°. When the anchor wing has the same lateral area, increasing the width of the anchor wing can effectively improve the bearing capacity. When the length of the anchor wing is greater than 1/2 of the anchor length, the bearing capacity obtained by increasing the anchor wing width is higher than that obtained by increasing the anchor wing length. The research findings can provide a reference for the optimization design of torpedo anchors. Based on the sediment rheological properties, Yu et al. [13] proposed a dynamic torpedo-anchor technology. This technology mainly uses high-frequency mechanical vibrations to fluidize the sediment near the anchor body, thereby reducing the resistance between the outer wall of the anchor body and the surrounding soil so that the anchor body can automatically and quickly penetrate the seabed. When the anchor body penetrates the preset depth, the vibration unit is shut down and the installation is completed. The dynamic torpedo anchor has a large tensile-strength ratio and can withstand the vertical pullout force. It is characterized by fast anchor placement and easy recovery. Raaj et al. [14] stated that soil properties and shear strength affected the pullout capacity and penetration depth. Chen et al. [15] reported that the pullout capacity and penetration depth of the torpedo anchor with vibrational shearing was not restrained by water depth and drop height. Based on the embedment depth, net weight, geometry, and soil properties, Wang et al. [16] proposed a model to predict the undrained monotonic holding capacity of the torpedo anchor, and the predicted results coincided well with the experimental results. With increasing fin length, Ads et al. [17] concluded that the penetration depth decreased and the maximum extraction resistance increased. Kim and Hossain [18] stated that the inclined pullout capacity of torpedo anchors depended on the anchor weight and anchor-soil contact area; thereby, the pullout capacity increased with increasing fin number. Chen et al. [19] stated that the maximum vertical pullout capacity increased exponentially with increasing embedment depth under no vibrations, whereas the pullout capacity increased linearly with increasing embedment depth under 200 Hz vibrations. Hossain et al. [20,21] reported that the pullout capacity of the torpedo anchors increased with increasing consolidation time, embedment depth, and undrained shear strength, and the rectangular fin and conical tip were more effective to improve the pullout capacity. Based on the centrifuge model test results, an analytical model to calculate the pullout capacity of the torpedo anchor was established according to the reverse end-bearing and frictional resistance. Based on the nonlinear regression analysis of the test results, Wang et al. [22] established a model to predict the maximum inclined force of torpedo anchor penetration into cohesive beds when the loading angle changed from 20° to 90°. With increasing embedment depth, relative density, and area of the bearing plate, Li et al. [23] reported that the bearing capacity of the end-bearing torpedo anchor remarkably improved, and the pullout capacity of the end-bearing anchor was significantly higher than that of the traditional anchor. Yi et al. [24] found that the failure envelope was mainly influenced by anchor incline and soil strength gradient, and a simple procedure was developed to predict the ultimate pullout capacity of the torpedo anchor. Raie and Tassoulas [25] noted that the dissipation of excess pore water pressure and the recovery of soil strength were essential factors to predict the pullout capacity of the torpedo anchor.

Compared with the torpedo anchor without an anchor wing, the end-bearing torpedo anchor forms pile end resistance using a bearing plate, thus further improving its bearing capacity. However, there are fewer studies on the bearing capacity of end-bearing torpedo anchors. In this paper, we conduct the pullout model test of torpedo anchors without anchor wings and end-bearing torpedo anchors and analyze the bearing characteristics of both types of torpedo anchors under vertical and inclined pullout loads. Based on the test results, we establish the V-H envelope of the torpedo-anchor bearing capacity using the regression analysis method. The research findings can provide a reference for the optimization design and bearing-capacity prediction of torpedo anchors.

2. Materials and Methods

2.1. Test Materials

The test soil waswashed kaolin, purchased from Jiashuo Building Materials Processing Co., Ltd., Lingshou County, Shijiazhuang City, Hebei Province, China. According to the standard for the geotechnical testing method (G/BT 50123-2019) [26], its basic physical properties are shown in Table 1.

Table 1. Main physical properties of kaolin.

G_s	ρ (g/cm^3)	w_L (%)	w_P (%)	I_P	S_u (kPa)	E (MPa)
2.60	1.60	42.3	21.3	21.0	1.63	0.82

The T-98 torpedo anchor wasselected as the model anchor. It is made of high-speed steel material, and its Young modulus and Poisson ratio was 200 GPa and 0.3, respectively. The anchor rod is 9.7 cm long, the anchor tip is 2.5 cm long, the anchor rod diameter (D) is 2 cm, and the anchor weighs 281.5 g. The anchor tip and the anchor rod are designed as two separate components, and the bearing plate can be assembled after detachment. A total of 12 types of bearing plates with a thickness of 2 mm were arranged for the test. The torpedo anchor without anchor wings (T type) is shown in Figure 1a, and the EN3 type torpedo-anchor bearing plate consists of 3 rectangular bearing plates with an angle of 120° for the center line, as shown in Figure 1b. The EN4-type torpedo-anchor bearing plate consists of 4 rectangular bearing plates with an angle of 90° for the center line, as shown in Figure 1c. The EC-type torpedo-anchor bearing plate was designed with a circular ring to ensure that the inner diameter of the bearing plate is the same as the outer diameter of the anchor rod and prevent the assembled bearing plate from any translation, as shown in Figure 1d. Table 2 lists the parameters of the torpedo-anchor bearing plate. As shown in the table, the size and weight of the rectangular bearing plate gradually increase with the increasing torpedo anchor number (1~4) when the type of bearing plate is fixed.

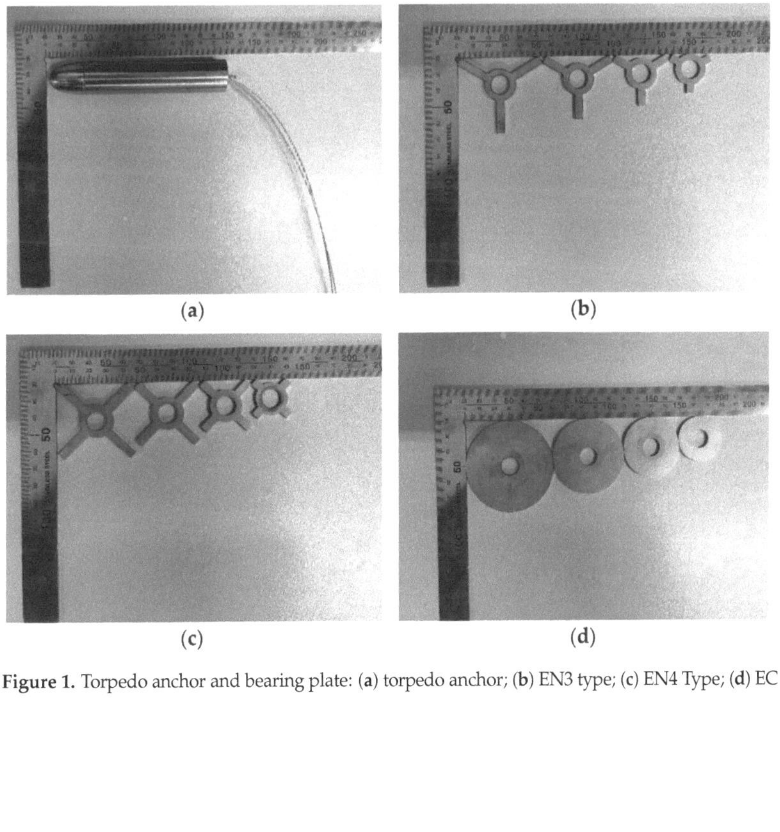

Figure 1. Torpedo anchor and bearing plate: (**a**) torpedo anchor; (**b**) EN3 type; (**c**) EN4 Type; (**d**) EC type.

Table 2. Parameters of torpedo-anchor bearing plate.

Label	EN3 Type Dimensions (mm)	EN3 Type Mass (g)	EN4 Type Dimensions (mm)	EN4 Type Mass (g)	EC Type Diameter (mm)	EC Type Mass (g)
a	3 × 5 × 5	5.59	4 × 5 × 5	6.50	30	12.96
b	3 × 5 × 10	7.61	4 × 5 × 10	8.67	40	25.10
c	3 × 5 × 15	9.22	4 × 5 × 15	10.81	50	40.74
d	3 × 5 × 20	10.76	4 × 5 × 20	12.92	60	59.59

2.2. Test Equipment

The pullout test device is made of a rigid frame, as shown in Figure 2. It mainly consists of a fixed pulley, motor, tension sensor, signal amplifier, signal acquisition box, and computer. The steel strand connected to the motor side passes through the pulley fixed at the beam, the other end is connected to the top of the torpedo anchor, and the middle of the steel strand is connected to the type-S tension sensor. The sensor is connected to the signal amplifier through a data cable and outputs the signal to the TWD information acquisition box. The measuring range of the tension sensor is ±10 kg, the motor speed is 1 cm/s, and the acquisition frequency of the TWD information acquisition box is 200 Hz. The test model box is made of plexiglass, as shown in Figure 3. The height of the model box is 75 cm, the outer diameter is 45 cm, and the wall thickness is 2 cm. There are 6 holes in the outer wall of the model box. The lines connecting the center of the model box with the center of each hole form an angle of 0°, 15°, 30°, 45°, 60°, and 75° with the horizontal line, respectively, so that the torpedo can be pulled in multiple directions. When the soil sample is still and saturated, each hole is blocked with a rubber plug to prevent water leakage.

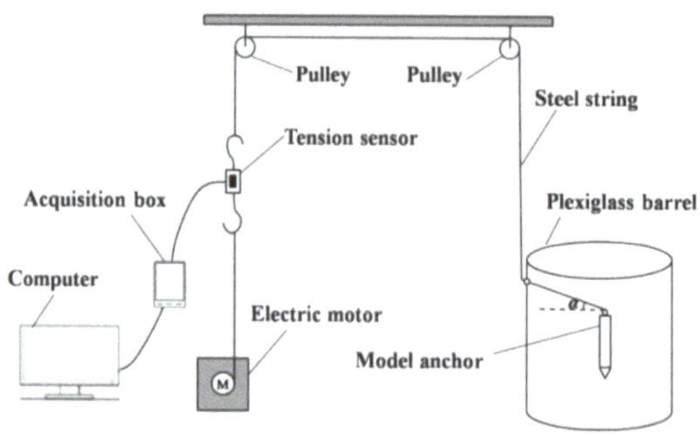

Figure 2. Schematic diagram of the model test device.

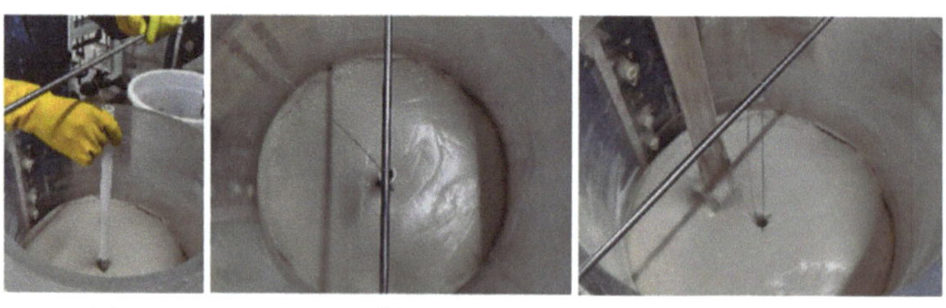

(a) **(b)** **(c)**

Figure 3. Flow chart of torpedo anchor installation: (**a**) installation; (**b**) vertically; (**c**) vibrating.

2.3. Test Methods

An amount of kaolin isslowly added into a proper amount of water, and we keep stirring to prevent kaolin from agglomerating and reduce bubble residues. Then, the model box is sealed with a plastic film to reduce soil moisture evaporation [23,26]. A steel strand hole is reserved at the tail of the torpedo anchor in the test, two steel strands are inserted into the hole, and the steel strand and the torpedo anchor can be firmly connected by tightening the limit screw. The diameter of the steel strands is 2 mm, and the lengths are 44 cm and 137 cm, respectively. The steel strand with a constant length of 44 cm is used to ensure the same embedded depth of the torpedo anchor and the same height, with the pullout hole at 0° with the outer wall of the model box. A PVC pipe with a length of 37 cm, a wall thickness of 3 mm, and the same outer diameter as the outer diameter of the torpedo anchor is used as the conduit. Pass the steel strand through the conduit and press the conduit down to ensure that the bottom of the conduit fits the top of the anchor rod. Slowly press the torpedo anchor into the soil using the conduit. In this process, the torpedo moves down at a low constant speed to reduce the disturbance to the surrounding soil. See Figure 3 for the installation process. When the torpedo anchor reaches the embedded depth, a steel rod is placed on the surface of the model box to limit the downward movement of the torpedo anchor so that the anchor is successfully placed. Then, withdraw the conduit vertically and slowly to reduce the disturbance to the soil above and the impact on the torpedo anchor. Vibrate the soil around the hole wall at a high frequency but low amplitude to accelerate the soil body flow and complete the collapse of the soil above the anchor top. In the process of vibration, make sure that vibration points are symmetrical and far away from the hole locations to reduce the impact on the anchor position. When the torpedo anchor is installed, keep it still for 12 h to improve the soil strength, and set a thin water level on the soil surface in this process to prevent the soil from cracking. Then, remove the suspended steel rod and connect the steel strand with a length of 137 cm to the tension sensor and the motor. Adjust the fixed pulley position of the model box to make the steel strand pass through the middle of the hole, and apply Vaseline around the hole to reduce the test error. When performing the vertical pullout test, make sure that the steel strand above the top of the torpedo anchor is vertically upward. When performing the oblique pullout test, make sure that the steel strand is laid through the hole center by adjusting the fixed pulley at the outer wall so that the pullout angle of the steel strand meets the test requirements. Start the motor and stop the test when the load reaches the peak strength or the anchor body is fully pulled out. During the test, the data acquisition system is used to record the value and time of the tension sensor, and the load-displacement curve can be obtained after conversion according to the uniform speed pulling of the motor.

3. Results and Discussion

3.1. Results and Analysis of Vertical Pullout Test of Torpedo Anchor

Under the action of vertical pullout load, the bearing capacity of the torpedo anchor without an anchor wing is mainly provided by the side frictional resistance of the anchor rod, and the bearing plate of the end-bearing torpedo anchor has a significant effect on the bearing capacity improvement of the torpedo anchor. To study the effect of the bearing-plate length on the bearing capacity of an end-bearing torpedo anchor, Figure 4 shows the load-displacement relation curve of the EN3-type torpedo anchor. The figure shows that when the displacement is small, the load-displacement curve of the torpedo anchor increases in an approximately linear mode, and the curve fluctuates and tends to be stable with the increase of the displacement. As for the primary cause, it is believed that the soil around the anchor is in the stage of elastic deformation when the curve increases linearly and the soil enters the stage of plastic deformation at the inflection point of the curve. During the upward movement of the torpedo anchor affected by the pullout load, the upward movement of the anchor tip leads to the formation of a cavity in the original position and the generation of suction, so the load is reduced. Through the release of active soil pressure, the soil on the cavity side moves towards the cavity and the cavity is gradually filled, thus gradually reducing the suction and increasing the pullout load. The torpedo anchor moves upward at a constant speed. Therefore, cavities appear and are filled now and then, so the load-displacement curve fluctuates. Under the action of the vertical pullout load, the bearing plate has a significant effect on the bearing capacity improvement of the torpedo anchor in comparison with the torpedo anchor without an anchor wing, and the bearing capacity rises with the increasing length of the bearing plate. When the length of the bearing plate is equal to the diameter of the anchor rod, the bearing capacity is increased by 0.78 times. When the length of the bearing plate is equal to the radius of the anchor rod, the bearing plate comes from the inside of the anchor rod more easily. Under such circumstances, the bearing capacity of the EN3-type torpedo anchor reaches 1.38 times that of the T-type torpedo anchor.

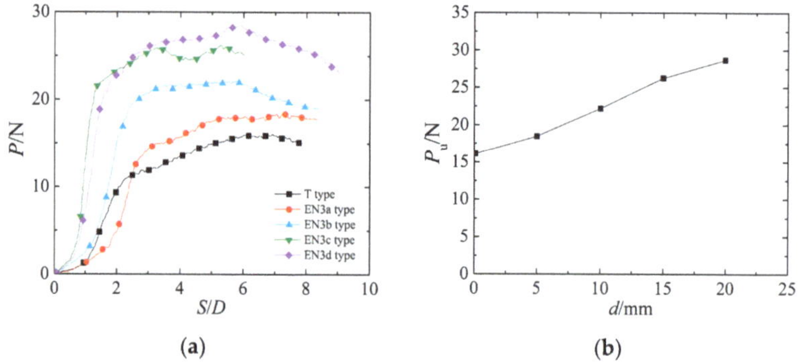

Figure 4. Relation curve between load and displacement of EN3-type torpedo anchor under vertical pullout load: (**a**) load-displacement curve; (**b**) bearing capacity–bearing-plate diameter curve.

Figure 5 shows the relation curve between the load and displacement of the EN4-type torpedo anchor. As shown in the figure, the findings are consistent with the previous study results. In detail, the bearing capacity of the torpedo anchor rises with the increasing length of the bearing plate and reaches the maximum value of 31.91 N when the length of the bearing plate is equal to the diameter of the torpedo anchor rod. The bearing capacity of the torpedo anchor is 24.59 N when the length of the bearing plate is equal to the radius of the torpedo anchor rod. In most cases, a longer bearing plate will provide a stronger bearing capacity. If the bearing plate is too long, it is difficult to install such a torpedo through structural design. Therefore, the length of the bearing plate involved in the test is limited to the range of 0.5 to 2 times the radius of the anchor rod. At the initial stage of pullout load for the torpedo anchor, the load increases linearly with the displacement. Under such circumstances, the soil is squeezed into the stage of elastic deformation by the anchor rod and the bearing plate, and the soil on the anchor side enters the stage of elastic deformation under the action of friction. However, the time nodes of entry into the stage of plastic deformation are not consistent. This finding is consistent with the conclusion of Richardson et al. [27].

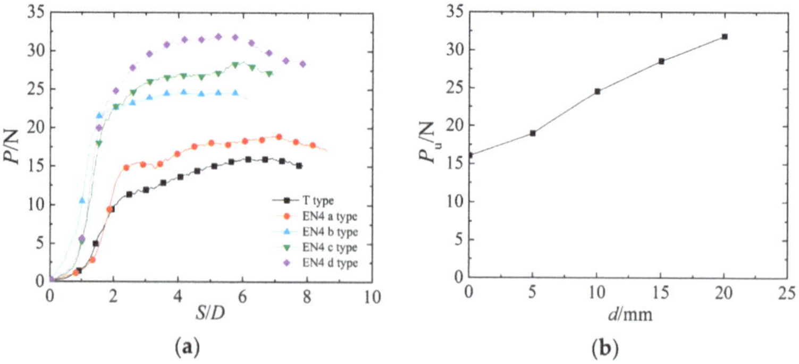

Figure 5. Relation curve between load and displacement of EN4-type torpedo anchor under vertical pullout load: (**a**) load-displacement curve; (**b**) bearing capacity–bearing-plate diameter curve.

In order to study the effect of the circular bearing-plate length on the bearing capacity of thetorpedo anchor, Figure 6 shows the load-displacement relation curve of the EC type of torpedo anchor. As shown in the figure, the bearing capacity of the torpedo anchor rises with the increasing radius of the bearing plate. When the radius of the bearing plate is equal to the diameter of the anchor rod, the bearing capacity of the end-bearing torpedo anchor is increased by 1.55 times compared with that of the torpedo anchor without the anchor wing. Figure 6b shows the relation curve between the bearing capacity and bearing plate radius of the EC torpedo anchor. As shown in the figure, a trend of approximate linear increase exists between the bearing capacity and the bearing-plate radius of the EC-type torpedo anchor, which is consistent with the above conclusion. When the bearing-plate radius is 20 mm, the bearing capacity of the torpedo anchor reaches 78.81 N, which ishigher than that of the T-type torpedo anchor. For the end-bearing torpedo anchor, the circular bearing plate forms a closed surface, which prevents the soil from flowing out of the bearing plate so that the bearing plate acts on the soil as a whole, further improving the bearing capacity. However, the circular bearing plate is an ideal type. In actual engineering practice, it is difficult for such a bearing plate to fully come from the anchor rod, while the EN3 and EN4 types of torpedo anchors can be constructed and installed more easily.

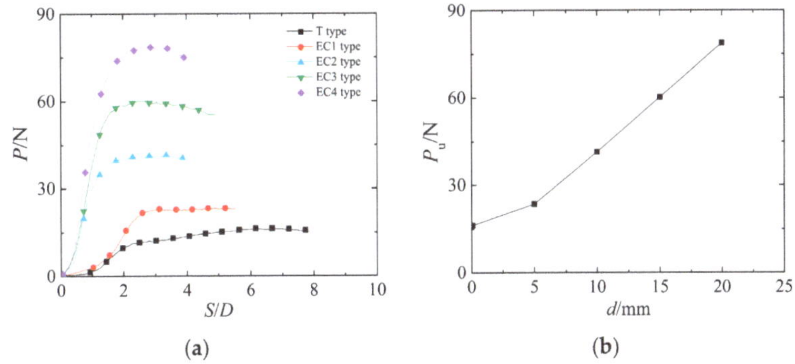

Figure 6. Relation curve between load and displacement of EC type torpedo anchor under vertical pullout load: (**a**) load-displacement curve; (**b**) bearing capacity–bearing-plate radius curve.

Figure 7 sums up the relation curves between the load and displacement of torpedo anchors with different types of bearing plates under vertical pullout load. As shown in the figure, the EC-type torpedo anchor can provide a greater bearing capacity than the EN3 and EN4 types of torpedo anchors, and the bearing-capacity difference gradually increases with the increasing length and diameter of the bearing plate. As for the primary cause, the EN3 and EN4 types of torpedo anchors have poor overall integrity with the increasing diameter of the bearing plate. Their bearing-capacity curve has a smaller slope than that of the EC-type torpedo anchors. The bearing plates of the EN3 and EN4 types oftorpedo anchors have the same effect on the bearing capacity of torpedo anchors, and the slope of the bearing-capacity curve of the EN4-type torpedo anchor is 1.23 times that of the EN3-type torpedo anchor, indicating that increasing the number and area of bearing plates can further improve the bearing capacity of torpedo anchors.

3.2. Results and Analysis of Inclined Pullout Test of Torpedo Anchor

Figure 8 shows the relation curve between the load, displacement, and pullout angle of a torpedo anchor without an anchor wing under an inclined pullout load. As shown in the figure, the bearing capacity of the torpedo anchor under an inclined pullout load is significantly increased compared with the bearing capacity under a vertical pullout load. The bearing capacity of the torpedo anchor without an anchor wing under different pullout angles is in the range of 18.25 N to 27.77 N and reaches the peak when the pullout angle is 45°. Based on comparisons, we discovered that the horizontal bearing capacity of the torpedo anchor without an anchor wing is approximately 1.18 times the vertical bearing capacity.

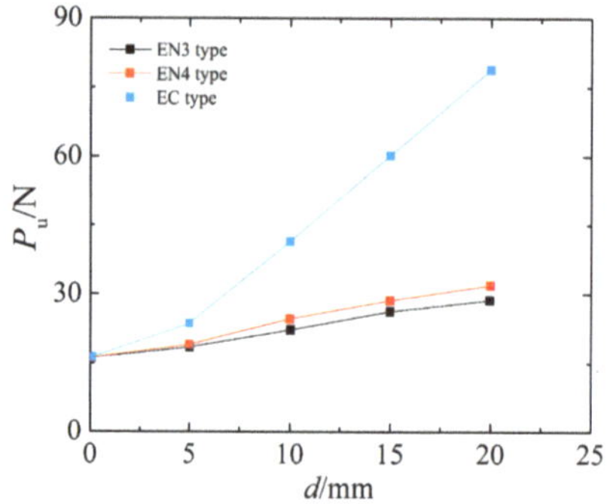

Figure 7. Relation curve between load and displacement of torpedo anchors with different types of bearing plates under vertical pullout load.

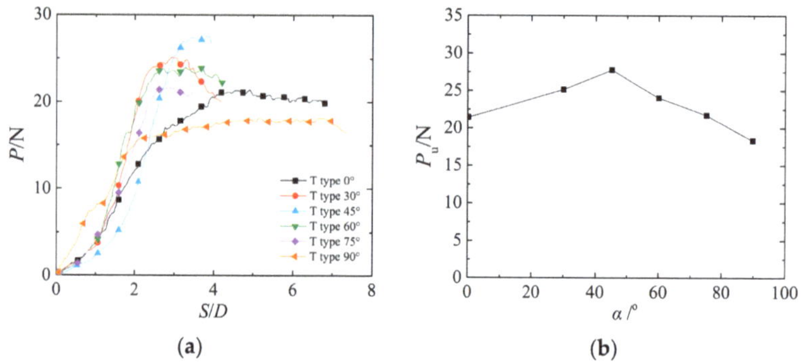

Figure 8. Relation curve between load, displacement, and pullout angle of T-type torpedo anchor under inclined pullout load: (**a**) load-displacement curve; (**b**) bearing capacity–pullout angle curve.

Figure 9 shows the relation curve between the load, displacement, and pullout angle of the EN3-type torpedo anchor under an inclined pullout load. As shown in the figure, the load-displacement curve of the torpedo anchor is approximately linear when there is a small displacement. Under such circumstances, it is believed that the soil around the anchor is in the stage of elastic deformation. When the pullout angle is in the range of 45° to 90°, the slope of the load-displacement curve is relatively large. When the pullout angle is 90°, the slope is the largest. Under the same test conditions, the horizontal bearing capacity of the EN3 torpedo anchor is 23.72 N, while the vertical bearing capacity is 21.86 N, indicating that the horizontal bearing capacity is 1.09 times the vertical bearing capacity. This fact is consistent with the research conclusion of O'Beirne [28]. For EN3-type torpedo anchors, the optimum pullout angle is 45°, and the corresponding bearing capacity is 28.03 N, which is about 1.28 times the vertical bearing capacity.

Figure 10 shows the relation curve between the load, displacement, and pullout angle of the EN4 torpedo anchor under an inclined pullout load. As shown in the figure, the load-displacement curve of the torpedo anchor under an inclined pullout load shows a trend of increasing first and then stabilizing. In the initial stage of load application, the curve increases in an approximately linear mode, and then the rate of increase gradually decreases and tends to be stable. When the torpedo anchor is subjected to vertical pullout, the anchor body moves from deep burial to shallow burial and then is gradually pulled out. Under such circumstances, the soil at the anchor top develops from local failure to structural failure, and the resistance of the soil at the anchor top towards the torpedo anchor decreases gradually. The bearing capacity of the EN4-type torpedo anchor also shows the trend of increasing first and then decreasing with the increasing pullout angle, and the optimum pullout angle is 45°. Under such circumstances, the corresponding bearing capacity reaches the maximum value of 30.08 N.

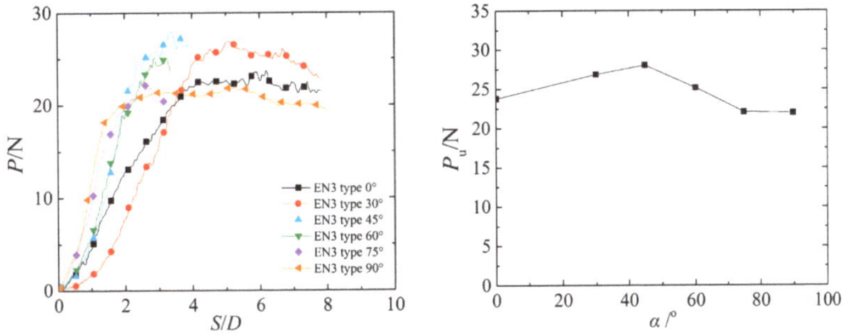

(**a**) (**b**)

Figure 9. Relation curve between load, displacement, and pullout angle of EN3-type torpedo anchor under inclined pullout load: (**a**) load-displacement curve; (**b**) bearing capacity–pullout angle curve.

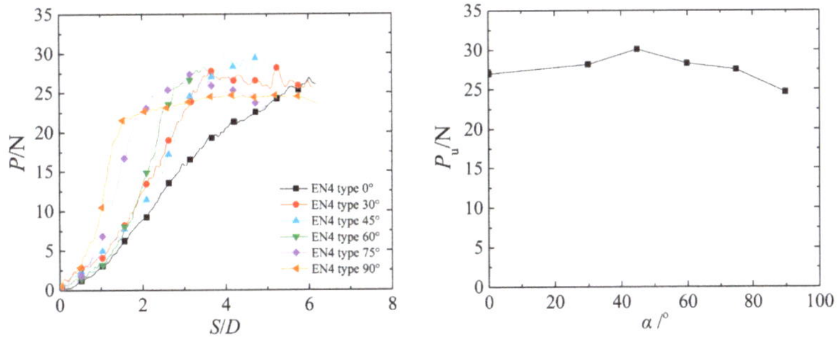

(**a**) (**b**)

Figure 10. Relation curve between load, displacement, and pullout angle of EN4-type torpedo anchor under inclined pullout load: (**a**) load-displacement curve; (**b**) bearing capacity–pullout angle curve.

Figure 11 shows the relation curve between the load, displacement, and pullout angle of the EC-type torpedo anchor under an inclined pullout load. As shown in the figure, the torpedo anchor has the smallest bearing capacity under the vertical load. When the pullout angle is 45°, the bearing capacity reaches the peak. Under the vertical pullout load, the corresponding displacement is only 1 D when the bearing capacity reaches the peak. As for the primary cause, the bearing plate, anchor top, and soil around the anchor enter the stage of elastic deformation when the torpedo anchor is under vertical pullout. With the gradual increase of displacement, the soil evolves from the elastic stage to the critical plastic stage, and then the bearing capacity reaches the peak. When the displacement continues to increase, the bearing capacity will not increase. However, the soil deformation continues, which corresponds to the second half of the load-displacement curve, and the bearing capacity reaches a stable trend or gradual downward trend. When the torpedo anchor is subjected to a horizontal pullout load, a large displacement is required to make the bearing capacity reach the peak load. The torpedo anchor starts to rotate from the vertical position. When it reaches a certain angle, the torpedo anchor provides the maximum bearing capacity at thattime. In other words, the maximum bearing-capacity equivalent to that of horizontal pullout is equivalent to the bearing capacity at a certain pullout angle. However, the peak horizontal bearing capacity is smaller than the peak bearing capacity at a specific angle mainly because the plastic deformation of the anchor top and anchor tip is the largest, and the plastic zone around the anchor expands with the rotation of the torpedo anchor when the torpedo anchor starts to rotate from the vertical position. When the torpedo anchor rotates to a certain angle, the soil around the anchor partially enters the stage of plastic deformation. Under such circumstances, the strength that the soil can provide is no longer the peak strength and maybe the residual strength. As shown in Figure 11, when the pullout angle changes from the vertical status to 45°, the bearing capacity of the torpedo anchor increases by 22.72% from 41.77 N to 51.26 N, indicating that the optimum pullout angle is 45°.

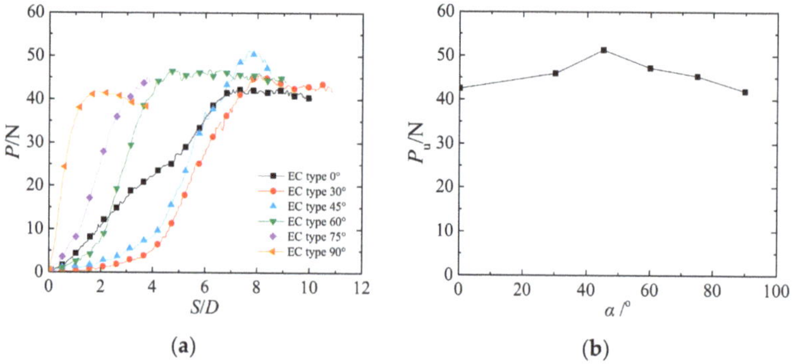

Figure 11. Relation curve between load, displacement, and pullout angle of EC-type torpedo anchor under inclined pullout load: (**a**) load-displacement curve; (**b**) bearing capacity–pullout angle curve.

Figure 12 sums up the relation curves between the bearing capacities of the T, EN3, EN4, and EC types of torpedo anchors and pullout angles. As shown in the figure, the bearing capacity of the torpedo anchors increases first and then decreases with the increasing pullout angle. When the pullout angle is 45°, the bearing capacity of all types of torpedo anchors reaches the peak. When the pullout angle is fixed, the bearing capacity of the T-typetorpedo anchor is the lowest, while the EC-type torpedo anchor has the largest bearing capacity, followed by the EN4-type torpedo anchor, indicating that the bearing capacity of the end-bearing torpedo anchor is significantly greater than that of the torpedo anchor without an anchor wing, and the bearing capacity of torpedo anchors rises with the increasing bearing-plate area.

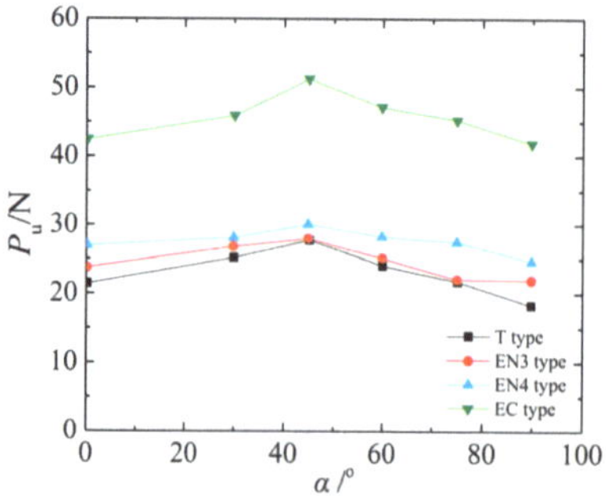

Figure 12. Relation curve between bearing capacity and pullout angle of torpedo anchor.

3.3. V-H Envelope Establishment

When studying the bearing capacity of torpedo anchors, the V-H bearing-capacity envelope is usually used to study the relations between the various components of bearing capacity [11]. The normalized V-H envelope is expressed as follows:

$$\left(\frac{P_{\mathrm{H}}}{P_{\mathrm{Hmax}}}\right)^{\mathrm{A}} + \left(\frac{P_{\mathrm{V}}}{P_{\mathrm{Vmax}}}\right)^{\mathrm{B}} = 1 \tag{1}$$

where P_{H} is the horizontal component of the bearing capacity of the torpedo anchor; P_{V} is the vertical component of the bearing capacity of the torpedo anchor; P_{Hmax} is the maximum horizontal bearing capacity, also the bearing capacity during horizontal pullout; P_{Vmax} is the maximum vertical bearing capacity, also the bearing capacity during vertical pullout; and A and B are the envelope coefficients.

Based on the model test results and fitting analysis, we obtained the V-H failure envelope diagram of the bearing capacities generated by different types of torpedo anchors. See Figure 13 for details. Table 3 lists the bearing-capacity envelope coefficients of the torpedo anchors. As shown in the table, the determination coefficients of the EN3, EN4, and EC types of torpedo anchors are higher than 0.990, indicating a good correlation. In order to simplify the prediction of torpedo anchor bearing capacity, the unified regression analysis wascarried out on the test results of different types of torpedo anchors; the values of envelope coefficients A and B are 2.753 and 4.522, respectively, and the determination coefficient reaches 0.930, indicating that the formula is suitable for predicting the bearing capacity of different types of torpedo anchors. Different from classic elliptic stresses yield function [29,30], the V-H failure envelope was generated by the bearing capacities of different types of torpedo anchors. Due to the determination coefficients of the T-type torpedo anchor being only 0.553, the V-H failure envelope with hyperelliptic shape was formed.

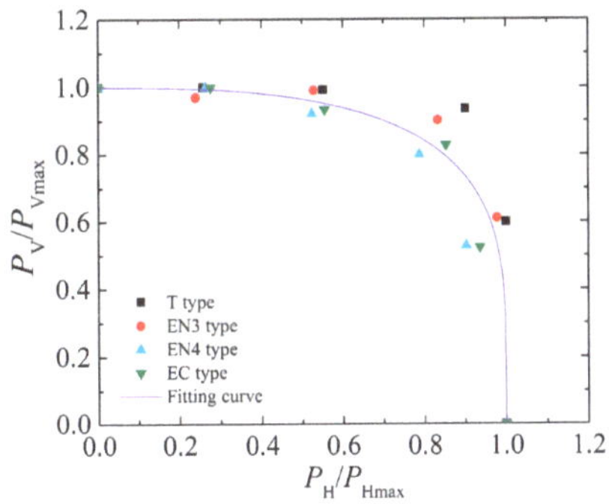

Figure 13. Bearing capacity V-H envelope.

Table 3. V-H envelope coefficient of bearing capacity.

Type	A	B	R^2
T	3.386	18.210	0.553
EN3	5.484	4.410	0.999
EN4	4.332	1.688	0.996
EC	10.484	1.090	0.995
Total	2.753	4.522	0.930

4. Conclusions

By conducting the pullout model test of torpedo anchors, we have studied the bearing characteristics of the T, EN3, EN4, and EC types of torpedo anchors under vertical loads and inclined loads and compared and analyzed the effect of the pullout angle and bearing-plate radius on the bearing capacity of torpedo anchors. Based on the model test results, we established the V-H envelope of bearing capacity for different types of torpedo anchors. The findings of this study are of certain reference value for predicting the bearing capacity of torpedo anchors in engineering practice. The main conclusions are as follows:

(1) When the displacement is small under a vertical pullout load, the load-displacement curve of torpedo anchors shows the trend of approximately linear increase. The curve fluctuates and tends to be stable with the increase of the displacement. Compared with the torpedo anchor without an anchor wing, the bearing capacity of the end-bearing torpedo anchor is increased significantly and rises with the increasing bearing plate length. The EC-type torpedo anchor provides a greater bearing capacity than the EN3 and EN4 types of torpedo anchors.
(2) When the pullout angle is in the range of 0° to 45° under an inclined pullout load, the slope of the load-displacement curve of torpedo anchors is smaller. When the pullout angle is in the range of 45° to 90°, the slope of the load-displacement curve is larger. When the pullout angle is 90°, the curve has the largest slope.
(3) With anincreasing pullout angle, the bearing capacity of the torpedo anchor increases first and then decreases. When the pullout angle is 45°, the bearing capacity of all types of torpedo anchors reaches the peak. When the pullout angle is fixed, the bearing capacity of the T-type torpedo anchor is the lowest, while the EC-type torpedo anchor has the largest bearing capacity, indicating that the bearing capacity of an end-bearing torpedo anchor is significantly greater than that of a torpedo anchor without an anchor wing, and the bearing capacity of torpedo anchors rises steadily with the increasing bearing-plate radius or area.
(4) Based on the bearing capacity model test results of torpedo anchors, we established the V-H envelope of torpedo-anchor bearing capacity. Through regression analysis, it is concluded that for the torpedo anchors of T, EN3, EN4, and EC types, the envelope coefficients A and B of bearing capacity are 2.753 and 4.522 respectively, and the determination coefficient reaches 0.930, indicating that the established V-H envelope formula is suitable for predicting the bearing capacity of torpedo anchors.

Author Contributions: Conceptualization, J.Z. and G.L.; methodology, J.Z. and J.L.; validation, Y.X. and H.K.; writing—original draft preparation, J.L.; writing—review and editing, G.L.; funding acquisition, G.L. All authors have read and agreed to the published version of the manuscript.

Funding: This research was funded by the Natural Science Basic Research Program of Shaanxi Province, grant number 2021JM-535 and the Special Fund for Scientific Research by Xijing University, grant number XJ18T01.

Institutional Review Board Statement: Not applicable.

Informed Consent Statement: Not applicable.

Data Availability Statement: Not applicable.

Conflicts of Interest: The authors declare no conflict of interest.

References

1. Meng, K.; Cui, C.Y.; Liang, Z.M.; Li, H.J.; Pei, H.F. A new approach for longitudinal vibration of a large-diameter floating pipe pile in visco-elastic soil considering the three-dimensional wave effects. *Comput. Geotech.* **2020**, *128*, 103840. [CrossRef]
2. Cui, C.Y.; Zhang, S.P.; Chapman, D.; Meng, K. Dynamic impedance of a floating pile embedded in poro-visco-elastic soils subjected to vertical harmonic loads. *Geomech. Eng.* **2018**, *15*, 793–803.
3. Cui, C.Y.; Meng, K.; Wu, Y.J.; Chapman, D.; Liang, Z.M. Dynamic response of pipe pile embedded in layered visco-elastic media with radial inhomogeneity under vertical excitation. *Geomech. Eng.* **2018**, *16*, 609–618.
4. Cui, C.Y.; Liang, Z.M.; Xu, C.S.; Xin, Y.; Wang, B.L. Analytical solution for horizontal vibration of end-bearing single pile in radially heterogeneous saturated soil. *Appl. Math. Model.* **2023**, *116*, 65–83. [CrossRef]
5. Cui, C.Y.; Meng, K.; Xu, C.S.; Wang, B.L.; Xin, Y. Vertical vibration of a floating pile considering the incomplete bonding effect of the pile-soil interface. *Comput. Geotech.* **2022**, *150*, 104894. [CrossRef]

6. Cui, C.Y.; Xu, M.Z.; Xu, C.S.; Zhang, P.; Zhao, J.T. An ontology-based probabilistic framework for comprehensive seismic risk evaluation of subway stations by combining Monte Carlo simulation. *Tunn. Undergr. Space Technol.* **2023**, *135*, 105055. [CrossRef]
7. Cui, C.Y.; Meng, K.; Xu, C.S.; Liang, Z.M.; Li, H.J.; Pei, H.F. Analytical solution for longitudinal vibration of a floating pile in saturated porous media based on a fictitious saturated soil pile model. *Comput. Geotech.* **2021**, *131*, 103942. [CrossRef]
8. Dong, Y.K.; Wang, D.; Randolph, M.F. Investigation of impact forces on pipeline by submarine landslide using material point method. *Ocean Eng.* **2017**, *146*, 21–28. [CrossRef]
9. Dong, Y.K.; Cui, L.; Zhang, X. Multiple-GPU parallelization of three-dimensional material point method based on single-root complex. *Int. J. Numer. Meth. Eng.* **2022**, *123*, 1481–1504. [CrossRef]
10. Cheng, Y.; Qiu, C.L. Numerical analysis of torpedo anchor's uplift bearing capacity by material point method. *J. Waterw. Harb.* **2021**, *42*, 114–122.
11. Yu, L.; Zhang, J.L.; Yang, Q.; Yang, G. Capacity of torpedo anchor in clay using finite element analysis. *Ocean Eng.* **2019**, *37*, 122–129.
12. Yu, L.; Yang, Q.; Zhang, J.L.; Yang, G. Simplified model for upper bound method to analyze horizontal bearing capacity of torpedo anchors. *Chin. J. Geotech. Eng.* **2020**, *42*, 773–781.
13. Yu, G.L.; Wang, W.K.; Wang, C. The structure and characteristics of powered torpedo anchor. *Ocean Eng.* **2018**, *36*, 143–148.
14. Raaj, S.K.; Saha, N.; Sundaravadivelu, R. Exploration of deep-water torpedo anchors—A review. *Ocean Eng.* **2023**, *270*, 113607. [CrossRef]
15. Chen, X.H.; Zhang, M.X.; Yu, G.L. A self-penetration torpedo anchor with vibrational shearing. *Ocean Eng.* **2021**, *236*, 109315. [CrossRef]
16. Wang, W.K.; Wang, X.F.; Yu, G.L. Vertical holding capacity of torpedo anchors in underwater cohesive soils. *Ocean Eng.* **2018**, *161*, 291–307. [CrossRef]
17. Ads, A.; Iskander, M.; Bless, S.; Omidvar, M. Visualizing the effect of Fin length on torpedo anchor penetration and pullout using a transparent soil. *Ocean Eng.* **2020**, *216*, 108021. [CrossRef]
18. Kim, Y.H.; Hossain, M.S. Numerical study on pull-out capacity of torpedo anchors in clay. *Géotechnique Lett.* **2016**, *6*, 275–282. [CrossRef]
19. Chen, X.H.; Yu, G.L.; Yue, S.L.; Zhou, H. Effect of frequency of minute-amplitude oscillatory shear loadings on ultimate yield stress of cohesive sediments. *Appl. Ocean Res.* **2021**, *113*, 102542. [CrossRef]
20. Hossain, M.S.; Kim, Y.; Gaudin, C. Experimental Investigation of installation and pullout of dynamically penetrating anchors in clay and silt. *J. Geotech. Geoenviron. Eng.* **2014**, *140*, 04014026. [CrossRef]
21. Hossain, M.S.; O'Loughlin, C.D.; Kim, Y. Dynamic installation and monotonic pullout of a torpedo anchor in calcareous silt. *Geotechnique* **2015**, *65*, 77–90. [CrossRef]
22. Wang, C.; Chen, X.H.; Yu, G.L. Maximum force of inclined pullout of atorpedo anchor in cohesive beds. *China Ocean Eng.* **2019**, *33*, 333–343. [CrossRef]
23. Li, G.; Zhang, J.L.; Liu, J. Model test of the pullout bearing capacity of end-bearing torpedo anchors. *J. Mar. Sci. Eng.* **2022**, *10*, 728. [CrossRef]
24. Yi, J.T.; Fu, Y.; Liu, C.F.; Zhang, Y.H.; Li, Y.P.; Zhang, X.Y. Pull-out capacity of an inclined embedded torpedo anchor subjected to combined vertical and horizontal loading. *Comput. Geotech.* **2020**, *121*, 103478. [CrossRef]
25. Raie, M.S.; Tassoulas, J.L. Simulation of torpedo anchor set-up. *Mar. Struct.* **2016**, *49*, 138–147. [CrossRef]
26. *G/BT 50123-2019*; Standard for Geotechnical Testing Method. Construction Ministry of PRC: Beijing, China, 2019.
27. Richardson, M. Dynamically Installed Anchors for Floating Offshore Structures. Ph.D. Thesis, The University of Western Australia, Perth, Australia, 2008.
28. O'Beirne, C.; O'Loughlin, C.D.; Wang, D.; Gaudin, C. Capacity of dynamically installed anchors as assessed through field testing and three-dimensional large-deformation finite element analyses. *Can. Geotech. J.* **2015**, *52*, 548–562. [CrossRef]
29. Kavvadas, M.; Amorosi, A. A constitutive model for structured soils. *Geotechnique* **2000**, *50*, 263–273. [CrossRef]
30. Savvides, A.A.; Papadrakakis, M. A computational study on the uncertainty quantifcation of failure of clays with a modifed Cam-Clay yield criterion. *SN Appl. Sci.* **2021**, *3*, 659. [CrossRef]

Journal of
Marine Science and Engineering

MDPI

Article

Three-Dimensional Ultrasonic Reverse-Time Migration Imaging of Submarine Pipeline Nondestructive Testing in Cylindrical Coordinates

Daicheng Peng [1], Fei Cheng [2,*], Xiaoyu She [1], Yunpeng Zheng [3], Yongjie Tang [4] and Zhuo Fan [3]

1 Key Laboratory of Exploration Technologies for Oil and Gas Resource, Ministry of Education, Yangtze University, Wuhan 430100, China; pengdc_geo@126.com (D.P.); shexiaoyu2000@163.com (X.S.)
2 College of Marine Science and Technology, China University of Geosciences, Wuhan 430074, China
3 Hubei Subsurface Multi-Scale Imaging Key Laboratory, School of Geophysics and Geomatics, China University of Geosciences, Wuhan 430074, China; zhengyp@cug.edu.cn (Y.Z.); zhuo@cug.edu.cn (Z.F.)
4 Department of Earth and Space Sciences, Southern University of Science and Technology, Shenzhen 518055, China; tyj_cug@126.com
* Correspondence: chengfly03@126.com; Tel.: +86-027-67886167

Abstract: Submarine pipelines are a safe and energy-efficient mode of gas transport. However, due to the complex manufacturing process and harsh operating environment, submarine pipelines are subject to fatigue cracks under long-term cyclic loading. A comprehensive and high-precision characterization strategy for submarine pipelines can effectively prevent potential safety hazards and have significant economic and social repercussions. As a matter of fact, pipeline defects cannot be reliably detected with current traditional 2D methods. On the other hand, in ultrasonic testing, cylindrical geometry increases the complexity of the 3D wave field in the submarine pipeline space and significantly influences the accuracy of the detection results. In this paper, we put forward a novel method for 3D ultrasonic image testing that is suitable for cylindrical coordinates. In order to accurately simulate the ultrasonic signal received from pipelines, we generalize the 3D staggered-grid finite-difference method from Cartesian coordinates to cylindrical ones and simulate the full wave field in the 3D pipeline space. Then, signal processing is performed on the ultrasound simulation records, and 3D reverse-time migration imaging of submarine pipeline defects can be effectively achieved using the reverse-time migration method and cross-correlation imaging conditions. The results obtained from simulations and real field data show that the proposed method provides high-quality 3D imaging of defects in pipelines, taking into account multiple scattering and mode conversion information at the bottom of the defects.

Keywords: submarine pipelines; nondestructive testing; 3D ultrasonic imaging; reverse-time migration; cylindrical coordinates

Citation: Peng, D.; Cheng, F.; She, X.; Zheng, Y.; Tang, Y.; Fan, Z. Three-Dimensional Ultrasonic Reverse-Time Migration Imaging of Submarine Pipeline Nondestructive Testing in Cylindrical Coordinates. *J. Mar. Sci. Eng.* **2023**, *11*, 1459. https://doi.org/10.3390/jmse11071459

Academic Editor: Sergei Chernyi

Received: 14 June 2023
Revised: 11 July 2023
Accepted: 20 July 2023
Published: 22 July 2023

1. Introduction

Submarine pipelines can connect subsea oil and gas resources with the entire onshore oil and gas production management system by the fastest, safest, and most economical route, which is called the "lifeline" of offshore oil and gas engineering [1]. However, due to the complex manufacturing processes and severe service environments [2], submarine pipelines are prone to crack voids, inclusions, and other defects, which greatly affects their mechanical properties and results in their premature failure. In real engineering applications, the initial imperfection is always introduced onto the pipes during the manufacture and installation procedures [3]. Submarine pipelines are subjected to fatigue loading in the harsh environment of the seabed operation conditions [4]. When the pipes are subjected to external pressure, failure first occurs in the cross-section with the most severe initial defects [5]. Additionally, once a crack develops, corrosion may promote its propagation, resulting in the cracking of the submarine pipeline [6]. When the pipeline is damaged and leakage occurs, it will not only cause economic losses, but also may bring about safety problems, endangering the safety of both people and the marine environment [7]. Therefore, before the submarine pipelines are put into service, a comprehensive high-precision inspection of them, including the rapid and accurate detection of defects, may effectively prevent safety hazards, which would guarantee a beneficial economic and social effect [8].

Several methods are available to inspect pipelines, including magnetic particle testing [9], magnetic flux leakage testing [10], acoustic wave testing [11], and penetration testing [12]. Ultrasonic testing may be used to detect the structure and shape of internal defects in submarine pipelines due to its high sensitivity, light equipment, and because it poses no harm to humans or the environment [13]. Ultrasonic testing can effectively detect the flaw size, the crack location [14], the elastic properties of materials [15], and the layup stacking sequence of composite materials [16,17] through the propagation of surface waves, guided waves, and body waves [18]. Then, by using ultrasonic imaging methods, defects in submarine pipelines can be visualized in an intuitive way [19–21]. The traditional ultrasonic methods of pipeline inspection include the synthetic aperture focusing technique (SAFT) [22–24], time-of-flight diffraction (TOFD) [25–27], and the total focusing method (TFM) [28,29]. Those methods have their own advantages and disadvantages. The SAFT is able to quickly provide media images; however, it has limitations in detecting vertical interfaces, lower boundaries of defects, and structures involving high-impedance contrasts. Additionally, there may be artifacts in SAFT-generated images due to surface waves, multiple reflections, and the mode conversions of the wave field originating at interfaces, resulting in wrong conclusions about the defect's location [30]. The TFM, on the other hand, only considers the direct ray path of ultrasound, without considering the mode conversions and multiple scattering arising from the interaction of ultrasonic waves with defects, which reduces the accuracy of the image [31,32]. Moreover, multiple wave reflections from the pipeline's lower side make the signals more complicated to analyze and lower the signal-to-noise ratio, often leading to artifacts in the reconstruction. As a consequence, it is difficult to image structures with vertical boundaries or complex geometry [33,34]. In order to circumvent such difficulties, modern ultrasonic testing techniques often involve full acoustic or elastic wave-equation modeling. In this paper, we address the 3D ultrasonic imaging of a pipeline using the reverse-time migration (RTM) technique, which has received much attention in the field of geophysics [35–39]. In recent years, RTM methods, originating from seismic imaging, have gained popularity in ultrasonic nondestructive testing applications [40–43]. The RTM method is a pre-stack imaging technique based on full-wave extrapolation. It does not suffer from the presence of oblique structures and offers a high resolution for the imaging of complex structures. In contrast to ray-based methods, RTM includes the effects of multiple scattering and mode conversions, as well as multiple wave reflections from the defect's lower side [33]. This allows one to gain more information, enabling the imaging of vertical interfaces and boundaries and providing higher quality images of interior defects [41,42]. However, most RTM studies on submarine pipelines are conducted using Cartesian coordinates [41,44], whereas, in any realistic case, the pipeline has an irregular cavity shape. In particular, for cylindrical structures, the pipeline's walls should be approximated as a staircase boundary using 3D Cartesian coordinates. Thus, it cannot accurately delineate the pipeline cavity, and the resulting grid scattering and dispersion analysis is affected by artifacts that reduce the imaging accuracy. To avoid those unwanted effects, we use cylindrical coordinates to model submarine pipelines in ultrasonic testing, which are more suitable for discretizing grids, and thus ensure higher accuracy [45]. In particular, using cylindrical coordinates makes the subdivision grids suitable to accurately represent the submarine pipeline cavity structure [46–48].

To address the shortcomings of traditional 2D methods and improve the accuracy of the ultrasonic characterization of submarine pipelines, we propose a 3D ultrasonic simulation and RTM imaging of submarine pipelines based on cylindrical coordinates. In our approach, the simulation of the 3D ultrasonic wave field and the wave field characteristic analysis of the submarine pipeline is performed by setting up a double free-surface and absorption boundary. Then, the ultrasonic RTM method is used to achieve high-quality 3D imaging of defects in pipelines. Numerical examples and real field data are used to prove the reliability and effectiveness of our method, paving the way for potential applications in practical ultrasonic testing.

2. Methodology

To perform accurate numerical simulations of ultrasonic waves in a 3D pipeline, we have developed a variable staggered-grid time-domain finite-difference numerical simulation method (FDM) in cylindrical coordinates. In the following paragraphs, we provide a brief description of the elastic wave equation, grid discretization, and boundary conditions in cylindrical coordinates.

2.1. Equations of Motion and Grid Discretization in Cylindrical Coordinates

For isotropic media, the first-order velocity-stress equation in the cylindrical coordinates (r, θ, z) can be expressed as follows [49]:

$$\begin{cases} \rho\frac{\partial v_r}{\partial t} = \frac{\partial \sigma_{rr}}{\partial r} + \frac{1}{r}\frac{\partial \tau_{r\theta}}{\partial \theta} + \frac{\partial \tau_{rz}}{\partial z} + \frac{\sigma_{rr}-\sigma_{\theta\theta}}{r} + f_r \\ \rho\frac{\partial v_\theta}{\partial t} = \frac{\partial \tau_{r\theta}}{\partial r} + \frac{1}{r}\frac{\partial \sigma_{\theta\theta}}{\partial \theta} + \frac{\partial \tau_{\theta z}}{\partial z} + \frac{2\tau_{r\theta}}{r} + f_\theta \\ \rho\frac{\partial v_z}{\partial t} = \frac{\partial \tau_{rz}}{\partial r} + \frac{1}{r}\frac{\partial \tau_{\theta z}}{\partial \theta} + \frac{\partial \sigma_{zz}}{\partial z} + \frac{2\tau_{rz}}{r} + f_z \end{cases} \tag{1}$$

$$\begin{cases} \frac{\partial \sigma_{rr}}{\partial t} = (\lambda+2\mu)\frac{\partial v_r}{\partial r} + \lambda\frac{v_r}{r} + \frac{\lambda}{r}\frac{\partial v_\theta}{\partial \theta} + \lambda\frac{\partial v_z}{\partial z} + g_{rr} \\ \frac{\partial \sigma_{\theta\theta}}{\partial t} = \lambda\frac{\partial v_r}{\partial r} + (\lambda+2\mu)\left(\frac{v_r}{r} + \frac{1}{r}\frac{\partial v_\theta}{\partial \theta}\right) + \lambda\frac{\partial v_z}{\partial z} + g_{\theta\theta} \\ \frac{\partial \sigma_{zz}}{\partial t} = \lambda\frac{\partial v_r}{\partial r} + \lambda\left(\frac{v_r}{r} + \frac{1}{r}\frac{\partial v_\theta}{\partial \theta}\right) + (\lambda+2\mu)\frac{\partial v_z}{\partial z} + g_{zz} \\ \frac{\partial \tau_{r\theta}}{\partial t} = \mu\left(\frac{1}{r}\frac{\partial v_r}{\partial \theta} - \frac{v_\theta}{r} + \frac{\partial v_\theta}{\partial r}\right) + g_{r\theta} \\ \frac{\partial \tau_{rz}}{\partial t} = \mu\left(\frac{\partial v_r}{\partial z} + \frac{\partial v_z}{\partial r}\right) + g_{rz} \\ \frac{\partial \tau_{\theta z}}{\partial t} = \mu\left(\frac{\partial v_\theta}{\partial z} + \frac{1}{r}\frac{\partial v_z}{\partial \theta}\right) + g_{\theta z} \end{cases} \tag{2}$$

where $f_i(i = r, \theta, z)$ denotes the point force source, $g_{ij}(i, j = r, \theta, z)$ represents the coupling, ρ represents the density, λ and μ are the Lame constants, $v_i(i = r, \theta, z)$ denotes the particle velocity in the i direction, $\sigma_{ii}(i = r, \theta, z)$ is the normal stress, and $\tau_{ij}(i, j = r, \theta, z)$ denotes the shear stress. Simulations of the wave field in time-domain FDM may be implemented by discretizing Equations (1) and (2) using a central differencing scheme with a staggered grid both in spatial and temporal domains [50]. As shown in Figure 1, within cell (i, j, k), the normal stresses $\sigma_{ii}(i = r, \theta, z)$ are located on the nodes (i, j, k), the shear stresses $\tau_{ij}(i, j = r, \theta, z)$ are located on the half nodes $(i + 1/2, j, k + 1/2)$, $(i, j + 1/2, k + 1/2)$, and $(i + 1/2, j + 1/2, k)$, and the velocity components $v_i(i = r, \theta, z)$ are located on the half nodes $(i + 1/2, j, k)$, $(i, j + 1/2, k)$, and $(i, j, k + 1/2)$. Then, according to the staggered grid with discretized points of velocity and stress components in Figure 1, Equations (1) and (2) are dissected in a finite-difference staggered-grid scheme. More details of the 3D time-domain FDM can be found in Liu et al. [49] for cylindrical coordinates.

2.2. Boundary Conditions

In finite-difference forward modeling of submarine pipelines, the boundary conditions are particularly important to properly simulate the propagation of ultrasonic waves. Figure 2 illustrates a schematic diagram of the implementation of boundary conditions. The absorbing boundary is placed along the R- and Z-direction of the model area, while the double free-surface boundary is along the inner and outer θ-direction of the pipeline. In our study, for the sake of computational efficiency, we introduce an improved vacuum formulation [51] into the cylindrical coordinates to set the double free-surface boundary condition in the θ-direction. At variance with the acoustic-elastic boundary method, which requires setting the free-surface boundary condition individually for each case, the improved vacuum formulation (IVF) is adaptable to an irregular free-surface. In order to eliminate interference from the region outside of the model, an absorbing boundary condition is assumed on the exterior of the model. We employ the split-field perfectly matched layer (S-PML) [52] to optimize absorption and minimize computational costs. As shown in Figure 2, the S-PML is placed in the R-direction and Z-direction of the pipe to absorb the body waves propagating toward the model boundary. The attenuation factor $d(x_i)$ of the absorbing boundaries is given by the following expression:

$$d(x_i) = \log\left(\frac{1}{R}\right)\frac{3V_p^{\max}}{2L}\left(\frac{x_i}{L}\right)^2 (i = r, z) \tag{3}$$

where R is the theoretical reflection coefficient, L is the thickness of the absorbing boundary, and x_i is the distance between grid points and model boundaries in the i direction.

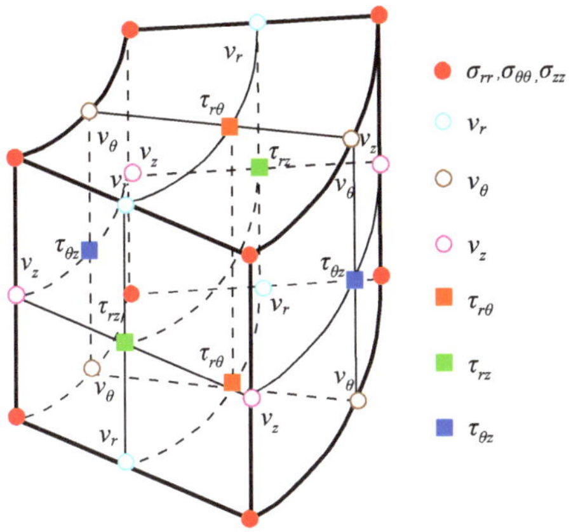

Figure 1. Staggered grid with discretized points of velocity and stress components. The $\sigma_{rr}, \sigma_{\theta\theta}, \sigma_{zz}$ denote the normal stresses, v_r, v_θ, v_z are the particle velocity components, and $\tau_{r\theta}$, τ_{rz}, $\tau_{\theta z}$ represent the shear stresses. The normal stresses are placed at the corner points around the staggered-grid cell, the particle velocity components are located at the cell edges, and the shear stresses are sampled at the center of the cell faces.

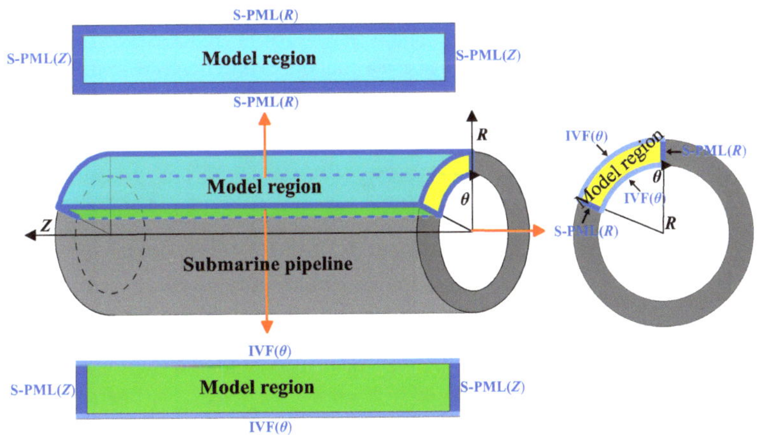

Figure 2. Schematic diagram of the model, illustrating the implementation of the boundary conditions. The absorbing boundary (S-PML) is placed along the R- and Z-direction of the model region, and the double free-surface boundary (IVF) is along the inner and outer θ-direction of the model region.

2.3. Reverse-Time Migration Method

Reverse-time migration (RTM), originally proposed by Whitemore, is based on the two-way wave equation, on which reverse-time extrapolation is performed on the time axis [35]. RTM was first applied to the field of seismic imaging, providing higher imaging accuracy than previous methods. Besides accuracy, RTM has no inclination limitation and may be applied to arbitrary complex velocity models [36,37]. The RTM algorithm consists of the following three steps: 1. forward propagation of the source wave fields; 2. backward propagation of the receiving wave fields; and 3. imaging using imaging conditions. In this study, we used cross-correlation imaging conditions, followed by source normalization, ultimately leading to the following expression [53]:

$$I(r,\theta,z) = \frac{\sum_{t=0}^{T} S(r,\theta,z,t)R(r,\theta,z,t)}{\sum_{t=0}^{T} S^2(r,\theta,z,t)} \tag{4}$$

where $I(r,\theta,z)$ represents the image result, $S(r,\theta,z,t)$ denotes the source field, and $R(r,\theta,z,t)$ is the field at the receiver. After that, the Laplace filtering method is employed to eliminate the low-frequency artifacts caused by the imaging conditions [54].

2.4. Signal Processing

The wavefields obtained from pipeline ultrasound inspection records are characterized by complex features, such that the preprocessing of records is required before imaging to enhance RTM accuracy. In this study, wavelet extraction and dynamic balance in the channel are employed. Wavelet extraction methods may be classified into two categories: deterministic [55] and statistical [56]. In this study, the statistical wavelet extraction method is employed to extract wavelets from reflected waves. To this aim, we have to first select the reference trace, and then search for wavelets of other tracks within the travel time range of the reflected wave using this reference track as a guide. Finally, we normalize the wavelets of all traces, and then stack all traces in the record.

2.5. Implementations

Figure 3 illustrates the steps of the proposed method. The first step is to construct a geophysical 3D submarine pipeline model and implement the staggered-grid finite-difference method in cylindrical coordinates. Then, the SPML-absorbing and the IVF double free-surface boundary conditions are implemented in cylindrical coordinates and a 3D ultrasonic wave field simulation of the submarine pipeline is performed. RTM calculations represent the third and final step. RTM itself involves three steps, i.e., the forward propagation of source wave fields, the backward propagation of receiver wave fields, and the imaging step using the imaging condition. The procedure is carried out as follows: At first, a source wavelet is placed on the pipeline surface and used to excite the propagation of ultrasonic waves. The source wave fields during the forward propagation from T = 0 to T = max are calculated. Then, the recorded signal at the boundaries is time-reversed and simultaneously propagated back into the simulation domain to obtain the source wave fields from T = max to T = 0. Then, signal processing is performed and used as a signal at the receiver position. The receiver wave fields are then calculated using the FDM. Finally, the source wave fields are cross correlated with the receiver wave fields at each time step to construct the image. The final ultrasonic 3D RTM imaging is obtained using Laplace filtering [54].

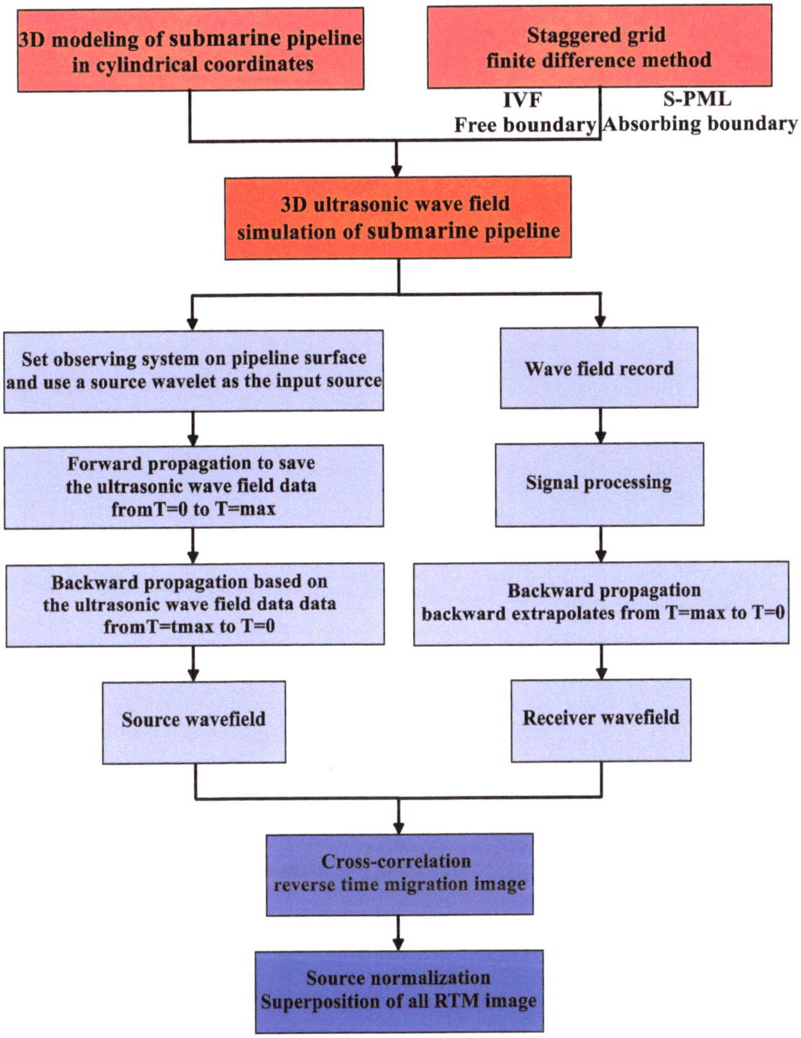

Figure 3. A flow-chart of implementation.

3. Numerical Simulation Results

3.1. Modeling and Survey Layout

3.1.1. Survey Layout

Our 3D cylindrical model of a submarine pipeline is shown in Figure 4. It includes a total of six survey lines arranged in the model region. Geophones are placed at the bottom, middle, and top of the pipeline's outer wall. The geophones are piezoelectric ceramic ultrasonic probes. Three survey lines are circumferentially placed (red lines in Figure 4), with a geophone spacing of 1.0 mm (Line-θ1, Line-θ2, and Line-θ3) and three are placed axially (blue lines in Figure 4), with a geophone spacing of 1.0 mm (Line-Z1, Line-Z2, and Line-Z3). The source is located at the center of the pipeline's outer wall.

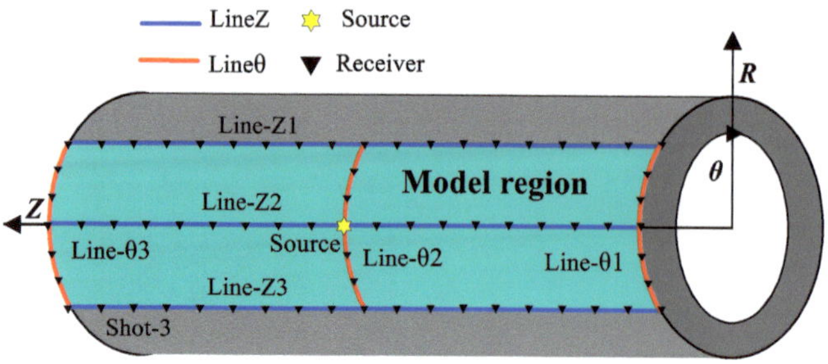

Figure 4. Schematic diagram of the ultrasonic testing and observation system.

3.1.2. Modeling

To analyze the propagation characteristics of the 3D wave field during the ultrasonic testing of pipelines, and to study the wave field characteristics resulting from different defects, we have designed three models (see Figure 5). Model-1 is a combined model of horizontal slag inclusion and hole defects, where the thickness of the slag is 2.0 mm, the width is 51.45 mm, the diameter of the hole is 2.4 mm, and the distance between the hole and the slag is 40 mm. Model-2 is a vertical slag inclusion model with a vertical thickness of 1.0 mm, a width of 51.45 mm, and a length of 60.0 mm. Model-3 describes another vertical slag inclusion with a slag width of 1.0 mm and a length of 60.0 mm. The model parameters are listed in Table 1.

Table 1. Parameters of the three models.

No.	$v_p(m/s)$	$\rho(kg/m^3)$
(1) Slag inclusion	1866.0	2466.0
(2) Hole	1400.0	1850.0
(3) Submarine pipeline	5600.0	7400.0

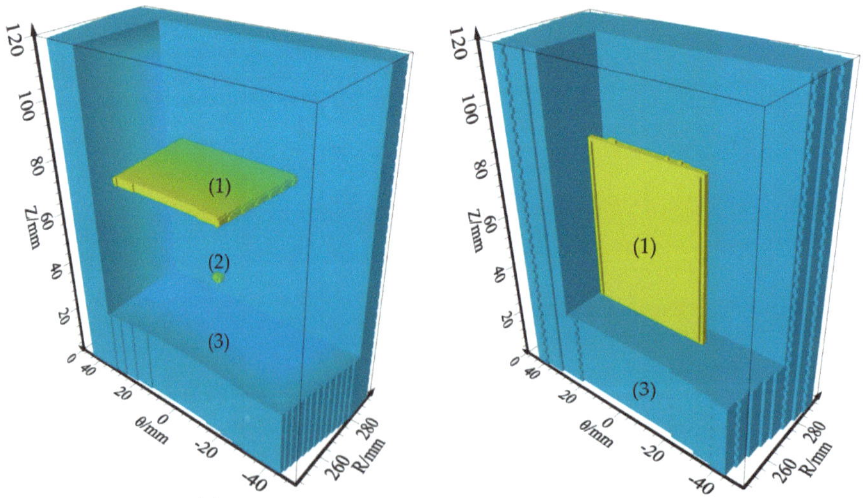

(a) (b)

Figure 5. *Cont.*

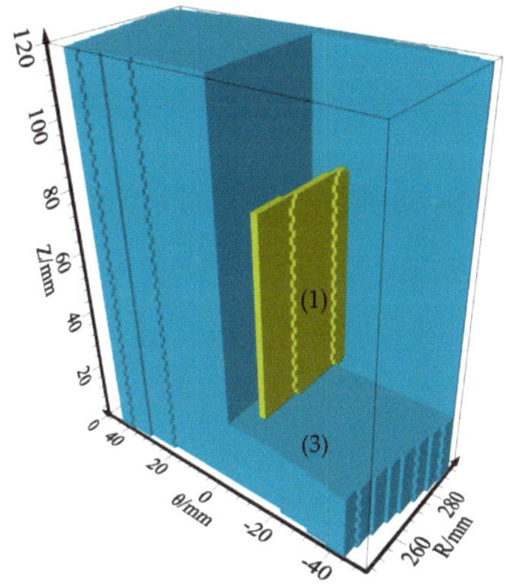

(c)

Figure 5. Submarine pipeline models developed to perform numerical simulations. (**a**) Combined model of a horizontal slag inclusion and a hole; (**b**) Vertical slag inclusion Model 1; and (**c**) Vertical slag inclusion 2. All parameters (1)–(3) in the figure are shown in Table 1.

3.2. Forward Modeling Results

In this study, we employ a spatial fourth-order and temporal second-order variable staggered-grid FDM. In our 3D cylindrical submarine pipeline model, the wall thickness is 45.0 mm, the inner diameter is 250.0 mm, the arc length is 102.97 mm (with a 20° rounding angle), and the axial length is 120.0 mm. The model size in the *R*-, θ-, and *Z*-directions is 45.0 mm × 102.97 mm (20°) × 120.0 mm. The radial step in the *R*-direction and the axial step in the *Z*-direction are $\Delta r = \Delta z = 0.2$ mm. The azimuthal step in the θ-direction is $\Delta\theta = 0.046°$. The corresponding arc length increases with the increase in wall thickness (0.2 mm at the inner wall and 0.236 mm at the outer wall). The time step of $\Delta t = 0.01$ μs. To ensure the consistency with the actual ultrasonic source, we assume a point-like force source. As shown in Figure 6, the source wavelet is a ricker wavelet with a frequency of 0.7 Mhz, a signal width of 0.33–1.16 Mhz, and a wavelet delay of 1.71 μs.

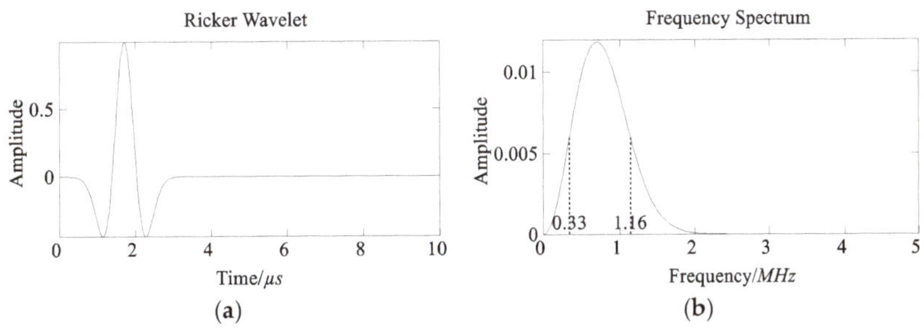

Figure 6. (**a**) Ricker wavelet waveform diagram and (**b**) Ricker wavelet spectrogram.

In order to visually analyze the propagation of the wavefields in Model-1, we consider the 3D wavefield snapshots at T = 12 μs, and the 2D wavefield snapshot profiles from three slice directions at T = 6, 12, and 18 μs. Figure 7a shows the 3D wavefield snapshot, where the yellow color indicates the defect's location; Figure 7b shows a slice diagram, where green indicates a ROZ slice located at the midpoint position in the θ-direction, purple denotes a ROθ slice located at the midpoint in the *Z*-direction, and orange indicates a θOZ slice located at the pipeline's outer wall; and Figure 7c shows snapshots of the 2D wavefield in three slices.

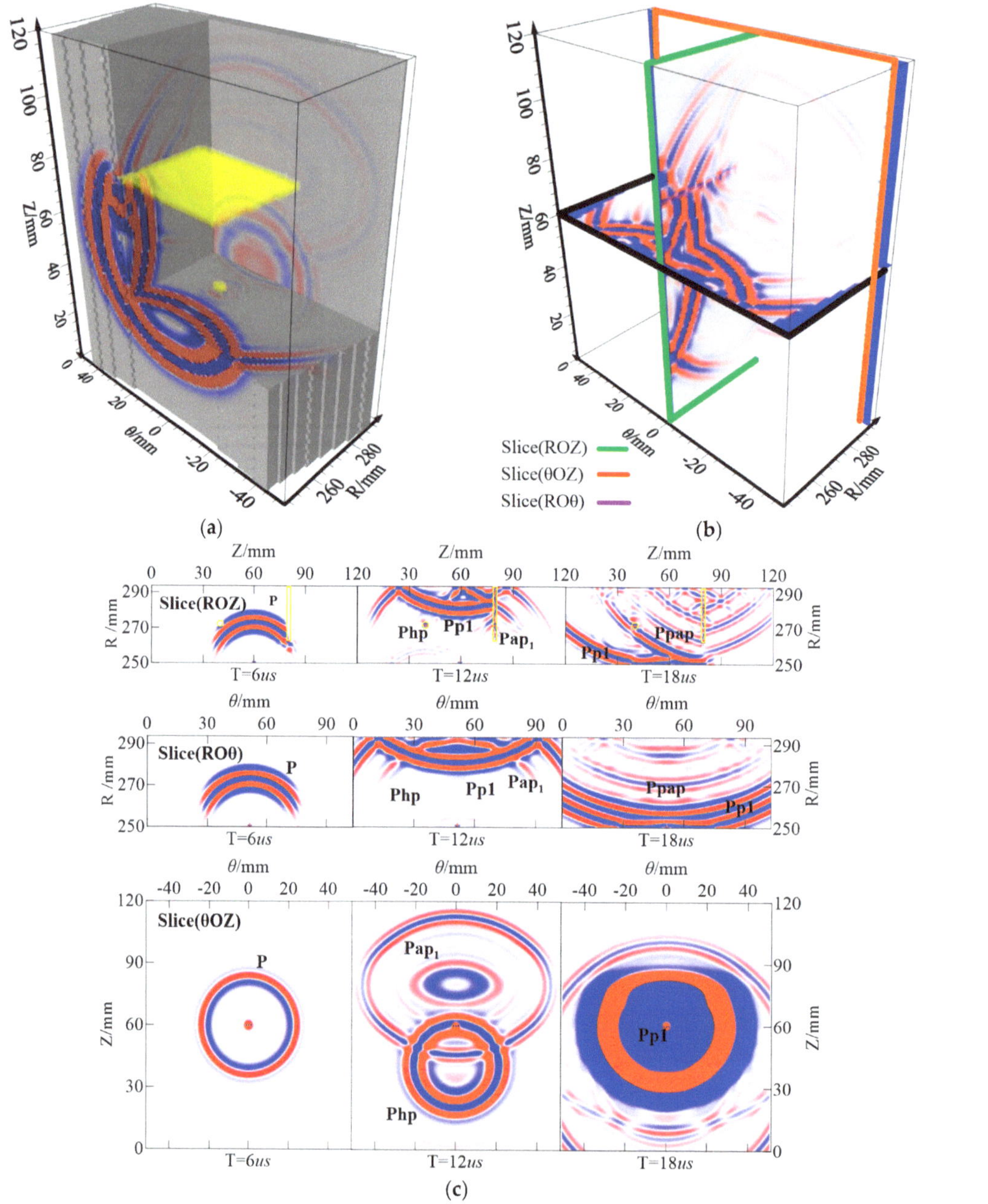

Figure 7. The radial component wave field snapshot (the yellow color indicates the defects location). (**a**) Show 3D snapshot at time T = 12 μs; (**b**) Slice diagram; and (**c**) Show 2D snapshot at time T = 6, 12, and 18 μs.

In Figure 7, at 6 μs, we only see the direct wave P, because the wavefront has not yet encountered the defect. At 12 μs, since the Fresnel condition is not satisfied, the diffraction wave Php is generated when the P wave passes through the hole. The diffraction wave Pap1 is produced when the P wave passes through the upper interface of the slag inclusion. At 18 μs, we see the reflection wave Pp1, which is generated when the P wave reaches the free-surface boundary of the pipeline's outer wall. When the reflected wave Pp1 reaches the slag inclusion defect, it generates a reflected diffracted wave Ppap. Overall, the kinematics and dynamics of each wave item are consistent with the wave field law, which confirms the accuracy of the wave field simulations.

3.2.1. Model-1

Based on the observation system in Figure 4, we were able to generate the wave field records of the two sets of survey lines through forward modeling. The simulation record of Model-1 is illustrated in Figure 8. Trace refers to the number of geophones in the survey line. As can be seen from the plots, the direct wave P, the reflected wave Pp1, the multiple Pp2, the diffraction wave Php, and the diffraction wave Pap1 carry most of the energy. Nevertheless, the diffraction waves Ppap generated by the Pp1 are also evident. According to the simulations, the arrival time of the direct wave P is 9.19 μs and the arrival times of the diffraction waves Pap1 and Php are 13.51 μs and 15.41 μs, respectively. The arrival time of the reflected wave Pp1 is 18.52 μs and the Ppap is 22.06 μs. This result confirms that the simulation results are accurate.

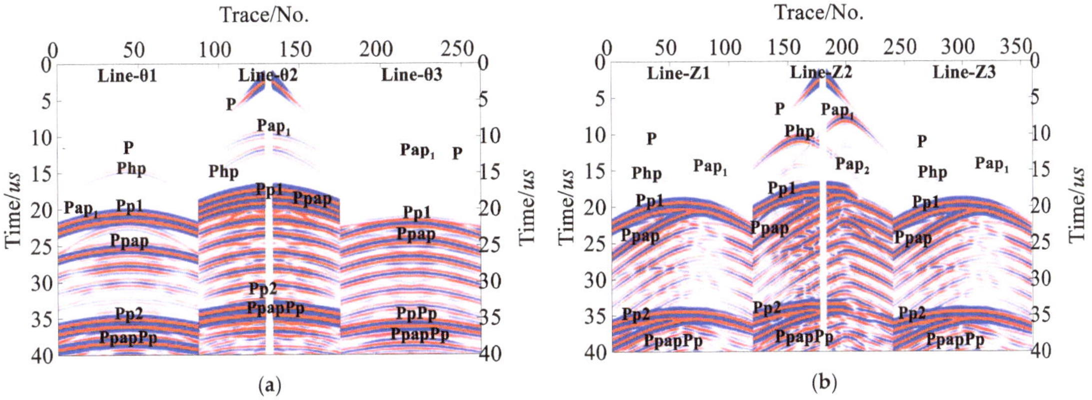

Figure 8. Synthetic *R*-component records of Model-1 (**a**) Line-θ1~θ3 and (**b**) Line-Z1~Z3.

3.2.2. Model-2

The results from Model-2 are consistent with those from Model-1. In particular, we see in Figure 9 that increasing the propagation length of the P wave leads to a reduced energy of the wave field and a corresponding reduction in the energy received by the geophone on both sides of the pipeline. Nevertheless, both the primary reflection waves Pmp and secondary reflection waves Pmp2 resulting from the vertical slag inclusion defect and the reflected waves Pmm reflected from the inner wall of the pipeline have high energy, which can be precisely identified. In addition, the diffraction waves PmD1 and PmD2 from the vertical slag defect endpoints are also clearly visible.

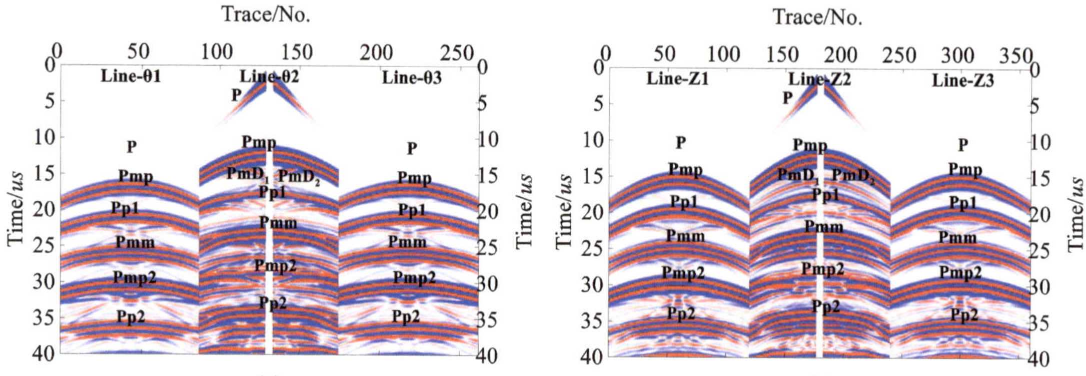

(a) (b)

Figure 9. Synthetic *R*-component records of Model-2 (**a**) Line-θ1~θ3 and (**b**) Line-Z1~Z3.

3.2.3. Model-3

In addition, the results of Model-3 are consistent with those of Model-1. We see in Figure 10 that the direct wave P and the reflected wave Pap from the vertical slag inclusion are weak in Line-θ1 and Line-θ3, while the reflected wave Pp1 and the reflected wave Ppap from the inner wall of the pipeline through the vertical slag are stronger. Nevertheless, the direct and reflected waves are clearly visible along Line-θ2, with energy decreasing with distance.

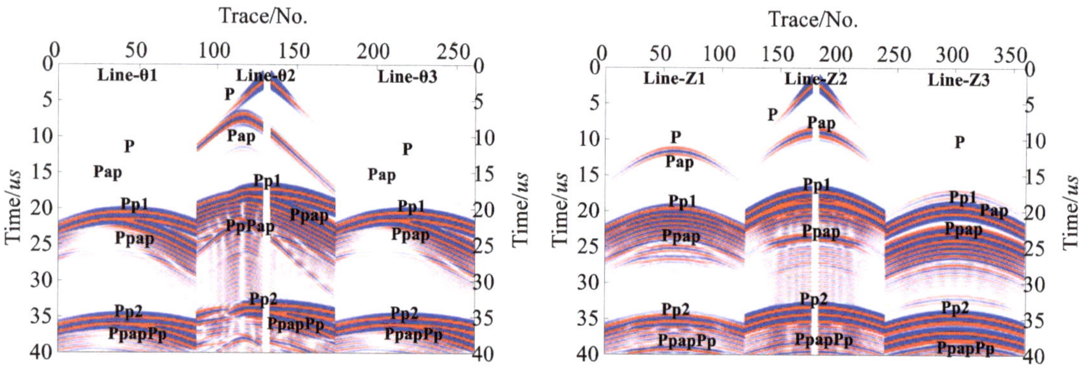

(a) (b)

Figure 10. Synthetic *R*-component records of Model-3 (**a**) Line-θ1~θ3 and (**b**) Line-Z1~Z3.

3.3. RTM Results

3.3.1. Signal Processing Results

The signal processing of wavefield records is one of the key steps in RTM imaging, and it ultimately determines the overall quality of the imaging. Here, we present the results from the signal processing of the original recordings of Model-3. Looking at Figure 10a, we can see that there is a large amount of information in the unprocessed original record, which interferes with the defect reflection wave. We thus start with removing the direct wave P (see Figure 11) and then proceed with the wavelet extraction method to suppress the free-surface boundary reflection wave. Finally, the dynamic balance in the track is used to obtain the final record, which is characterized by an improved signal-to-noise ratio.

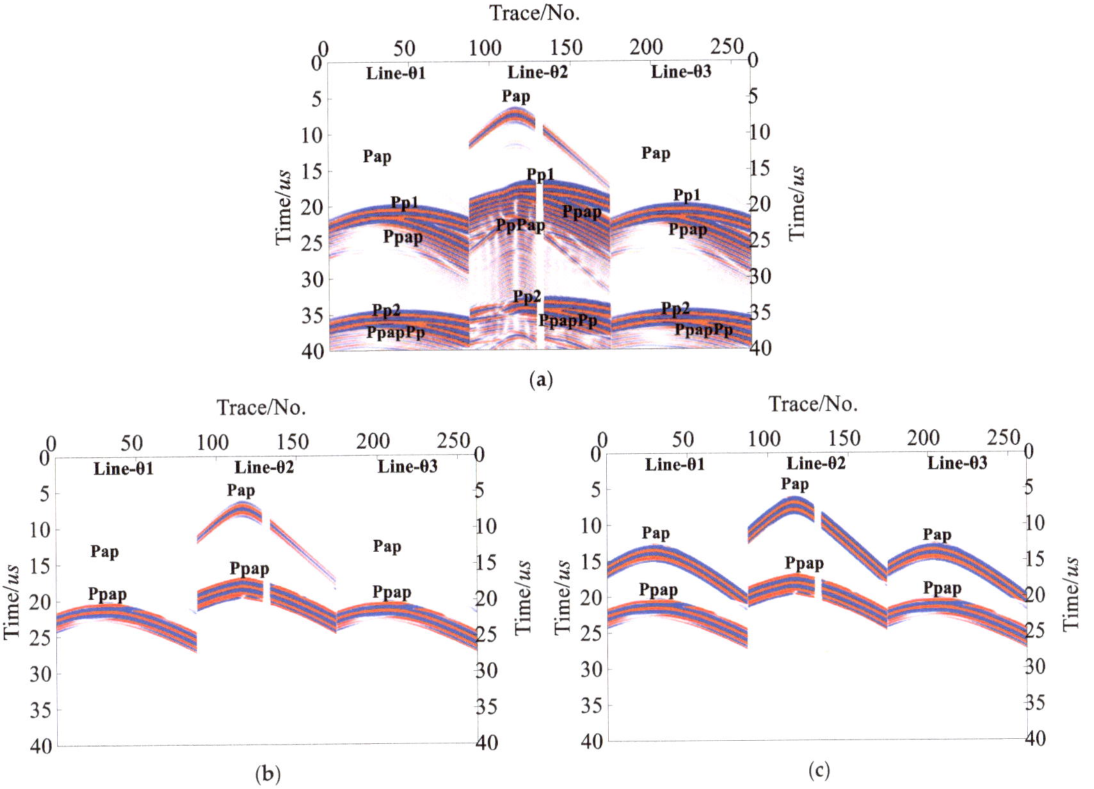

Figure 11. Signal processing results. (**a**) Record after direct wave removal; (**b**) Record after reflected wave suppression; and (**c**) Record after dynamic balance within the track.

3.3.2. Imaging Results

Figure 12 illustrates the results of RTM imaging for Model-2 after signal processing. As shown in Figure 12a, the unprocessed RTM image has a very low signal-to-noise ratio and contains many artifacts, which makes it impossible to identify the slag inclusion defect accurately. From Figure 12b, we can see that removing the direct wave allows us to eliminate the yellow-dashed-frame artifact on the outer wall of the pipeline. After the suppression of the free-surface boundary reflection wave, we obtain the image shown in Figure 12c, where the red-dotted-frame artifact has also been removed. Finally, using dynamic balance within the track, we improve the signal-to-noise ratio and obtain the final image of Figure 12d. When compared with the initial model in Figure 12a, it is evident that the position and shape of the vertical slag inclusion defect are essentially the same, and no other artifacts are present.

Figure 13 shows the three-dimensional ultrasonic RTM images obtained in cylindrical coordinates denoised with a Laplace filter and with the noise of the receiver point removed. The results for a hole defect are shown in Figure 13a, where the location and shape are clearly visible. The same is true for the boundaries and the position of the slag inclusion interface, as shown in Figure 13b. Figure 13c shows that, for a vertical slag inclusion defect, the upper and lower boundaries can be well imaged, as well as its bottom boundary. The position of the boundaries corresponds to those of the real model, and, although the lack of reflection point removes the information in the central region, the four boundaries are sufficient for determining the location of the slag inclusions.

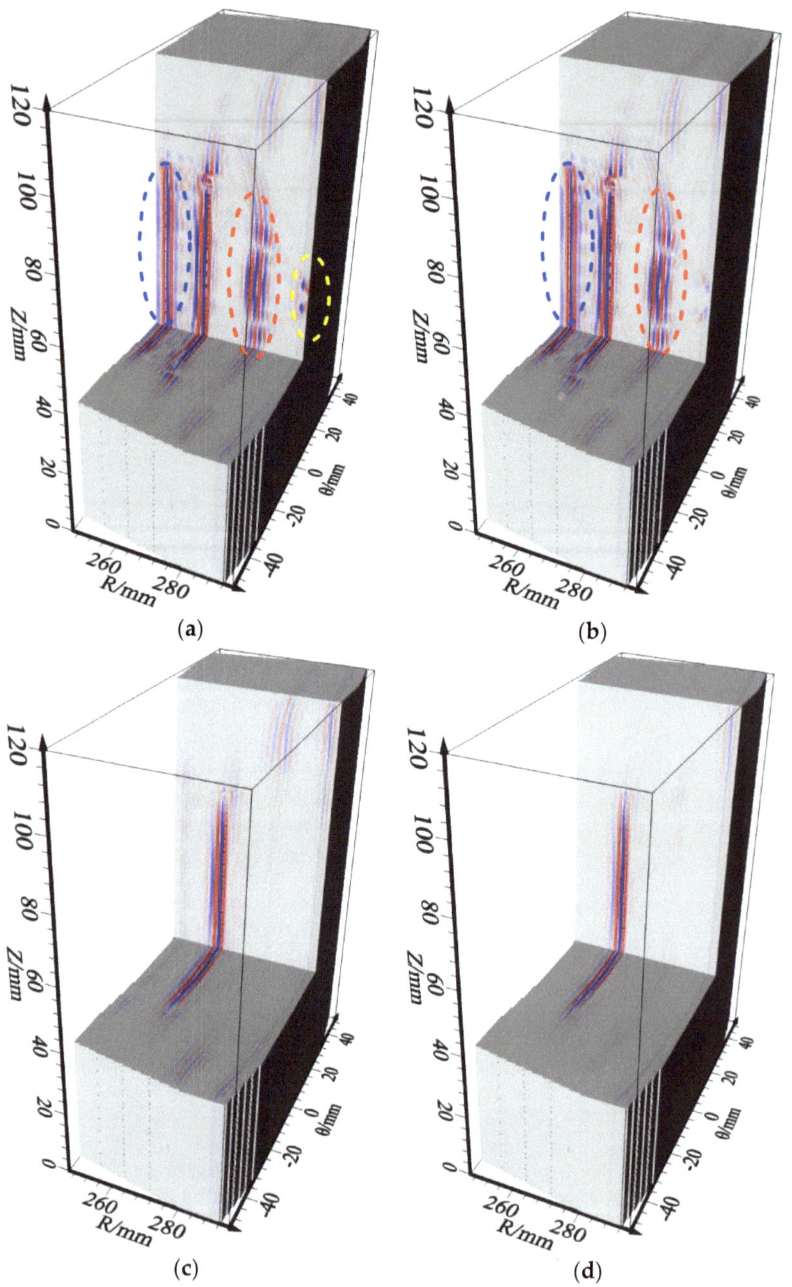

Figure 12. Imaging results: (**a**) Unprocessed imaging; (**b**) Imaging after direct wave removal; (**c**) Imaging after reflected wave suppression; and (**d**) Imaging after dynamic balance within the track.

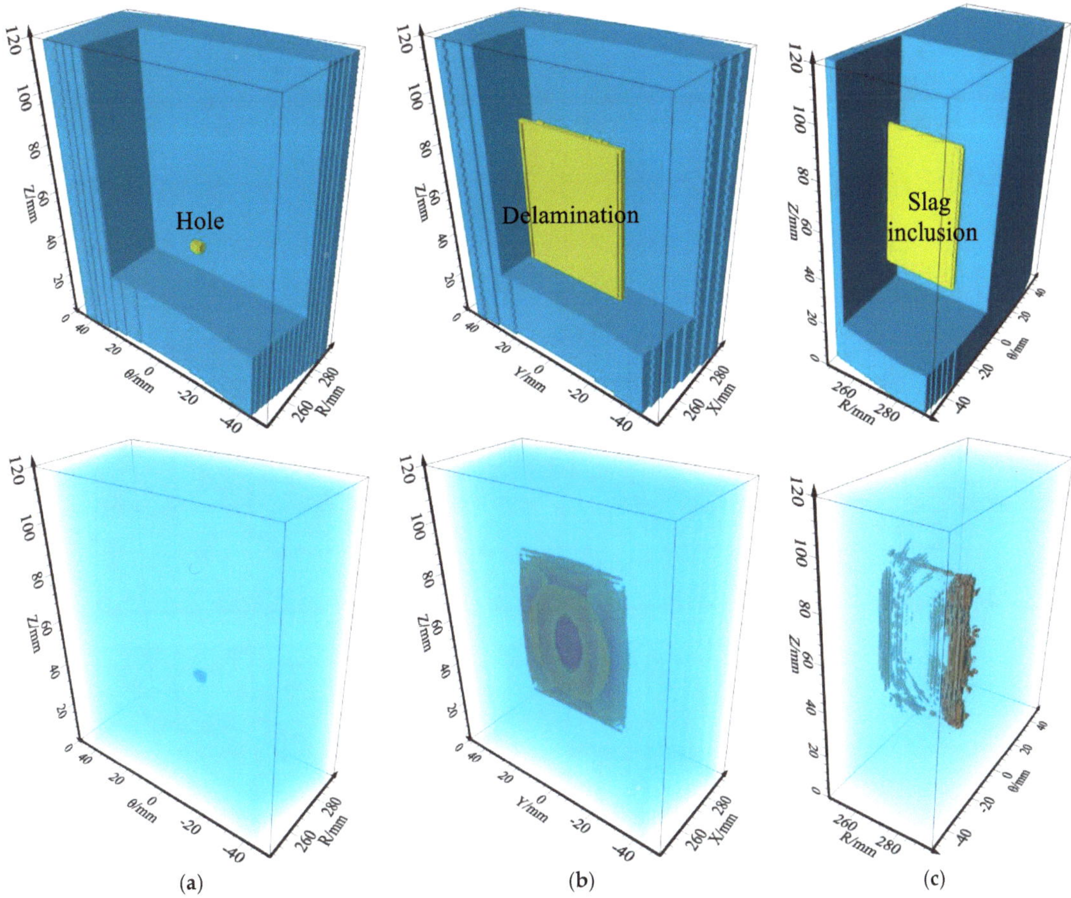

Figure 13. Three-dimensional ultrasonic RTM imaging results: (**a**) Hole defect; (**b**) Lamination defect; and (**c**) Slag inclusion.

4. Laboratory Experiment and Results

4.1. Experimental Setup and Observing System

Ultrasonic seismic physics simulation laboratory equipment is used to collect the actual data underwater for pipelines with defects. Figure 14 illustrates the major components of the data acquisition equipment. The ultrasonic data acquisition system consists of a computer, an ultrasonic pulse transmitter, a water pool, high-speed data-acquisition, and a probe-motion double-3D-coordinate automatic positioning control system. The experimental procedure is as follows: The computer sets the parameters for the sampling points, the starting and ending position of the receiver probe, and so on. Then, the 3D positioning device moves between the sampling start and end points. Once the ultrasonic receiving transducer reaches a sampling point, the ultrasonic pulse generator transmits a fixed-length synchronous signal. Finally, the ultrasonic signal received by the ultrasonic receiving transducer is sent to the computer for processing.

The experimental specimen is shown in Figure 14c. The submarine pipeline length is 550 mm, the outside diameter is 219 mm, and the wall thickness is 45 mm. In order to simulate slag inclusions in actual an ultrasonic testing, cement is injected into the crack to simulate a low velocity body. The slag inclusion on the outer wall of the pipeline measures 60 mm in arc length, 3 mm in width, and 22.5 mm in depth. As shown in Figure 14d, for this slag entrapment model, we designed an observation system on the outer wall of the pipeline, with a total of seven lines. The spacing between each line is 1 cm, the length of each line is 150 mm, the channel spacing is 1 mm, and the sampling time is 60 μs. The source is located at the end point, and the offset distance is 2 cm. The ultrasonic pulse generation receiver frequency is 0.5 MHz.

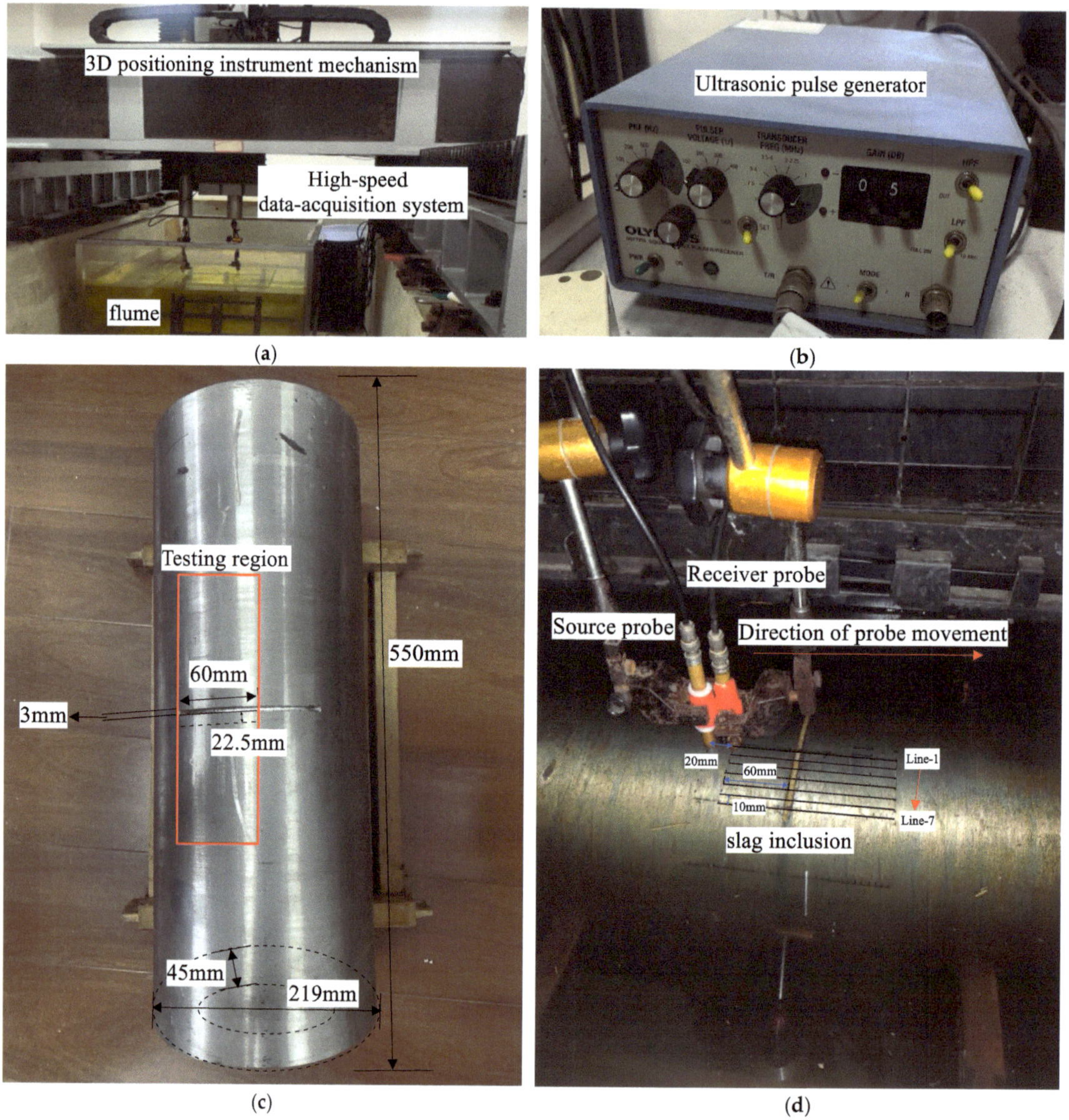

Figure 14. Ultrasonic experimental setup and observing system. (**a**) Three-dimensional positioning instrument mechanism; (**b**) Ultrasonic pulse generator; (**c**) Specimen; and (**d**) Observing system.

4.2. Results

Figure 15 illustrates the actual filtered data. The direct wave P can be clearly seen, as well as the surface wave R, and slag inclusions diffraction wave Pap. Since water is used as a coupling agent, the ultrasonic excitation propagates in the pipeline, and also in the water, thus producing a direct wave Pw, whose speed is much slower than the propagation in the steel pipe (medium speed), but with greater energy.

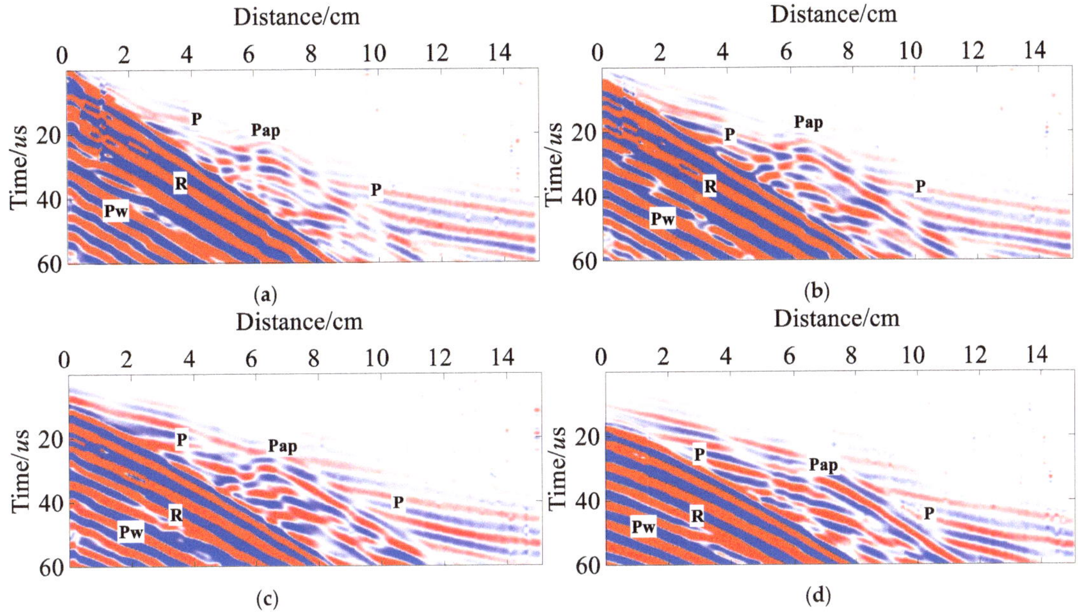

Figure 15. Ultrasonic test data. (**a**–**d**) Line-1~Line-4.

We have also constructed a pipeline model with the same slag inclusion defect, matching the size and physical parameters of the test. The simulation data (Figure 16a) are compared to the experimental data (Figure 16b). The figure illustrates that the diffracted wave Pap position of the defect is basically the same. In the experimental data, we can see the effects of t noise and absorption by the water layer, resulting in a low excitation frequency, incomplete wave field separation, and R energy covering part of the effective wave field. However, the diffraction wave Pap of the slag inclusions can still be identified, thereby confirming the reliability of our cylindrical FDM simulations in providing theoretical guidance for the ultrasonic nondestructive testing of pipeline defects.

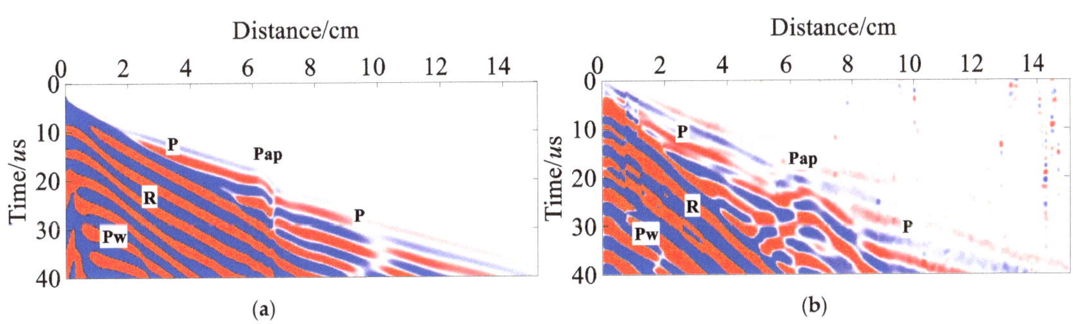

Figure 16. Comparison of test and simulation records. (**a**) Simulation record and (**b**) Test record.

Based on the actual ultrasound Line-1~Line-7 data, cross-correlation imaging conditions are used for RTM imaging after data processing. The imaging results after noise processing are shown in Figure 17b. The image clearly shows the lower interface and vertical boundaries of the defect, and the wave field energy is concentrated at the real cement-filled defects. The results confirm the reliability of the method proposed in this paper, and that ultrasonic testing of submarine pipelines is feasible.

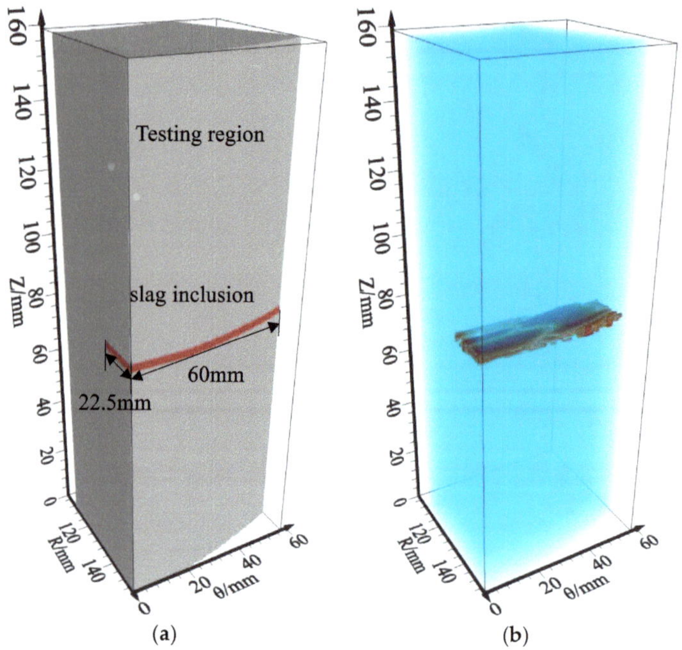

Figure 17. The test results of 3D ultrasonic RTM imaging. (**a**) Test Model and (**b**) 3D ultrasonic RTM imaging.

5. Conclusions

In this study, we have proposed a staggered-grid FDM in cylindrical coordinates to match the natural cylindrical symmetry of submarine pipeline cavities. A realistic pipeline model has been designed by including a double free-surface boundary and an absorbing boundary. Using this scheme, we have simulated the ultrasonic wave fields resulting from three types of defects in pipelines, i.e., a hole, a vertical slag inclusion defect, and a horizontal slag inclusion. Numerical and experimental examples are provided to verify the reliability and accuracy of the method. Signal processing has been performed using the ultrasonic information at hand. In particular, by using cross-correlation imaging conditions, 3D RTM imaging of a pipeline's space in cylindrical coordinates has been realized. Compared to traditional 2D ultrasonic testing methods, our scheme is capable of providing accurate, high-quality 3D imaging of pipeline defects with high resolution and accuracy. The improvement in accuracy comes from taking into account the converted wave on the inner wall of the pipelines and the information coming from the multiple waves reflection at the bottom of the defects. Numerical and experimental results indicate that the method is effective, and that it may be potentially applied to the practical ultrasonic nondestructive testing of submarine pipelines.

Author Contributions: Conceptualization, D.P. and F.C.; methodology, D.P., X.S., F.C. and Y.Z.; software, D.P., Y.Z. and Y.T.; validation, D.P., X.S. and Y.Z.; formal analysis, D.P. and F.C.; investigation, Y.Z. and Z.F.; resources, D.P.; data curation, D.P., F.C., Y.Z. and Z.F.; writing—original draft preparation, D.P.; writing—review and editing, X.S., Y.Z. and Y.T.; visualization, D.P.; supervision, F.C.; project administration, D.P. and F.C.; funding acquisition, F.C. All authors have read and agreed to the published version of the manuscript.

Funding: This research was funded by the National Natural Science Foundation of China (No. 41704146).

Institutional Review Board Statement: Not applicable.

Informed Consent Statement: Not applicable.

Data Availability Statement: The data presented in this study are available on request from the corresponding author.

Conflicts of Interest: The authors declare no conflict of interest. The funders had no role in the design of the study; in the collection, analyses, or interpretation of data; in the writing of the manuscript, or in the decision to publish the results.

References

1. Du, F.; Li, C.; Wang, W. Development of Subsea Pipeline Buckling, Corrosion and Leakage Monitoring. *J. Mar. Sci. Eng.* **2023**, *11*, 188. [CrossRef]
2. Dong, Y.; Wang, D.; Randolph, M.F. Investigation of impact forces on pipeline by submarine landslide using material point method. *Ocean Eng.* **2017**, *146*, 21–28. [CrossRef]
3. Li, R.; Chen, B.Q.; Guedes Soares, C. Design Equation of Buckle Propagation Pressure for Pipe-in-Pipe Systems. *J. Mar. Sci. Eng.* **2023**, *11*, 622. [CrossRef]
4. Dong, Y.; Ji, G.; Fang, L.; Liu, X. Fatigue Strength Assessment of Single-Sided Girth Welds in Offshore Pipelines Subjected to Start-Up and Shut-Down Cycles. *J. Mar. Sci. Eng.* **2022**, *10*, 1879. [CrossRef]
5. Mahmutoglu, Y.; Turk, K. Received signal strength difference based leakage localization for the underwater natural gas pipelines. *Appl. Acoust.* **2019**, *153*, 14–19. [CrossRef]
6. Kakaie, A.; Soares, C.G.; Ariffin, A.K.; Punurai, W. Fatigue Reliability Analysis of Submarine Pipelines Using the Bayesian Approach. *J. Mar. Sci. Eng.* **2023**, *11*, 580. [CrossRef]
7. Dong, Y.; Liao, Z.; Wang, J.; Liu, Q.; Cui, L. Potential failure patterns of a large landslide complex in the Three Gorges Reservoir area. *Bull. Eng. Geol. Environ.* **2023**, *82*, 41. [CrossRef]
8. Hong, X.; Huang, L.; Gong, S.; Xiao, G. Shedding damage detection of metal underwater pipeline external anticorrosive coating by ultrasonic imaging based on HOG + SVM. *J. Mar. Sci. Eng.* **2021**, *9*, 364. [CrossRef]
9. Sheng, H.; Wang, P. Evaluation of Pipeline Steel Mechanical Property Distribution Based on Multimicromagnetic NDT Method. *IEEE Trans. Instrum. Meas.* **2023**, *72*, 6001715. [CrossRef]
10. Wu, D.; Liu, Z.; Wang, X.; Su, L. Composite magnetic flux leakage detection method for pipelines using alternating magnetic field excitation. *NDT E Int.* **2017**, *91*, 148–155. [CrossRef]
11. Abou-Khousa, M.A.; Rahman, M.S.U.; Donnell, K.M.; Al Qaseer, M.T. Detection of Surface Cracks in Metals using Microwave and Millimeter Wave Nondestructive Testing Techniques—A Review. *IEEE Trans. Instrum. Meas.* **2023**, *72*, 8000918. [CrossRef]
12. Yao, Y.; Tung ST, E.; Glisic, B. Crack detection and characterization techniques—An overview. *Struct. Control Health Monit.* **2014**, *21*, 1387–1413. [CrossRef]
13. Felice, M.V.; Fan, Z. Sizing of flaws using ultrasonic bulk wave testing: A review. *Ultrasonics* **2018**, *88*, 26–42. [CrossRef]
14. Vogelaar, B.; Golombok, M. Quantification and localization of internal pipe damage. *Mech. Syst. Signal Process.* **2016**, *78*, 107–117. [CrossRef]
15. Barros, B.; Conde, B.; Cabaleiro, M.; Riveiro, B. Deterministic and probabilistic-based model updating of aging steel bridges. *Structures* **2023**, *54*, 89–105. [CrossRef]
16. Morokov, E.; Levin, V.; Chernov, A.; Shanygin, A. High resolution ply-by-ply ultrasound imaging of impact damage in thick CFRP laminates by high-frequency acoustic microscopy. *Compos. Struct.* **2021**, *256*, 113102. [CrossRef]
17. Morokov, E.; Titov, S.; Levin, V. In situ high-resolution ultrasonic visualization of damage evolution in the volume of quasi-isotropic CFRP laminates under tension. *Compos. Part B Eng.* **2022**, *247*, 110360. [CrossRef]
18. Zhu, W.; Xiang, Y.; Zhang, H.; Zhang, M.; Fan, G.; Zhang, H. Super-resolution ultrasonic Lamb wave imaging based on sign coherence factor and total focusing method. *Mech. Syst. Signal Process.* **2023**, *190*, 110121. [CrossRef]
19. Drinkwater, B.W.; Wilcox, P.D. Ultrasonic arrays for non-destructive evaluation: A review. *NDT E Int.* **2006**, *39*, 525–541. [CrossRef]
20. Portzgen, N.; Gisolf, D.; Blacquiere, G. Inverse wave field extrapolation: A different NDI approach to imaging defects. *IEEE Trans. Ultrason. Ferroelectr. Freq. Control* **2006**, *54*, 118–127. [CrossRef]

21. Bai, Z.; Chen, S.; Jia, L.; Zeng, Z. Phased array ultrasonic signal compressive detection in low-pressure turbine disc. *NDT E Int.* **2017**, *89*, 1–13. [CrossRef]
22. Langenberg, K.; Berger, M.; Kreutter, T.; Mayer, K.; Schmitz, V. Synthetic aperture focusing technique signal processing. *NDT Int.* **1986**, *19*, 177–189. [CrossRef]
23. Ni, C.Y.; Chen, C.; Ying, K.N.; Dai, L.N.; Yuan, L.; Kan, W.W.; Shen, Z.H. Non-destructive laser-ultrasonic Synthetic Aperture Focusing Technique (SAFT) for 3D visualization of defects. *Photoacoustics* **2021**, *22*, 100248. [CrossRef] [PubMed]
24. Seo, H.; Pyun, D.K.; Jhang, K.Y. Synthetic aperture imaging of contact acoustic nonlinearity to visualize the closing interfaces using tone-burst ultrasonic waves. *Mech. Syst. Signal Process.* **2019**, *125*, 257–274. [CrossRef]
25. Silk, M.G. The use of diffraction-based time-of-flight measurements to locate and size defects. *Br. J. Non-Destr. Test.* **1984**, *26*, 208–213.
26. Sun, X.; Lin, L.; Jin, S. Resolution Enhancement in Ultrasonic TOFD Imaging by Combining Sparse Deconvolution and Synthetic Aperture Focusing Technique (Sparse-SAFT). *Chin. J. Mech. Eng.* **2022**, *35*, 94. [CrossRef]
27. Yang, F.; Shi, D.; Lo, L.-Y.; Mao, Q.; Zhang, J.; Lam, K.-H. Auto-Diagnosis of Time-of-Flight for Ultrasonic Signal Based on Defect Peaks Tracking Model. *Remote Sens.* **2023**, *15*, 599. [CrossRef]
28. Holmes, C.; Drinkwater, B.W.; Wilcox, P.D. Post-processing of the full matrix of ultrasonic transmit–receive array data for non-destructive evaluation. *NDT E Int.* **2005**, *38*, 701–711. [CrossRef]
29. He, H.; Sun, K.; Sun, C.; He, J.; Liang, E.; Liu, Q. Suppressing artifacts in the total focusing method using the directivity of laser ultrasound. *Photoacoustics* **2023**, *31*, 100490. [CrossRef]
30. Müller, S.; Niederleithinger, E.; Bohlen, T. Reverse time migration: A seismic imaging technique applied to synthetic ultrasonic data. *Int. J. Geophys.* **2012**, *2012*, 128465. [CrossRef]
31. He, J.; Leckey, C.A.; Leser, P.E.; Leser, W.P. Multi-mode reverse time migration damage imaging using ultrasonic guided waves. *Ultrasonics* **2019**, *94*, 319–331. [CrossRef]
32. Yang, X.; Wang, K.; Xu, Y.; Xu, L.; Hu, W.; Wang, H.; Su, Z. A reverse time migration-based multistep angular spectrum approach for ultrasonic imaging of specimens with irregular surfaces. *Ultrasonics* **2020**, *108*, 106233. [CrossRef] [PubMed]
33. Rao, J.; Wang, J.; Kollmannsberger, S.; Shi, J.; Fu, H.; Rank, E. Point cloud-based elastic reverse time migration for ultrasonic imaging of components with vertical surfaces. *Mech. Syst. Signal Process.* **2022**, *163*, 108144. [CrossRef]
34. Nguyen, L.T.; Modrak, R.T. Ultrasonic wavefield inversion and migration in complex heterogeneous structures: 2D numerical imaging and nondestructive testing experiments. *Ultrasonics* **2018**, *82*, 357–370. [CrossRef] [PubMed]
35. Whitmore, N.D. Iterative depth migration by backward time propagation. In *SEG Technical Program Expanded Abstracts 1983*; Society of Exploration Geophysicists: Houston, TX, USA, 1983; pp. 382–385. [CrossRef]
36. Baysal, E.; Kosloff, D.D.; Sherwood JW, C. Reverse time migration. *Geophysics* **1983**, *48*, 1514–1524. [CrossRef]
37. Chang, W.F.; McMechan, G.A. 3-D elastic prestack, reverse-time depth migration. *Geophysics* **1994**, *59*, 597–609. [CrossRef]
38. Ma, X.; Li, H.; Gui, Z.; Peng, X.; Li, G. Frequency-Domain Q-Compensated Reverse Time Migration Using a Stabilization Scheme. *Remote Sens.* **2022**, *14*, 5850. [CrossRef]
39. Fang, J.; Shi, Y.; Zhou, H.; Chen, H.; Zhang, Q.; Wang, N. A High-Precision Elastic Reverse-Time Migration for Complex Geologic Structure Imaging in Applied Geophysics. *Remote Sens.* **2022**, *14*, 3542. [CrossRef]
40. Fink, M. Time reversal of ultrasonic fields. I. Basic principles. *IEEE Trans. Ultrason. Ferroelectr. Freq. Control* **1992**, *39*, 555–566. [CrossRef]
41. Ji, K.; Zhao, P.; Zhuo, C.; Chen, J.; Wang, X.; Gao, S.; Fu, J. Ultrasonic full-matrix imaging of curved-surface components. *Mech. Syst. Signal Process.* **2022**, *181*, 109522. [CrossRef]
42. Liu, H.; Qi, Y.; Chen, Z.; Tong, H.; Liu, C.; Zhuang, M. Ultrasonic inspection of grouted splice sleeves in precast concrete structures using elastic reverse time migration method. *Mech. Syst. Signal Process.* **2021**, *148*, 107152. [CrossRef]
43. Zhang, Y.; Gao, X.; Zhang, J.; Jiao, J. An Ultrasonic Reverse Time Migration Imaging Method Based on Higher-Order Singular Value Decomposition. *Sensors* **2022**, *22*, 2534. [CrossRef] [PubMed]
44. Rao, J.; Saini, A.; Yang, J.; Ratassepp, M.; Fan, Z. Ultrasonic imaging of irregularly shaped notches based on elastic reverse time migration. *NDT E Int.* **2019**, *107*, 102135. [CrossRef]
45. Jia, D.; Zhang, W.; Wang, Y.; Liu, Y. A new approach for cylindrical steel structure deformation monitoring by dense point clouds. *Remote Sens.* **2021**, *13*, 2263. [CrossRef]
46. Ren, Y.; Wang, J.; Yang, Z.; Xu, X.; Chen, L. Pre-stack elastic reverse time migration in tunnels based on cylindrical coordinates. *J. Rock Mech. Geotech. Eng.* **2022**, *14*, 1933–1945. [CrossRef]
47. Zheng, Y.; Cheng, F.; Liu, J.; Fan, Z.; Han, B.; Wang, J. Elastic full-wave field simulation in 3D tunnel space with the variable staggered-grid finite-difference method in cylindrical coordinates. *J. Appl. Geophys.* **2023**, *213*, 105013. [CrossRef]
48. Nguyen, L.T.; Kocur, G.K.; Saenger, E.H. Defect mapping in pipes by ultrasonic wavefield cross-correlation: A synthetic verification. *Ultrasonics* **2018**, *90*, 153–165. [CrossRef]
49. Liu, Q.H.; Sinha, B.K. A 3D cylindrical PML/FDTD method for elastic waves in fluid-filled pressurized boreholes in triaxially stressed formations. *Geophysics* **2003**, *68*, 1731–1743. [CrossRef]
50. Virieux, J. P-SV wave propagation in heterogeneous media: Velocity-stress finite-difference method. *Geophysics* **1986**, *51*, 889–901. [CrossRef]

51. Zeng, C.; Xia, J.; Miller, R.D.; Tsoflias, G.P. An improved vacuum formulation for 2D finite-difference modeling of Rayleigh waves including surface topography and internal discontinuities. *Geophysics* **2012**, *77*, T1–T9. [CrossRef]
52. Liu, Q.H. Perfectly matched layers for elastic waves in cylindrical and spherical coordinates. *J. Acoust. Soc. Am.* **1999**, *105*, 2075–2084. [CrossRef]
53. Chattopadhyay, S.; McMechan, G.A. Imaging conditions for prestack reverse-time migration. *Geophysics* **2008**, *73*, S81–S89. [CrossRef]
54. Zhang, Y.; Sun, J. Practical issues in reverse time migration: True amplitude gathers, noise removal and harmonic source encoding. *First Break* **2009**, *27*. [CrossRef]
55. Velis, D.R.; Ulrych, T.J. Simulated annealing wavelet estimation via fourth-order cumulant matching. *Geophysics* **1996**, *61*, 1939–1948. [CrossRef]
56. Buland, A.; Omre, H. Bayesian wavelet estimation from seismic and well data. *Geophysics* **2003**, *68*, 2000–2009.

Journal of
Marine Science and Engineering

MDPI

Article

Ontology Framework for Sustainability Evaluation of Cement–Steel-Slag-Stabilized Soft Soil Based on Life Cycle Assessment Approach

Chunyang Yu [1], Jia Yuan [1], Chunyi Cui [1,*], Jiuye Zhao [1], Fang Liu [2] and Gang Li [2]

1 Department of Transportation Engineering, Dalian Maritime University, Dalian 116026, China; yuchunyang@dlmu.edu.cn (C.Y.); yuanjia@dlmu.edu.cn (J.Y.); zhaojiuye@dlmu.edu.cn (J.Z.)
2 Shaanxi Key Laboratory of Safety and Durability of Concrete Structures, Xijing University, Xi'an 710123, China; liufang_winter@163.com (F.L.); t_bag945@126.com (G.L.)
* Correspondence: cuichunyi@dlmu.edu.cn

Abstract: Steel slag has become a promising supplementary cementitious material for soft soil stabilization. However, there is a lack of research on the integrated assessment of cement–steel-slag-stabilized soft soils (SCSs) from the performance, environmental, and economic perspectives. In this study, an ontology framework for the sustainable evaluation of SCSs was developed based on the life cycle assessment (LCA) approach, which combined a knowledge base with semantic web rules to achieve an automated decision design for soft soil stabilization, considering comprehensive benefits. The ontology framework was applied to a marine soft soil stabilization case to verify its scientificity and practicability and to evaluate the influence of the fineness, carbonation degree, and substitution ratio of steel slag on the sustainability of SCSs. The results show that, when compared to pure-cement-stabilized soil (S-C), using 10% and 20% of fine steel slag carbonated for 18 h (FSS-C-18h) as cement substitutes can significantly reduce carbon emissions and costs while achieving a similar strength performance as S-C, demonstrating the feasibility of steel slag as a sustainable supplementary cementitious material for soft soil stabilization.

Keywords: ontology; steel slag; stabilized soils; foundation; LCA; sustainability evaluation

check for updates

Citation: Yu, C.; Yuan, J.; Cui, C.; Zhao, J.; Liu, F.; Li, G. Ontology Framework for Sustainability Evaluation of Cement–Steel-Slag-Stabilized Soft Soil Based on Life Cycle Assessment Approach. *J. Mar. Sci. Eng.* **2023**, *11*, 1418. https://doi.org/10.3390/jmse11071418

Academic Editor: Dmitry A. Ruban

Received: 6 June 2023
Revised: 13 July 2023
Accepted: 13 July 2023
Published: 14 July 2023

1. Introduction

Soft soils are widely distributed in coastal areas and are characterized by a high natural water content, low bearing capacity, and high compressibility, and cannot be used directly in engineering [1]. However, the treatment of the soft soil foundation is a very challenging task, which requires certain methods such as preloading, chemical stabilization, and electro-osmosis [2,3]. The chemical stabilization method is widely used to achieve the stabilization of soft soil by mixing some cementitious materials (such as cement, lime, water glass, ion-exchange-class, and polymer-class stabilized materials, etc.) into the soft soil [4–6].

Cement is widely used in the field of soft soil stabilization due to its wide range of sources and stable properties. However, cement has problems such as high energy consumption, high carbon emission, consumption of non-renewable resources, and environmental pollution [7,8]. Replacing cement with low-carbon supplementary cementitious materials (SCMs) such as steel slag, fly ash, and silica fume is considered as the most promising strategy for sustainable soft soil stabilization design [9–11].

Steel slag is the solid waste the metallurgical industry produces, accounting for about 12–15% of steel output [12,13]. Steel slag is a potential SCM due to its similar composition to cement [14,15]. The low reactivity and poor volume stability of steel slag can be improved by grinding and carbonation [16,17]. Most of the existing studies on the partial replacement of cement by steel slag for soft soil stabilization have focused on the effect of steel slag on the mechanical properties of stabilized soil [18,19]. However, there is a lack of scientific quantitative research on the environmental and economic impact of cement–steel-slag-stabilized soils (SCSs).

Life cycle assessment (LCA) is a scientific method to qualitatively and quantitatively assess the potential environmental impact of a product or process during its life cycle [20,21], and LCA is considered one of the essential tools for sustainability assessment. Ghasemi et al. [22] performed the LCA study to compare the environmental impact using slurry and wet carbonation processes for converter slag and found that the net avoided global warming potential (GWP) of the slurry and wet routes were 525.56 and 426.67 $kgCO_2eq/MWh$, respectively. Li et al. [23] evaluated the environmental impact of steel slag aggregates and steel slag blocks. The GWP results showed that steel slag blocks could achieve negative carbon emissions. However, LCA involves complex logical relationships between products, activities, and the environment, the results of which are often difficult to understand and to use to share inventory information between different disciplines [24].

Ontology can standardize concepts, terms, and their relationships, provide methodologies for building knowledge frameworks, and is widely used for information retrieval, integration, decision making, and knowledge sharing between different domains by combining knowledge terms and predefined rules [25,26]. Zhang et al. [27] proposed an ontology-based semantic representation method for the product life cycle, which implemented a formalized and shared product life cycle design. Hou et al. [28] developed the concrete structures design ontology with embodied energy and carbon as sustainable evaluation indices. Meng et al. [29] proposed a multi-objective design method for an energy pile system based on ontology from the perspective of technology, economy, and sustainability. Based on the Monte Carlo simulation approach, Cui et al. [30] established a comprehensive seismic risk assessment ontology framework for the subway station. Ontology has obvious advantages in solving multi-domain, multi-objective problems due to its shareability, interoperability, and reusability. However, little research has been carried out to develop ontology frameworks for soft soil stabilization.

This study intends to propose a sustainable evaluation framework for SCSs based on LCA and ontology. The framework takes unconfined compressive strength (UCS), GWP, and cost as indicators, and conducts a multi-objective decision-making study on stabilized soils by combining a knowledge base with semantic web rules, with a view to obtaining a design method for soft soil stabilization with optimal overall benefits. This paper is organized as follows: In Section 2, the LCA method is applied to quantify the GWP and costs at each stage of the life cycle and to propose sustainable environmental and economic indicators. Section 3 presents the development of the ontology framework for the evaluation of stabilized soils (OntoESS). Section 4 conducts a case study of marine soft soil stabilization to verify the practicality of the OntoESS framework and to investigate the effects of steel slag fineness, carbonation degree, and substitution ratio on the sustainability of stabilized soils and to compare it with the sustainability of pure-cement-stabilized soils (S-C). The main conclusions and next steps of this study are presented in Section 5.

2. Life Cycle Assessment Methodology

2.1. Goal and Scope Definition

The goal of the life cycle assessment (LCA) in this study is to assess the potential environmental and economic impacts of SCSs and compare them with pure-cement-stabilized soil (S-C). The global warming potential (GWP, measured in terms of carbon dioxide equivalent), which is most important to industries, is selected as the life cycle environmental indicator [31]. The GWP is the mass of CO_2 for whom the greenhouse effect of various greenhouse gases corresponds to the same effect in a 100-year time frame. CO_2 is used as the reference gas because it has the greatest impact on global warming [32].

The functional unit is defined as "the cement–steel slag stabilizer required to stabilize 1 m^3 soft soils". The system research boundary belongs to the "cradle to gate" as shown in Figure 1, including the production of raw materials, the transportation of raw materials to the factory, and the stabilizer preparation process (grinding and carbonation), without considering the transportation of stabilizers to the construction site and the compaction process with soft soil, as the carbon emissions of SCSs and S-C at these stages can be considered to be similar. The input of the LCA includes energy (diesel, coal, and electricity) and raw materials (cement and steel slag). The output of the LCA only considers greenhouse gases represented by CO_2.

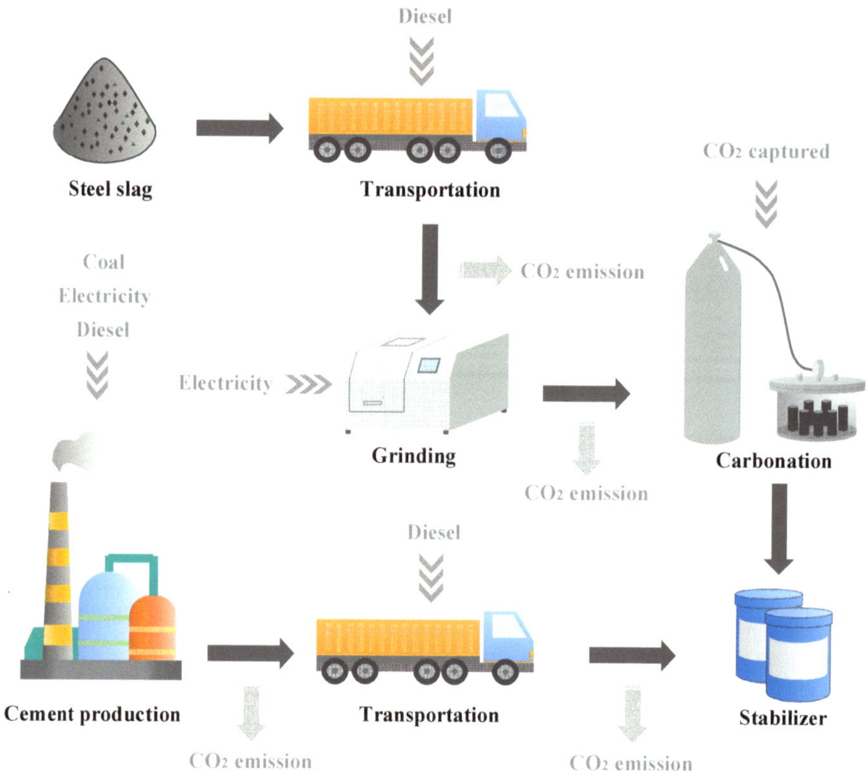

Figure 1. System boundary for cement–steel slag stabilizer production (cradle to gate).

2.2. Life Cycle Inventory

To assess carbon emissions at all stages within the system boundary, it is necessary to collect all input inventory data. In this study, the raw materials required and the measured energy consumption are used as the basis for the life cycle inventory analysis, and the carbon emission factors are taken from the inventory of carbon and energy (ICE) and from relevant literature [33,34].

Steel slag is mostly disposed of in landfills as a waste product from the metallurgical industry [35], and, in most cases, is considered as unintended residual waste. Therefore, according to the recommendations of ISO 14040 [36], the environmental impact of steel slag production is not allocated. The electricity consumption in the preparation of the stabilizer comes from the grinding of the steel slag, which generates carbon emissions that should be converted using the average carbon emission factors for China's regional power grids published by the National Development and Reform Commission [34]. The carbonation of steel slag is a self-heating reaction without an external heat source, so there is no energy consumption. The carbon emission factors of cement production, electricity consumption, and heavy diesel truck transportation are shown in Table 1.

Table 1. Carbon emission factors of materials, transportation, and electricity.

Cement Production ($CO_{2\text{-}mf}$)	Electricity ($CO_{2\text{-}ef}$)	Transportation ($CO_{2\text{-}tf}$)
($kgCO_2$-eq/t)	($kgCO_2$-eq/kWh)	($kgCO_2$-eq/t-km)
735	0.7769	0.162

2.3. Life Cycle Environmental Impact Assessment

According to the system boundary, carbon emissions during the life cycle include the three stages of raw material production, raw material transportation, and stabilizer preparation. Moreover, the resource utilization of steel slag can avoid carbon emissions caused by landfills, herein referred to as the transportation of steel slag to landfills. Total carbon emissions can be calculated based on Equation (1):

$$CO_2 = CO_{2-m} + CO_{2-t} + CO_{2-p} - CO_{2-a} \tag{1}$$

where CO_{2-m} is the carbon emissions of the raw material production process ($kgCO_2$-eq), CO_{2-t} is the carbon emissions of the raw material transportation process, CO_{2-p} is the carbon emissions of the stabilizer preparation process, and CO_{2-a} is the carbon emissions of avoiding landfill.

Carbon emissions from material production only consider cement, calculated by Equation (2):

$$CO_{2-m} = W_c \times CO_{2-\mathrm{mf}} \tag{2}$$

where W_c is the weight of cement (t), and $CO_{2-\mathrm{mf}}$ is the carbon emission factor for cement production ($kgCO_2$-eq/t); see Table 1 for details.

CO_2 emissions from material transportation can be obtained by Equation (3):

$$CO_{2-t} = W_c \times D_c \times CO_{2-tf} + W_s \times D_s \times CO_{2-tf} \tag{3}$$

where D_c and D_s represent the distance from the cement and steel slag to the plant (km), respectively. $CO_{2-\mathrm{tf}}$ represents the carbon emissions of fuel consumed when transporting each unit of the ith material in 1 km ($kgCO_2$-eq/t-km); see Table 1 for details.

Carbon emissions from the stabilizer preparation include the electricity consumption for the grinding of the steel slag, and the CO_2 *uptake* during carbonation, as shown in Equation (4):

$$CO_{2-p} = E_g \times CO_{2-ef} - W_s \times CO_2\ uptake \times 1000 \tag{4}$$

where E_g is the electricity consumption per hour (kWh) for the grinding processes, and CO_{2-ef} is the electricity carbon emission factor ($kgCO_2$-eq/kWh); see Table 1 for details. W_s is the weight of the steel slag (t). CO_2 *uptake* (%) represents the ratio of CO_2 absorption to the amount of steel slag, measured by the rate of mass weight gain of the steel slag after carbonation. CO_2 *uptake* can be calculated by Equation (5):

$$CO_2\ uptake = \frac{m_{s-carbonated} - m_{s-initial}}{m_{s-initial}} \times 100\% \tag{5}$$

where $m_{s\text{-}initial}$ is the weight of the dried steel slag before the carbonation reaction, and $m_{s\text{-}carbonated}$ is the weight of the dried carbonated steel slag.

Carbon emissions of steel slag from avoiding landfill can be calculated based on Equation (6):

$$CO_{2-a} = W_s \times D_a \times CO_{2-tf} \tag{6}$$

where D_a represent the distance from the steel slag to the landfill (km).

2.4. Life Cycle Economic Impact Assessment

The production cost within the system boundary is used as the life cycle economic indicator, including raw material, transportation, stabilizer preparation, and avoiding landfill costs, which are converted into US dollars based on market prices and exchange rates during the study period [37]. (In January 2023, the median exchange rate of the Chinese RMB in the interbank foreign exchange market was US $1 to RMB 6.7626.) Raw material costs include the purchase of cement and steel slag. Transportation costs include the transportation of cement and steel slag to the plant. The cost of stabilizer preparation is the electricity consumption for the steel-slag-grinding process. The cost of avoiding landfill only considers the transportation of steel slag to the landfill. Unit prices of materials, transportation, and electricity are shown in Table 2. Labor and mechanical costs are not considered in this study.

Table 2. The unit price of materials, transportation, and electricity.

Cement ($Cost_{mc}$)	Steel Slag ($Cost_{ms}$)	Transportation ($Cost_{tf}$)	Electricity ($Cost_{ef}$)
(USD/t)	(USD/t)	(USD/t-km)	(USD/kWh)
67.8	10	0.14	0.145

The total cost can be obtained by Equation (7):

$$Cost = Cost_m +_t + Cost_p - Cost_a \tag{7}$$

where $Cost_m$ is the cost of raw materials, $Cost_p$ is the cost of stabilizer preparation, $Cost_t$ is the cost of material transportation, and $Cost_a$ is the cost of avoiding landfill.

The raw material cost can be expressed as Equation (8):

$$Cost_m = W_c \times Cost_{mc} + W_s \times Cost_{ms} \tag{8}$$

where W_c and W_c represent the weight of the cement and steel slag (t), respectively. $Cost_{mc}$ and $Cost_{ms}$ are the price per ton of cement and steel slag (USD/t), respectively; see Table 2 for details.

The material transportation cost is calculated by Equation (9):

$$Cost_t = W_c \times D_c \times Cost_{tf} + W_s \times D_s \times Cost_{tf} \tag{9}$$

where D_c and D_s represent the distance from the cement and steel slag to the plant (km), respectively. $Cost_{tf}$ represents the cost of fuel consumed when transporting each unit of the ith material in 1 km (USD/t-km); see Table 2 for details.

The stabilizer preparation cost is calculated by Equation (10):

$$Cost_p = E_g \times Cost_{ef} \tag{10}$$

where E_g is the electricity consumption (kWh) for the grinding processes, and $Cost_{ef}$ refers to the unit price of electricity (USD/kWh); see Table 2 for details.

The cost of avoiding landfill for steel slag is calculated according to Equation (11):

$$Cost_a = W_s \times D_a \times Cost_{tf} \tag{11}$$

where D_a represents the distance from the steel slag to the landfill (km).

2.5. Sustainability Index

In order to realize the wide application of steel slag partially replacing cement in soft soil stabilization, the sustainability efficiency of SCSs needs to be studied, considering their mechanical properties, carbon emission, and cost. Damineli et al. [38] proposed the carbon emissions required to achieve a strength of 1 MPa as an indicator for material evaluation. Based on this method, the SCS sustainability indices are obtained by Equations (12) and (13). The unconfined compressive strength (UCS) of SCSs and the CO_2

and cost calculated from Equations (1) and (7) are normalized using S-C as the basis, as shown in Equations (14)–(16). Obviously, a lower index indicates better sustainability of SCSs. Taking S-C with a sustainability index of 1 as the benchmark, if both sustainability indices of an SCS scheme are less than 1, it indicates that the scheme has better sustainability than S-C and can be used as a preliminary screening design scheme to provide a decision reference for designers.

$$SUI_{environment} = \frac{N_{CO_2}}{N_{UCS}} \quad (12)$$

$$SUI_{economic} = \frac{N_{Cost}}{N_{UCS}} \quad (13)$$

$$N_{CO_2} = \frac{CO_2}{R_{CO_2}} \quad (14)$$

$$N_{Cost} = \frac{Cost}{R_{Cost}} \quad (15)$$

$$N_{UCS} = \frac{UCS}{R_{UCS}} \quad (16)$$

where $SUI_{environment}$ is the sustainability environment index of SCSs, and $SUI_{economic}$ is the sustainability economic index of SCSs. N_{CO_2}, N_{Cost}, and N_{UCS} are the normalized values of CO_2, cost, and UCS based on S-C. R_{CO_2}, R_{Cost}, and R_{UCS} are the CO_2, cost, and UCS of the reference stabilized soil S-C.

3. Design and Development of Ontology Framework

3.1. System Framework

The open-source software Protégé is used to develop the ontology framework for the evaluation of stabilized soils (OntoESS). The OntoESS framework consists of the database layer, the knowledge base layer, and the user layer, as illustrated in Figure 2. The database layer contains knowledge content such as the stabilizer production process, material properties, energy information, etc., which can be obtained through books and literature, the life cycle inventory (LCI) database, stabilized soil tests, etc. The knowledge base layer is the core of the ontology framework. Domain knowledge is transformed into the ontology model and semantic web rule language (SWRL) through the rule editor, and stored in the knowledge base in the form of the OWL file [39,40]. The predetermined rules can reason new facts through the reasoner. In the user layer, engineers can define semantic query web rule language (SQWRL) query reasoning results based on design requirements to obtain design schemes and optimization directions that meet the requirements [41]. In order to ensure the correctness of the ontology logic construction, knowledge reasoning and consistency checking are realized by the Pellet inference engine.

3.2. The Development of OntoESS

Before the establishment of the ontology model, domain terms need to be collected to form knowledge items, such as involving the steel slag preparation process, including "carbonation, CO_2 captured, etc.". Furthermore, there are complicated semantic relationships among products, activities, and environments in life cycle analysis. To provide an explicit description, life cycle semantics is used in the construction of ontology models, such as the introduction of the terms "process and flow" in ISO 14040. The key concepts and relationships in OntoESS are established using the Unified Modeling Language (UML), as illustrated in Figure 3.

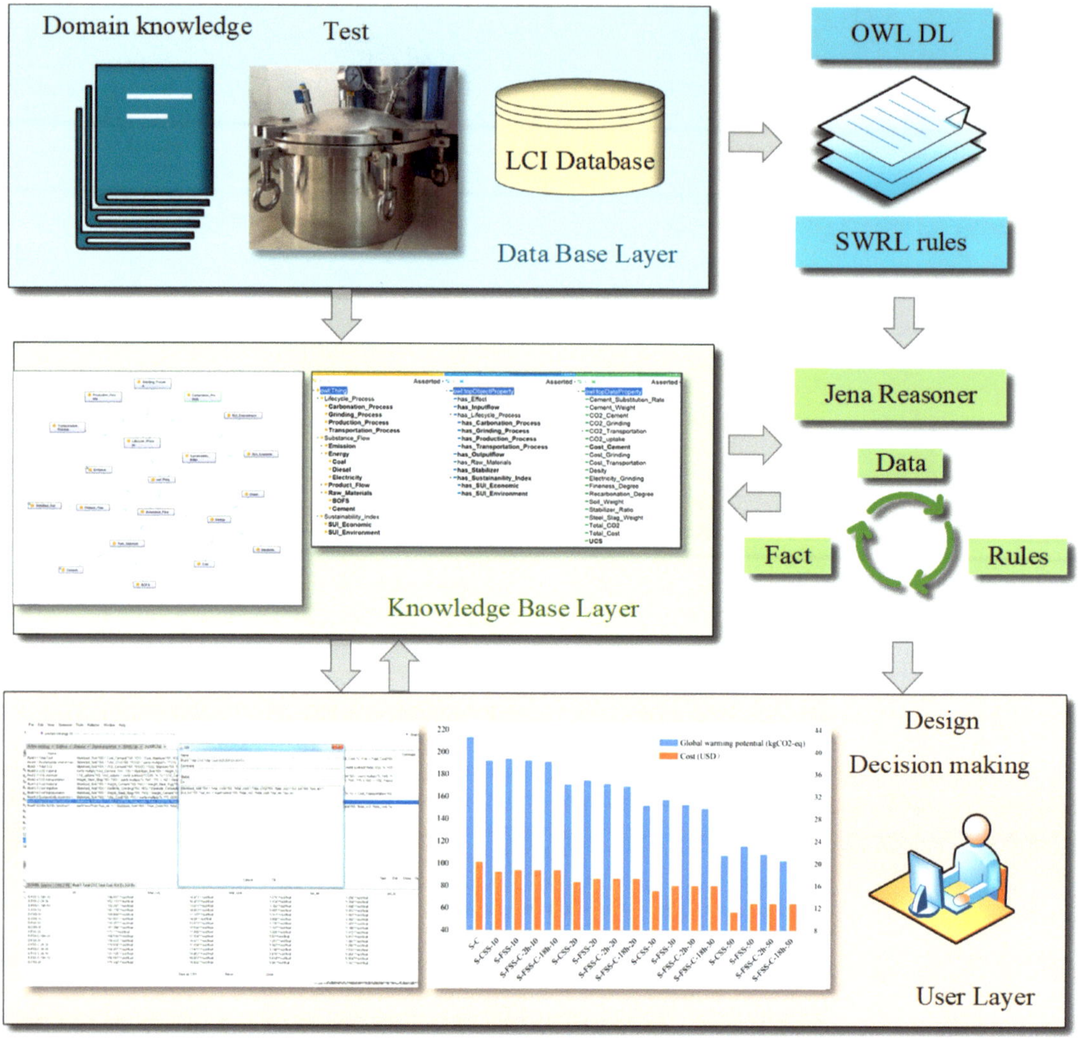

Figure 2. The developed ontology framework.

3.2.1. Define Hierarchy and Classes

The ontology model based on life cycle semantics is shown in Figure 4. The top-level classes involving stabilizer production mainly include the elementary flow, process, and product flow classes. The elementary flow class has two subclasses, the resource and emission classes. Resources are energy (diesel, coal, etc.) and raw materials (cement, steel slag) that enter the production system from the natural environment. Emissions are substances released from the production system to the air, water, or soil, such as CO_2. The process is used to describe various activities in the product life cycle, including the production process, transportation process, etc. The product flow represents the output of a process or production system, e.g., stabilizers.

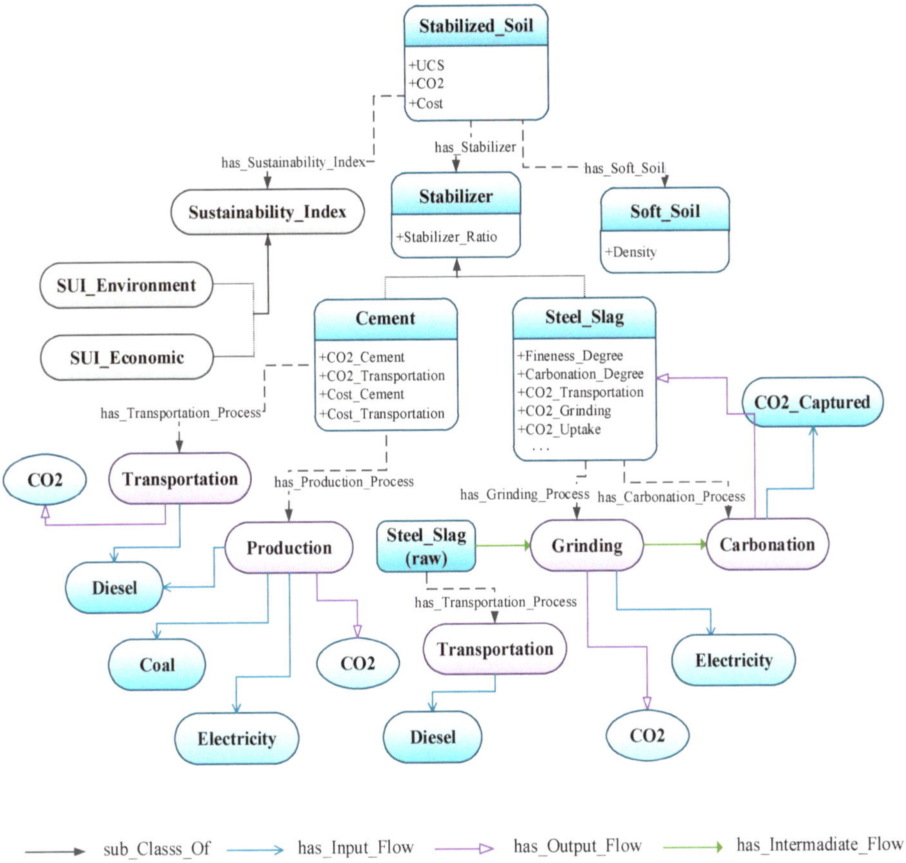

Figure 3. UML class diagram of the key concepts in the ontology.

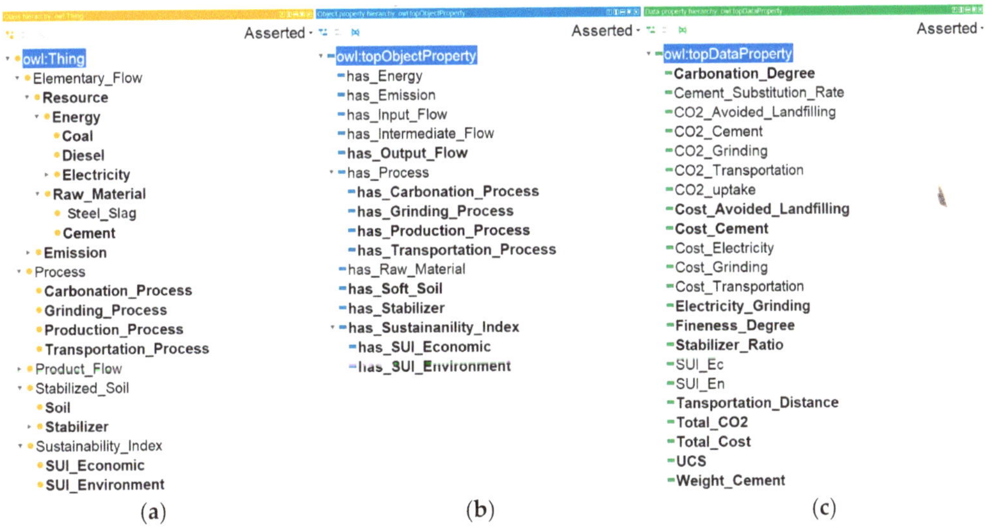

Figure 4. Development of ontology in Protégé-OWL 5.2: (**a**) classes, (**b**) object properties, (**c**) data properties.

3.2.2. Define Properties of Classes

Properties of ontology include object properties and data properties. The object properties are used to define the relationships between different classes; for example, the relationship between cement and production is defined as "has _ Production _ Process". The property "has _ Input _ Flow" is set between the process and the material or product entering the process. "has _ Output _ Flow" is placed between the process and the material or product leaving the process. "has _ Intermediate _ Flow" represents the relationship between adjacent processes in the production system. Data properties are used to qualitatively or quantitatively describe the instance properties, such as material weight, strength, carbon emission, etc.

3.2.3. Define SWRL Rules

To enable a comprehensive decision analysis of SCSs, the ontology framework needs to have strong reasoning and computational capabilities. Therefore, SWRL rules are used for function strengthening to calculate carbon emissions, production costs, and sustainability indicators for SCSs. SWRL rules can be combined with the elements defined in the ontology. Generally, SWRL rules consist of class atoms, individual property atoms, data-valued property atoms, and built-in atoms. The atoms are connected by '^', the reasoning and the result are connected by '->', and '?' is used to represent variables. For example, the SWRL rules for calculating the total carbon emissions of SCSs are shown in Table 3, and the SWRL rules for Equations (1)–(16) are shown in Tables S1–S3.

Table 3. The SWRL rules of calculating the total CO_2 emission.

Rule 1	Calculating total carbon emission of SCSs: $CO_2 = CO_{2-m} + CO_{2-t} + CO_{2-p} - CO_{2-a}$ Stabilized_Soil(?SS)^CO2_Cement(?SS,?CO2C)^CO2_Stabilizer(?SS, ?CO2S)^CO2_Transportation(?SS,?CO2T)^ CO2_Avoided_Landfilling(?SS,?CO2a)^swrlb:add(?x, ?CO2C, ?CO2S, ?CO2T)^swrlb:subtract(?total_CO2, ?x, ?CO2a) -> Total_CO2 (?SS, ?total_CO2)

3.2.4. Define SQWRL Rules

SQWRL is an ontology rule query language based on the SWRL extension. After SWRL rule reasoning, engineers can query the relevant information of SCSs according to the design requirements, such as the cost and sustainability indices of each scheme. See Tables S1–S3 for details of the SQWRL rules.

4. Case Study

4.1. Case Study Description

A case study was conducted to demonstrate the practicality of OntoESS in the sustainable evaluation of stabilized soils. The soft soils in this case were sampled from a marine soft soil foundation in Dalian, Liaoning Province, China. The raw materials of the stabilizer were 42.5R Portland cement and steel slag (type for basic oxygen furnace slag). The physical properties of soft soils are shown in Table 4 [18]. According to the unified soil classification system ASTM-2487, the soil was classified as low plasticity clay (CL). Table 5 shows the chemical composition of soft soil, cement, and steel slag [18]. The transportation distances of the raw materials are listed in Table 6.

Table 4. Basic physical properties of soft soil.

Parameter		Value
Initial water content (%)		61.12
Liquid limit (%)		38.2
Plastic limit (%)		19.2
Plasticity index		18.7
Clay fraction (%)	<0.002 mm	5.1
Silt fraction (%)	0.002–0.075 mm	55.2
Sand fraction (%)	0.075–2 mm	39.7
Optimum water content (%)		11.7
Maximum dry density (g/cm^3)		1.92

Table 5. Chemical compositions of soft soil, cement, and steel slag (w/w%).

Sample	CaO	SiO_2	Fe_2O_3	Al_2O_3	MgO	MnO	TiO_2	Others
Soft soil	5.6	43.8	8.65	22.50	6.41	-	-	10
Cement	61.12	21.46	2.88	5.25	2.08	-	-	2.5
Steel slag	39.02	14.45	23.42	3.83	7.27	7.31	1.35	1.1

Table 6. Material transportation information.

Material	Transport Distance (km)	
	Plant	Landfill
Cement	21	-
Steel slag	58	35

To investigate the influence of the steel slag fineness, carbonation degree, and substitution rate on the sustainability of SCSs, four kinds of steel slag were prepared, as shown in Figure 5. A ball mill with a capacity of 20−25 t/h and a power of 1500 kW was used for steel slag grinding. The specific surface area of coarse steel slag (CSS) increased from 117.3 m^2/kg to 747.2 m^2/kg after grinding with a ball mill for 1 h, consuming about 75 kwh/t of electrical energy. Only fine steel slag (FSS) was used for carbonation as the finer particle size was found to be more conducive to the carbonation reaction [42]. The FSS mixed with a certain amount of water was uniformly placed in the carbonation chamber and carbonated by a concentration of 99.9% CO_2 under a room temperature and pressure environment (temperature 25 °C, pressure 0.2 MPa). According to the measured weight gain rate of the steel slag after carbonation, the CO_2 *uptake* for fine steel slag carbonated for 2 h (FSS-C-2h) and 18 h (FSS-C-18h) was 5% and 9%, respectively.

Four types of steel slag were mixed with cement at different substitution ratios to form the composite stabilized material, with S-C as the control group, to investigate the effect of steel slag on the strength of the stabilized soil. The total amount of stabilizer ($m_{cement+steel\ slag}/m_{dry\ soil}$) was controlled to be 15%, and the proportion of steel slag replacing cement was 10%, 20%, 30%, and 50%, respectively. Based on the deep mixing method widely used for foundation treatment [43,44], the water content of the specimen preparation was set at 1.5 times the liquid limit, i.e., 57.2%, to ensure the fluidity of the stabilized soil. The specimens were maintained at room temperature for 60 days after demoulding. Design parameters and UCS_{60d} of SCSs and S-C are given in Table 7.

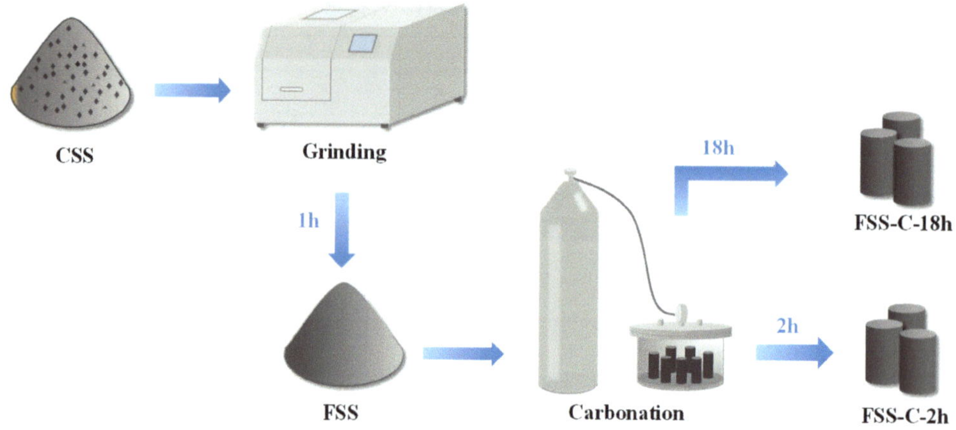

Figure 5. The preparation of the four types of steel slag.

Table 7. Design parameters and UCS_{60d} of SCSs and S-C.

Designated Name	Steel Slag Substitution Ratio (%)	Binder Mixtures (kg)		UCS_{60d} (kPa)
		Cement	Steel Slag	
S-C	0	288	0	2070.2
S-CSS	10	259.2	28.8	1919.92
	20	230.4	57.6	1727.86
	30	204.6	86.4	1331.41
	50	144	144	793.66
S-FSS	10	259.2	28.8	2032.2
	20	230.4	57.6	1681.93
	30	204.6	86.4	1214.95
	50	144	144	988.44
S-FSS-C-2h	10	259.2	28.8	1949.23
	20	230.4	57.6	1651.62
	30	204.6	86.4	1404.69
	50	144	144	916.29
S-FSS-C-18h	10	259.2	28.8	2267.32
	20	230.4	57.6	1975.27
	30	204.6	86.4	1345.3
	50	144	144	842.16

From the strength results in Table 7, with the increase in substitution ratio, the strength generally shows a downward trend. The strength of S-FSS-C-18h with 50% steel slag substitution (S-FSS-C-18h-50) is only 40.7% of that of S-C. Previous studies have found that there is an interaction between steel slag and cement, which can stimulate each other's activity under a certain ratio; however, excess steel slag can hinder the hydration of the cement [45,46]. It is noteworthy that low-content carbonated steel slag can significantly improve the strength of stabilized soil; the strength of S-FSS-C-18h-10 reaches 110% of S-C. This is probably due to the formation of $CaCO_3$ dispersed in the cement paste after carbonation of the steel slag to form additional nucleation sites, which promote the hydration of the cement [47].

4.2. The Application of OntoESS

4.2.1. Impact Analysis of Steel Slag Preparation

According to the specimen design in Table 7, individuals of stabilized soil are created in the ontology model. Information such as material amounts, transport distance, and power consumption are input into each individual as data attributes. The calculation rules of Equations (1)–(16) were invoked via the SWRL Tab plug-in and the reasoner is run to perform the rule inference. After the system runs, the individual's carbon emissions and cost results at each stage are automatically generated, as shown in Figure 6.

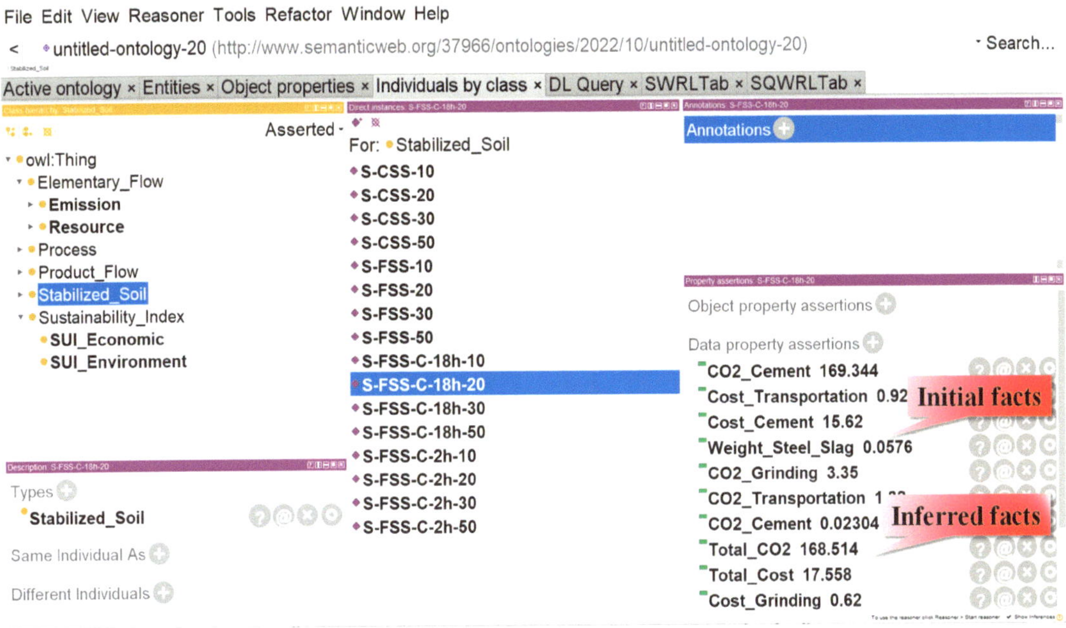

Figure 6. The interface of new fact.

To compare the environmental impacts of the four steel slag preparation processes, S-CSS-20, S-FSS-20, S-FSS-C-2h-20, and S-FSS-C-18h-20 were used as examples; according to the above method, the carbon emissions of the four equivalent steel slags during transport, grinding, carbonation, and avoided landfilling were queried, and the results are shown in Table 8. Electricity consumption during the grinding process is the main contributor to the GWP of steel slag; FSS has the highest GWP of the four kinds of steel slag, reaching 3.56 kgCO_2-eq. Carbonated fine steel slag (FSS-C-2h, FSS-C-18h) can reduce GWP due to the absorption of a certain amount of CO_2, of which FSS-C-18h achieved a negative GWP −1.62 kgCO_2-eq due to the longer carbonation time. Fuel consumption from the transportation process and avoided landfilling has little contribution to GWP, which is mainly related to transport distance. From the environmental impact perspective, the adverse effects of the steel slag transportation and preparation process can be almost offset by the *CO_2 uptake* and avoided landfilling.

Table 8. GWP of the four types of steel slag (kgCO_2-eq).

Types	Transportation	Avoided Landfilling	Grinding	CO_2 Uptake	Total
CSS	0.54	−0.3265	-	-	0.2135
FSS	0.54	−0.3265	3.35	-	3.5635
FSS-C-2h	0.54	−0.3265	3.35	−2.88	0.6835
FSS-C-18h	0.54	−0.3265	3.35	−5.814	−1.6205

4.2.2. Impact Analysis of SCSs

Engineers can also use the SQWRL plug-in to query the specified results according to the design requirements. The carbon emissions, costs, and sustainability indices of each scheme are shown in Figures 7–9.

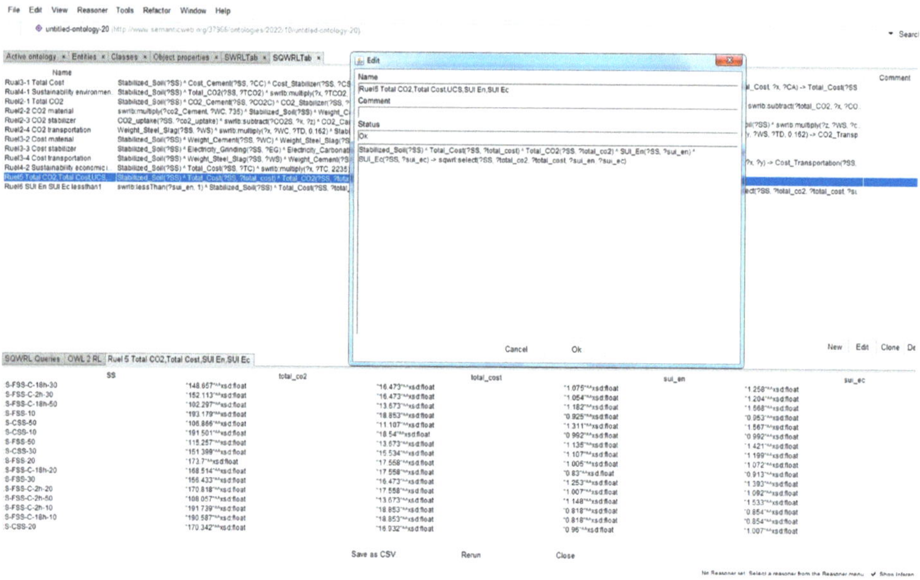

Figure 7. The execution and results after running the SQWRL rules of the query SCS indicators.

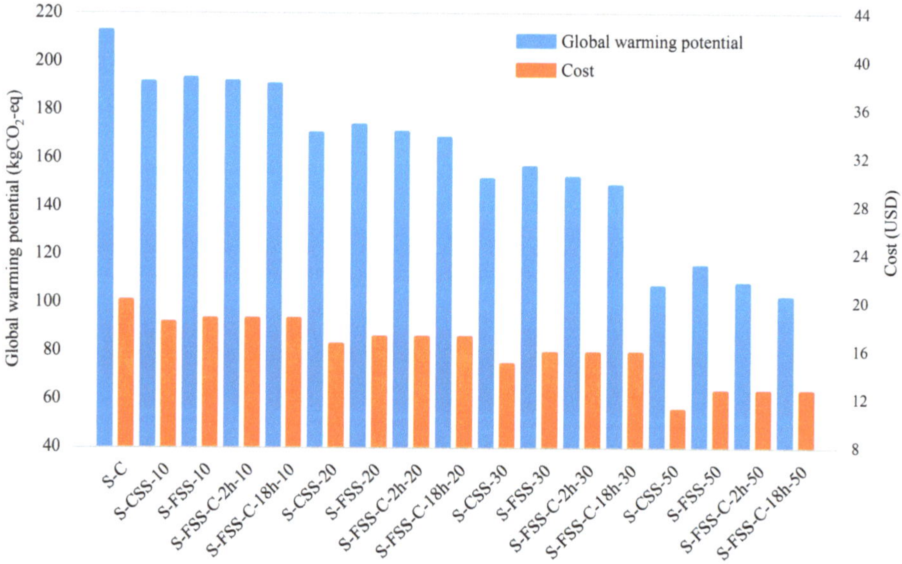

Figure 8. Querying results of GWP, Cost.

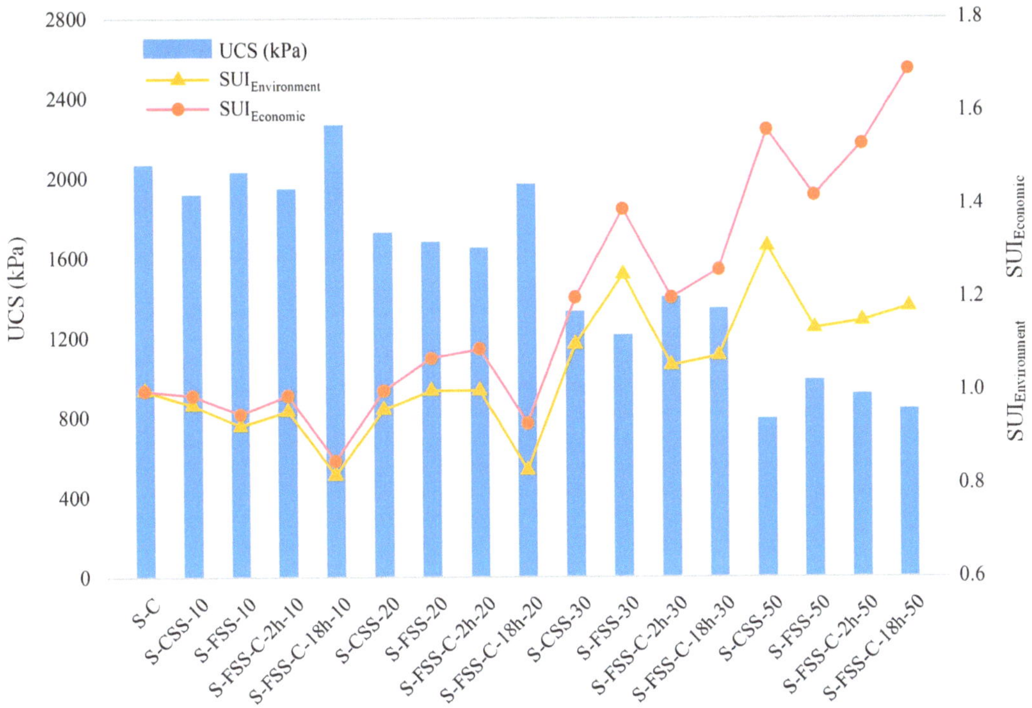

Figure 9. Querying results of $SUI_{Environment}$, $SUI_{Economic}$, and UCS_{60d}.

Figure 8 presents the GWP and cost of the S-C and SCSs. The difference in GWP from the steel slag preparation is not evident in the stabilized soil, because more than 90% of the GWP from stabilized soil is from cement production, which results from the high energy consumption of the cement production processes. Meanwhile, the cement cost reaches more than 70% in each scheme due to the high cement price. With the steel slag substitution ratio increase, the GWP and cost of SCSs decreased significantly. When the substitution ratio is 50%, the GWP and cost of S-FSS-C-18h are 48.1% and 70% of S-C, respectively.

Figure 9 shows the UCS_{60d} and sustainability indices of each scheme. Steel slag as an SCM can effectively reduce the GWP and production cost of the stabilized soil by partially replacing cement. However, cement is the main source of its strength [48], and excess steel slag can cause insufficient strength of the stabilized soil, thus leading to a high sustainability index of the stabilized soil, which is not conducive to engineering applications. We set the strength of the stabilized soil to meet the strength grade 1.0 as required by GJ/T 526-2018 [49], which means that the strength of the specimen should be greater than 1 MPa and, therefore, the steel slag substitution ratio should not exceed 50%. The schemes with both $SUI_{Environment}$ and $SUI_{Economic}$ of less than 1 can be preliminarily selected by running the SQWRL rule. The results are shown in Figure 10. The stabilized soil with low-content steel slag has better sustainability than S-C, and the strength is not less than 90% of S-C. Among them, S-FSS-C-18h-10 and S-FSS-C-18h 20 have the best sustainability with a GWP reduction of 10.4% and 20.4% compared to S-C, respectively.

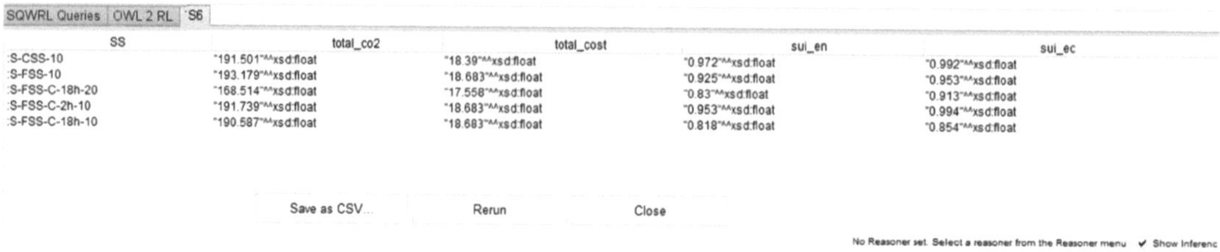

SQWRL Queries | OWL 2 RL | S6

ss	total_co2	total_cost	sui_en	sui_ec
:S-CSS-10	"191.501"^^xsd:float	"18.39"^^xsd:float	"0.972"^^xsd:float	"0.992"^^xsd:float
:S-FSS-10	"193.179"^^xsd:float	"18.683"^^xsd:float	"0.925"^^xsd:float	"0.953"^^xsd:float
:S-FSS-C-18h-20	"168.514"^^xsd:float	"17.558"^^xsd:float	"0.83"^^xsd:float	"0.913"^^xsd:float
:S-FSS-C-2h-10	"191.739"^^xsd:float	"18.683"^^xsd:float	"0.953"^^xsd:float	"0.994"^^xsd:float
:S-FSS-C-18h-10	"190.587"^^xsd:float	"18.683"^^xsd:float	"0.818"^^xsd:float	"0.854"^^xsd:float

Save as CSV... Rerun Close

No Reasoner set. Select a reasoner from the Reasoner menu ✔ Show Inferenc

Figure 10. Query results for schemes with $SUI_{Environment}$ and $SUI_{Economic}$ of less than 1.

4.2.3. Sensitivity Analysis

While this case demonstrates the viability of steel slag as a soft soil stabilizer, the fuel consumption for material transportation can also significantly affect the sustainability of the stabilized soil. Improper selection of steel slag suppliers may be contrary to the design intention to reduce the carbon footprint and cost. Therefore, a sensitivity analysis of steel slag transportation for S-FSS-C-18h-10 and S-FSS-C-18h-20 is selected to determine the maximum transport distance that meets sustainable design requirements and inform the designer's selection of suppliers.

The pure cement stabilizer production (21 km transport distance from the cement production site to the stabilizer plant) is taken as the base scenario to assess the environmental and cost impacts of increasing the transport distance from the steel slag production site to the stabilizer plant. The transport distance to achieve the GWP and production cost of a pure cement stabilizer is used as the maximum transport distance for steel slag (D_{smax}), calculated according to Equations (17) and (18). The SWRL calculation rules are shown in Tables S1–S3. The results are shown in Figure 11. Arriving at the same GWP as S-C, the steel slag transport distances ($D_{smax\text{-}GWP}$) for S-FSS-C-18h-10 and S-FSS-C-18h-20 are 4794.55 km and 4791.55 km, respectively, indicating that the effect of steel slag transport on the GWP is not significant and that considerable environmental benefits can be achieved even over long distances. The effect of steel slag transport distance on the cost of stabilizer production cannot be ignored. To achieve lower costs than S-C, it is recommended that the transport distance of steel slag ($D_{smax\text{-}Cost}$) for S-FSS-C-18h-10 and S-FSS-C-18h-20 is controlled to within 379 km.

$$D_{smax-GWP} = \frac{212.66 - (CO_2 - CO_{2-ts})}{W_s \times 0.162} \quad (17)$$

$$D_{smax-Cost} = \frac{18.85 - (Cost - Cost_{ts})}{W_s \times 0.14} \quad (18)$$

where $D_{smax-GWP}$ is the maximum transport distance of steel slag to reach the benchmark scenario GWP, and $D_{smax-Cost}$ is the maximum transport distance of steel slag to reach the benchmark scenario cost. CO_2 and $Cost$ are the carbon emission and cost of SCSs in the initial scenario, respectively. CO_{2-ts} and $Cost_{ts}$ are the carbon emission and cost of steel slag transportation in the initial scenario, respectively.

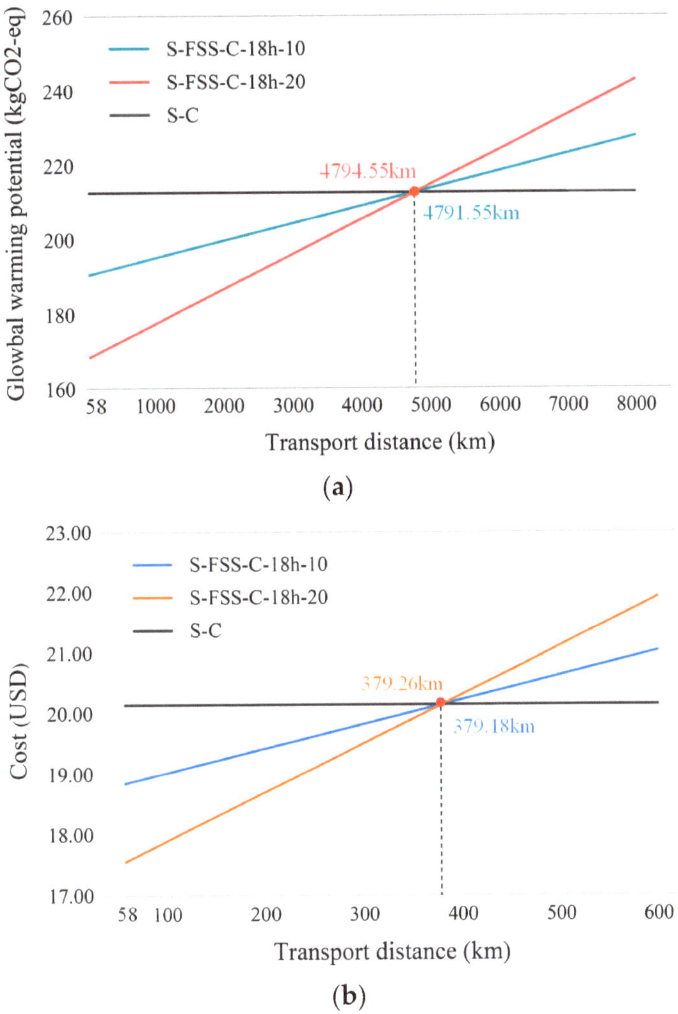

Figure 11. Sensitivity analysis for the steel slag transport distance: (**a**) global warming potential, (**b**) production cost.

5. Discussion

Cement is the most commonly used soft soil stabilization material for foundation treatment in coastal engineering, but the environmental problems caused by the high carbon emission of its production have gradually attracted the attention of all countries. The use of steel slag to partially replace cement to stabilize soft soil has caused widespread concern among researchers. The ontology framework proposed in this study fills the gap in the research on the sustainability evaluation of cement steel-slag-stabilized soils, because most of the current evaluation studies on stabilized soils only focus on one or two aspects of engineering performance, the environment, and the economy [50,51], failing to integrate the indicators from the perspective of sustainability, and the results of their evaluations are difficult to use for providing references for designers.

From the study results, the steel slag treated by grinding and carbonation can obtain better sustainability than S-C by replacing the cement to stabilize soft soil with a lower content, demonstrating the feasibility of steel slag for soft soil stabilization. However, long transport distances for steel slag should be avoided so that the cost advantages of SCSs are not masked. It should be noted that the preparation of stabilizers in this study was carried out at the laboratory scale. In the future, the influence of the equipment and labor factors required for the mass production and on-site construction of stabilizers should be considered. In addition, durability is also a focus of future research on the sustainability of stabilized soils. Based on the flexibility and reusability of the ontology framework, it can be improved and extended according to new materials and processes, and tools with unified interfaces can be developed for different projects and phases, so as to realize the efficient and collaborative design, operation, and maintenance management of stabilized soils in the whole life cycle.

6. Conclusions

In this study, an ontology framework was developed to evaluate the sustainability of stabilized soil with a cement–steel slag blend (SCSs). Firstly, a quantitative approach for sustainability evaluation indicators of SCSs based on LCA was proposed. Then, an SCS ontology model was developed, and related domain knowledge and basic data are integrated into the knowledge base. According to the semantic web rules, the reasoning and query of evaluation indices were further realized. The ontology framework proposed in this study can clearly describe the logical relationship between production activities and the environment during the life cycle of SCSs, which helps designers to clarify the influence of materials and processes on the sustainability of stabilized soils, and then obtain the optimal design and optimization direction from a macro perspective.

The practicability of the proposed ontology framework was verified by a case of marine soft soil stabilization. The case study found that the four types of steel slag can achieve better sustainability than pure-cement-stabilized soil (S-C) at a lower content. The stabilization of soft soils with FSS-C-18h with 10% and 20% substitution rates represented the best stabilization scheme, achieving similar strengths to S-C while significantly reducing carbon emissions and costs. From the sensitivity analysis of the transport distance of steel slag, even if the transport distance of steel slag is significant, SCSs are still favorable to the environment. However, the transport distance of steel slag greatly affects the cost, which must be considered when selecting suppliers.

Supplementary Materials: The following supporting information can be downloaded at: https://www.mdpi.com/article/10.3390/jmse11071418/s1, Table S1: The SWRL rules of calculating CO_2 emission; Table S2: The SWRL rules of calculating cost; Table S3: The SWRL rules of calculating the Sustainability Index; Table S4: The SQWRL rules of selecting all the information; Table S5: The SQWRL rules with Sustainability Index less than 1; Table S6: The SWRL rules of calculating D_{smax}.

Author Contributions: Conceptualization, F.L.; methodology, C.Y.; software, J.Y.; validation, J.Z. and G.L.; resources, J.Z.; data curation, F.L.; writing—original draft preparation, C.Y.; writing—review and editing, J.Y.; supervision, C.C.; funding acquisition, C.C. All authors have read and agreed to the published version of the manuscript.

Funding: This work was supported by the National Key Research and Development Program of China (Grant No. 2021YFB2601102), the Natural Science Foundation of Liaoning Province of China (Grant No. 2022-MS-153), and the Cultivation Program for the Excellent Doctoral Dissertation of Dalian Maritime University (0143210270).

Institutional Review Board Statement: Not applicable.

Informed Consent Statement: Not applicable.

Conflicts of Interest: The authors declare no conflict of interest.

References

1. Zentar, R.; Wang, H.; Wang, D. Comparative study of stabilization/solidification of dredged sediments with ordinary Portland cement and calcium sulfo-aluminate cement in the framework of valorization in road construction material. *Constr. Build. Mater.* **2021**, *279*, 122447. [CrossRef]
2. Xue, Z.J.; Tang, X.W.; Yang, Q.; Tian, Z.; Zhang, Y.; Xu, W. Mechanism of electro-osmotic chemical for clay improvement: Process analysis and clay property evolution. *Appl. Clay Sci.* **2018**, *166*, 18–26. [CrossRef]
3. Dong, Y.K.; Cui, L.; Zhang, X. Multiple-GPU for three dimensional MPM based on single-root complex. *Int. J. Numer. Methods Eng.* **2022**, *123*, 1481–1504. [CrossRef]
4. Gilazghi, S.T.; Huang, J.; Rezaeimalek, S.; Bin-Shafique, S. Stabilizing sulfate-rich high plasticity clay with moisture activated polymerization. *Eng. Geol.* **2016**, *211*, 171–178. [CrossRef]
5. Onyelowe, K.C.; Ebid, A.M.; Nwobia, L.I.; Obianyo, I.I. Shrinkage limit Multi-AI-Based predictive models for sustainable utilization of activated rice husk ash for treating expansive pavement subgrade. *Transp. Infrastruct. Geotechnol.* **2021**, *9*, 835–853. [CrossRef]
6. Diaz-Loya, I.; Juenger, M.; Seraj, S.; Minkara, R. Extending supplementary cementitious material resources: Reclaimed and remediated fly ash and natural pozzolans. *Cem. Concr. Compos.* **2019**, *101*, 44–51. [CrossRef]
7. Ali, M.B.; Saidur, R.; Hossain, M.S. A review on emission analysis in cement industries. *Renew. Sustain. Energy Rev.* **2011**, *15*, 2252–2261. [CrossRef]
8. Onyelowe, K.C.; Ebid, A.M.; Onyia, M.E.; Amanamba, E.C. Estimating the swelling potential of non-carbon-based binder (NCBB)-treated clayey soil for sustainable green subgrade using AI (GP, ANN and EPR) techniques. *Int. J. Low-Carbon Technol.* **2022**, *17*, 807–815. [CrossRef]
9. Anastasiou, E.; Georgiadis Filikas, K.; Stefanidou, M. Utilization of fine recycled aggregates in concrete with fly ash and steel slag. *Constr. Build. Mater.* **2014**, *50*, 154–161. [CrossRef]
10. Verian, K.P.; Behnood, A. Effects of deicers on the performance of concrete pavements containing air-cooled blast furnace slag and supplementary cementitious materials. *Cem. Concr. Compos.* **2018**, *90*, 27–41. [CrossRef]
11. Onyelowe, K.C.; Ebid, A.M.; Egwu, U.; Onyia, M.E.; Onah, H.N.; Nwobia, L.I.; Onwughara, I.; Firooz, A.A. Erodibility of Nanocomposite-Improved Unsaturated Soil Using Genetic Programming, Artificial Neural Networks, and Evolutionary Polynomial Regression Techniques. *Sustainability* **2022**, *14*, 7403. [CrossRef]
12. Shao, X.; Mehdizadeh, H.; Li, L.F.; Ling, T.C. Life cycle assessment of upcycling waste slag via CO_2 pre-treatment: Comparative study of carbonation routes. *J. Clean. Prod.* **2022**, *378*, 134115. [CrossRef]
13. Jia, R.Q.; Liu, J.X.; Jia, R.Q. A study of factors that influence the hydration activity of mono-component CaO and bi-component $CaO/Ca_2Fe_2O_5$ systems. *Cem. Concr. Res.* **2017**, *91*, 123–132. [CrossRef]
14. Liu, S.H.; Li, L.H. Influence of fineness on the cementitious properties of steel slag. *J. Therm. Anal. Calorim.* **2014**, *117*, 629–634. [CrossRef]
15. Wang, D.; Chang, J.; Ansari, W.S. The effects of carbonation and hydration on the mineralogy and microstructure of basic oxygen furnace slag products. *J. CO2 Util.* **2019**, *34*, 87–98. [CrossRef]
16. Nielsen, P.; Boone, M.A.; Horckmans, L.; Snellings, R.; Quaghebeur, M. Accelerated carbonation of steel slag monoliths at low CO_2 pressure—Microstructure and strength development. *J. CO2 Util.* **2020**, *36*, 124–134. [CrossRef]
17. Ghouleh, Z.; Guthrie, R.I.L.; Shao, Y. High-strength KOBM steel slag binder activated by carbonation. *Constr. Build. Mater.* **2015**, *99*, 175–183. [CrossRef]
18. Cui, C.Y.; Yu, C.Y.; Zhao, J.Y.; Zheng, J.J. Steel Slag/Precarbonated Steel Slag as a Partial Substitute for Portland Cement: Effect on the Mechanical Properties and Microstructure of Stabilized Soils. *KSCE J. Civ. Eng.* **2022**, *26*, 3803–3814. [CrossRef]
19. Yu, C.Y.; Cui, C.Y.; Wang, Y.; Zhao, J.Y.; Wu, Y.J. Strength performance and microstructural evolution of carbonated steel slag stabilized soils in the laboratory scale. *Eng. Geol.* **2021**, *295*, 106410. [CrossRef]
20. Khoo, H.H.; Bu, J.; Wong, R.L.; Kuan, S.Y.; Sharratt, P.N. Carbon capture and utilization: Preliminary life cycle CO_2, energy, and cost results of potential mineral carbonation. *Energy Procedia* **2011**, *4*, 2494–2501. [CrossRef]
21. Van den Heede, P.; De Belie, N. Environmental impact and life cycle assessment (LCA) of traditional and 'green' concretes: Literature review and theoretical calculations. *Cem. Concr. Compos.* **2012**, *34*, 431–442. [CrossRef]
22. Ghasemi, S.; Costa, G.; Zingaretti, D.; Bäbler, M.U.; Baciocchi, R. Comparative Life-cycle Assessment of Slurry and Wet Accelerated Carbonation of BOF Slag. *Energy Procedia* **2017**, *114*, 5393–5403. [CrossRef]
23. Li, L.F.; Jiang, Y.; Pan, S.Y.; Ling, T.C. Comparative life cycle assessment to maximize CO_2 sequestration of steel slag products. *Constr. Build. Mater.* **2021**, *298*, 123876. [CrossRef]
24. Bertin, B.; Scuturici, V.M.; Pinon, J.M.; Risler, E. A semantic approach to life cycle assessment applied on energy environmental impact data management. In Proceedings of the 2012 Joint EDBT/ICDT Workshops, Berlin, Germany, 26–30 March 2012; pp. 87–94. [CrossRef]
25. Zhang, J.S.; Li, H.J.; Zhao, Y.H.; Ren, G.Q. An ontology-based approach supporting holistic structural design with the consideration of safety, environmental impact and cost. *Adv. Eng. Softw.* **2018**, *115*, 26–39. [CrossRef]
26. Zhong, B.T.; Wu, H.T.; Li, H.; Sepasgozar, S.; Luo, H.B.; He, L. A scientometric analysis and critical review of construction related ontology research. *Autom. Constr.* **2019**, *101*, 17–31. [CrossRef]

27. Zhang, Y.Z.; Luo, X.F.; Buis, J.J.; Sutherland, J.W. LCA-oriented semantic representation for the product life cycle. *J. Clean. Prod.* **2015**, *86*, 146–162. [CrossRef]
28. Hou, S.J.; Li, H.Z.; Rezgui, Y. Ontology-based approach for structural design considering low embodied energy and carbon. *Energy Build.* **2015**, *102*, 75–90. [CrossRef]
29. Meng, K.; Cui, C.Y.; Li, H.J.; Liu, H.L. Ontology-Based Approach Supporting Multi-Objective Holistic Decision Making for Energy Pile System. *Buildings* **2022**, *12*, 236. [CrossRef]
30. Cui, C.Y.; Xu, M.Z.; Xu, C.S.; Zhang, P.; Zhao, J.T. An ontology-based probabilistic framework for comprehensive seismic risk evaluation of subway stations by combining Monte Carlo simulation. *Tunn. Undergr. Space Technol.* **2023**, *135*, 105055. [CrossRef]
31. Bala, A.; Raugei, M.; Benveniste, G.; Gazulla, C.; Fullana, P.; Pere, A. Simplified tools for global warming potential evaluation: When 'good enough' is best. *Int. J. Life Cycle Assess.* **2010**, *15*, 489–498. [CrossRef]
32. IPCC. Climate Change 2022 Mitigation of Climate Change, Retrieved December 18, 2022. Available online: https://www.ipcc.ch/ (accessed on 13 July 2023).
33. Jones, G.H.C. Embodied Carbon—The ICE Database. 2023. Available online: https://circularecology.com/embodied-carbon-footprint-database.html (accessed on 13 July 2023).
34. *GB/T51366-2019*; Standard for Building Carbon Emissions Calculation. China Architecture and Building Press: Beijing, China, 2019.
35. Li, L.F.; Ling, T.C.; Pan, S.Y. Environmental benefit assessment of steel slag utilization and carbonation: A systematic review. *Sci. Total Environ.* **2022**, *806*, 150280. [CrossRef] [PubMed]
36. *ISO 14040*; Environmental Management-Life Cycle Assessment: Principles and Framework. ISO: Geneva, Switzerland, 2006. Available online: https://www.iso.org/standard/37456.html (accessed on 13 July 2023).
37. CCement. Spot Prices of Lead Industry in China. 2023. Available online: https://www.price.ccement.com/ (accessed on 13 July 2023).
38. Damineli, B.L.; Kemeid, F.M.; Aguiar, P.S.; John, V.M. Measuring the eco-efficiency of cement use. *Cem. Concr. Compos.* **2010**, *32*, 555–562. [CrossRef]
39. Zhang, F.; Ma, Z.M.; Li, W.J. Storing OWL ontologies in object-oriented databases. *Knowl.-Based. System.* **2015**, *76*, 240–255. [CrossRef]
40. Chen, R.C.; Huang, Y.H.; Bau, C.T.; Chen, S.M. A recommendation system based on domain ontology and SWRL for anti-diabetic drugs selection. *Expert Syst. Appl.* **2012**, *39*, 3995–4006. [CrossRef]
41. Ma, Z.L.; Liu, Z. Ontology- and freeware-based platform for rapid development of BIM applications with reasoning support. *Autom. Constr.* **2018**, *90*, 1–8. [CrossRef]
42. Song, Q.F.; Guo, M.Z.; Wang, L.; Ling, T.C. Use of steel slag as sustainable construction materials: A review of accelerated carbonation treatment. *Resour. Conserv. Recycl.* **2021**, *173*, 10574. [CrossRef]
43. Gullu, H.; Canakci, H.; Al Zangan, I.F. Use of cement based grout with glass powder for deep mixing. *Constr. Build. Mater.* **2017**, *137*, 12–20. [CrossRef]
44. Hessouh, J.J.M.M.; Eslami, J.; Beaucour, A.L.; Noumowe, A.; Mathieu, F.; Gotteland, P. Physical and mechanical characterization of deep soil mixing (DSM) materials: Laboratory vs construction site. *Constr. Build. Mater.* **2023**, *368*, 130436. [CrossRef]
45. Chen, Z.M.; Li, R.; Zheng, X.M.; Liu, J.X. Carbon sequestration of steel slag and carbonation for activating RO phase. *Cem. Concr. Res.* **2021**, *139*, 106271. [CrossRef]
46. Liu, Q.; Liu, J.X.; Qi, L.Q. Effects of temperature and carbonation curing on the mechanical properties of steel slag-cement binding materials. *Constr. Build. Mater.* **2016**, *124*, 999–1006. [CrossRef]
47. Meng, T.; Qiang, Y.J.; Hu, A.F.; Xu, C.T.; Lin, L. Effect of compound nano-$CaCO_3$ addition on strength development and microstructure of cement-stabilized soil in the marine environment. *Constr. Build. Mater.* **2017**, *151*, 775–781. [CrossRef]
48. Coffetti, D.; Crotti, E.; Gazzaniga, G.; Carrara, M.; Pastore, T.; Coppola, L. Pathways towards sustainable concrete. *Cem. Concr. Res.* **2022**, *154*, 106718. [CrossRef]
49. *GJ/T 526-2018*; Stabilizer for Soft Soil. Ministry of Housing and Urban-Rural Development of the People's Republic of China: Beijing, China, 2018.
50. Ghadir, P.; Zamanian, M.; Mahbubi-Motlagh, N.; Siberian, L.J.; Ranjbar, N. Shear strength and life cycle assessment of volcanic ash-based geopolymer and cement stabilized soil: A comparative study. *Transp. Geotech.* **2021**, *31*, 100639. [CrossRef]
51. MolaAbasi, H.; Kharazmi, P.; Khajeh, A.; Saberian, M.; Chenari, R.; Harandi, M.; Li, J. Low plasticity clay stabilized with cement and zeolite: An experimental and environmental impact study. *Resour. Conserv. Recycl.* **2022**, *184*, 106408. [CrossRef]

MDPI AG
Grosspeteranlage 5
4052 Basel
Switzerland
Tel.: +41 61 683 77 34

Journal of Marine Science and Engineering Editorial Office
E-mail: jmse@mdpi.com
www.mdpi.com/journal/jmse

MDPI

www.ingramcontent.com/pod-product-compliance
Lightning Source LLC
Chambersburg PA
CBHW041419210926
51850CB00013B/3312
9783725831241